Linear and Nonlinear Instabilities in Mechanical Systems

Linear and Nonlinear Instabilities in Mechanical Systems

Analysis, Control and Application

Hiroshi Yabuno
University of Tsukuba, Japan

WILEY

This edition first published 2021

Registered Offices
John Wiley & Sons, Inc., 111 River Street, Hoboken, NJ 07030, USA
John Wiley & Sons Ltd, The Atrium, Southern Gate, Chichester, West Sussex, PO19 8SQ, UK

Editorial Office
111 River Street, Hoboken, NJ 07030, USA

For details of our global editorial offices, customer services, and more information about Wiley products visit us at www.wiley.com.

Wiley also publishes its books in a variety of electronic formats and by print-on-demand. Some content that appears in standard print versions of this book may not be available in other formats.

Library of Congress Cataloging-in-Publication Data

Names: Yabuno, Hiroshi, 1961- author.
Title: Linear and nonlinear instabilities in mechanical systems : analysis, control and application / Hiroshi Yabuno, University of Tsukuba.
Description: First edition. | Hoboken, NJ : John Wiley & Sons, Inc., 2021. | Includes bibliographical references and index.
Identifiers: LCCN 2020040692 (print) | LCCN 2020040693 (ebook) | ISBN 9781119066538 (cloth) | ISBN 9781119066552 (adobe pdf) | ISBN 9781119066545 (epub)
Subjects: LCSH: Machinery, Dynamics of–Mathematics. | Stability–Mathematical models. | Control theory.
Classification: LCC TJ173 .Y33 2021 (print) | LCC TJ173 (ebook) | DDC 621.8/11–dc23
LC record available at https://lccn.loc.gov/2020040692
LC ebook record available at https://lccn.loc.gov/2020040693

Cover Design: Wiley
Cover Image: Hiroshi Yabuno with permission from Central Japan Railway company

Set in 9.5/12.5pt STIXTwoText by SPi Global, Chennai, India
Printed and bound by CPI Group (UK) Ltd, Croydon, CR0 4YY

10 9 8 7 6 5 4 3 2 1

To Yuko and Ayaka

Contents

Preface

Mechanical systems have been often analyzed using linearized mathematical models in order to obtain the solution through the application of the well-established linear theory. As they become lighter, faster, and more flexible, the systems are easily enforced in a nonlinear regime. The linearized mathematical models and the linear theoretical analysis have turned out to be not applicable for the complete interpretation and description of the complex dynamics produced by nonlinear effects. In such a context, the aim of this book is to provide fundamental methods for analyzing nonlinear instability phenomena as well as linear instability ones. Additional relevant goals of this book are to seek positive exploitations of nonlinear effects to develop innovative high-performance mechanical systems, to gain insights into the physical meaning (phenomenological understanding) of analytical results obtained by mathematical procedures as well as the dynamical systems theory (Thompson and Stewart (2002)) through real phenomena observation. The mathematical descriptions and the consequent interpretation of linear and nonlinear instability phenomena from a physical point of view are emphasized throughout this book for engineering applications of the dynamical systems theory. For each subject discussed in the book, several videos showing the real instability phenomena produced in mechanical systems are supplied to bridge the theoretical results and the corresponding phenomena.

Key properties of linear systems are (i) the principle of superposition from a mathematical point of view and (ii) the uniqueness of the equilibrium state from a physical point of view. In single-degree-of-freedom conservative systems, since the potential energy curve is expressed by a quadratic function with respect to the displacement and has only one extremum, it indicates that there is only one equilibrium state. In nonlinear systems, there can exist multiple equilibrium states, and the number of such equilibrium states may be changed by acting on the system parameters, according to the so called control parameter in the dynamical systems theory. A linearized spring–mass–damper system in resonance – subject to an external or forced periodic excitation-exhibits a response amplitude determined by the magnitude of the damping effect since the inertia and restoring forces are cancelled out each other. Thus, the applied excitation balances the remaining force, that is, the damping effect. On the other hand, the occurrence of self-excited oscillation due to negative damping can be theoretically predicted from the linearized mathematical model, but the theoretical result shows that the amplitude infinitely grows with time. This result is caused by the fact that there is no term balancing the negative damping effect because of the above cancellation of the inertia and restoring forces. In the situation when the amplitude becomes not so small after the growth, the nonlinear effects neglected in linearization of the mathematical model turn out to affect the dynamics. Therefore, to elucidate the behavior in the nonlinear regime, accounting for the nonlinear effects on the mathematical model and the nonlinear analysis is essential, and the finite response amplitude can be realized by the balance of the negative damping force to the nonlinear effects. Such phenomena other than self-excited oscillation occur in a variety of practical systems as described in literature (for example, Lacarbonara (2013); Nayfeh and Mook (2008); Shaw and Balachandran (2008); Thompson and Stewart (2002); Thomsen (2003); Troger and Steindl (2012)) and include buckling and parametric excitation, which are dealt with in this book. While the critical load to produce buckling is obtained by the linear analysis in the linearized model, the postbuckling state, i.e. the state in the case when the compressive force is above the critical load, is obtained by nonlinear analysis in the nonlinear mathematical model. Similarly, in parametric resonance, only the resonance region with respect to excitation frequency and amplitude is obtained by the linear analysis of the linearized mathematical model, whereas in order to obtain the steady-state amplitude after the growth of response amplitude, nonlinear analysis has to be carried out for the mathematical model accounting for the nonlinear effects.

To conduct the nonlinear analysis, mathematical models with different accuracy are introduced depending on the behavior of interest. For example, let us consider the motion of a pendulum (Meirovitch 1975). The restoring force due to gravity is proportional to $\sin\theta$, where θ is the angle of the pendulum. Taylor expansion of $\sin\theta$ is $\sin\theta = \theta - \frac{1}{3!}\theta^3 + \frac{1}{5!}\theta^5 + \cdots$. If only the behavior at small amplitudes is of interest, since the nonlinear terms with respect to θ can be truncated, the linearization of $\sin\theta \approx \theta$ is suitable. On the other hand, in the analysis of the rotary motion, no approximation is applicable. Furthermore, in the analysis of parametric resonance produced by the periodically vertical excitation, at least the cubic nonlinear term in addition to the linear term is kept in the mathematical model, as mentioned in Chapter 11 for detail. Dynamical behaviors often consist of components changing with different time scales. A familiar example of coexisting time scales is the resonance in a linear spring–mass–damper system subject to external excitation mentioned above. The time history of the response is characterized by the rapid periodic oscillation with the same frequency as the excitation frequency and the growth of the amplitude, which can be regarded as the fast and slow dynamics, respectively. The slow dynamics often reveal directly the essential characteristics of system dynamics as stability. The equations governing the slow dynamics, which is also called amplitude equation, can be extracted from the equations of the original system through singular perturbation methods as the method of multiple scales dealt with in Chapter 9. This can be regarded as a reduction method for the original dynamics while another reduction is the center manifold reduction (Chapter 8), in which the lower dimensional system is approximately obtained in the neighborhood of the critical condition of stability by focusing on the subspaces classified by the difference of flow speed. Based on the special strategies mentioned above, this book provides analytical approaches for linear and nonlinear instability phenomena. This book consists of the following chapters, which are related as shown in the "Reading Paths of the Chapters" and the level is advanced step by step and straightforward.

- From Chapters 1 through 4, to prepare the reader understanding the nonlinear analysis, the underlying linear theory is sufficiently treated so as to smoothly lead to the nonlinear theory in subsequent chapters. Chapter 1 introduces the definitions of equilibrium states and their stability as some of the most important concepts to analyze dynamical systems. Chapter 2 is devoted to the analysis of linear dynamical systems and clarifies the local stability of equilibrium states by relating it to the eigenvalues of the linear operator. Chapters 3 and 4 present the linear stability of nonconservative systems due to circulatory force and mathematically characterize the mechanism to produce the self-excited oscillation using the nonorthogonality of each modal vector.
- Chapter 5 discusses static instabilities based on the linear theory introduced in Chapters 1 and 2. Through the static unstable dynamics of a two-link model for a slender straight elastic rod subject to compressive force and spring mass damper models representing MEMS (Micro-Electro-Mechanical Systems) actuator and switch, the mathematical descriptions of the static instabilities are interpreted from a physical point of view showing how they are obtained by linear analysis of the linearized models. In some conditions, linear analysis results in unacceptable phenomena from a physical point of view and the necessity of nonlinear analysis to a suitable nonlinear mathematical model is indicated.
- Chapter 6 is concerned with dynamic instabilities. First, we deal with a belt-driven mass–spring–damper system. The produced dynamics instability is caused by a negative damping characteristic due to the Coulomb friction between belt and mass. In the remaining part of the chapter, the dynamic instabilities of a wing, a railway vehicle, a rotor, and a fluid-conveying pipe are introduced. The resonance mechanisms caused by the circulatory forces are clarified based on the theory of Chapters 3 and 4.
- Chapter 7 introduces local bifurcations that are produced in the neighborhood of the trivial equilibrium state. It is analytically shown, by taking into account the effects of nonlinearity in a system, that, as the parameters in the system are varied, the stability of equilibrium states may be changed or the equilibrium states may be created or destroyed. Such changes depending on the parameters are called bifurcations. The newly generated equilibrium states created by nonlinear effects often solve the unacceptable phenomena obtained by the linear theory.
- Chapter 8 is devoted to reduction methods of nonlinear dynamical systems. By using center manifold theory, the decrease of the dimension is achieved. By using a nonlinear coordinate transformation, the nonlinear terms in the original nonlinear dynamical system are eliminated as many as possible to obtain a reduced equivalent dynamical system.
- Chapter 9 deals with the method of multiple scales as one of singular perturbation methods to extract the equations governing the slowly time varying components in nonlinear phenomena, which is sometimes called amplitude equation. The method is utilized in the nonlinear analysis in Chapters 10 through 13 and Chapter 15.

- Chapter 10 analyzes the nonlinear effect on the self-excited oscillations caused from negative damping force and circulatory force, which are analyzed in the linear theory in Chapter 6. While the amplitude of self-excited oscillation is predicted to grow by the linear theory, it is shown by nonlinear analysis that the nonlinear effects retain the growth of amplitude.
- Chapter 11 is concerned with parametric resonance produced in the system with time-dependent perturbed stiffness. The vertically excited pendulum, which is a conventional system to produce the parametric resonance, is introduced to show the fundamental nonlinear characteristics of parametric resonance through bifurcation analysis.
- Chapter 12 considers the stabilization of statically unstable equilibrium state by high-frequency excitation. A corresponding conventional model is an inverted pendulum and the upright position is the statically unstable equilibrium state. By using the method of multiple scales, it is clarified that the high-frequency excitation stabilizes the upright position without any feedback control.
- Chapter 13 is concerned with atomic force microscope (AFM) that is an instrument to measure the profile of sample surfaces with atomic scale by detecting the variation in the dynamics of a force sensing cantilever probe. In order to detect the natural frequency of the probe and to obtain AFM images without contact of the probe to a sample, the self-excited oscillation and its amplitude control are proposed based on the theory mentioned in Chapter 8.
- Chapter 14 introduces another application of self-excited oscillation. First, it is theoretically shown that weakly coupled resonators can be regarded as an ultrasensitive mass sensor. In this method, the accuracy of detection of the eigenmode of the resonators affects the sensitivity of mass measurement. Even in the viscous environment, the exploitation of self-excited oscillation achieves the very high sensitive measurement.
- Chapter 15 introduces the utilization of pitchfork bifurcation phenomena for the motion control of an underactuated manipulator. By perturbing the pitchfork bifurcations produced in a free arm by high-frequency excitation, the stable equilibrium states are shifted and the motion control is carried out without feedback control. This is an example of how the positive utilization of nonlinear phenomena enables a motion control which cannot be achieved from the viewpoint of linear theory.
- Chapter 16 collects the videos of practical systems corresponding to the phenomena mentioned in previous chapters. These videos based on the author's personal research are referred to in each chapter. The analytical results and the corresponding phenomena are bridged by using video of various experiments in real systems. It helps the reader to develop a real initiation of the instability mechanisms without being hindered by the mathematical complex expressions.

Reading paths of the chapters

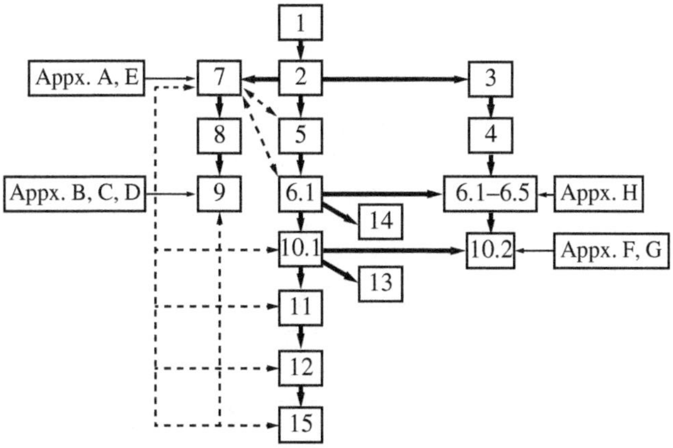

This is a textbook on basic nonlinear dynamics and control for senior under graduate students and first-year graduate students in engineering courses, and can be also used for self-study or reference by engineers who attempt nonlinear analysis and design for practical engineering systems. It is designed to be self-sufficient as possible, and no prior knowledge of nonlinear dynamics is required.

The minimal prerequisite for reading this book is a working knowledge of elemental linear algebra, analysis, and physics, which are taught at junior or senior level courses and are here used throughout. The advanced mathematics and physics are provided where needed.

Several important topics and theories are left out of this book. In particular, the book does not contain materials related to chaos. That is due to the following reasons: (i) practical real systems easily exhibit nonlinear behaviors other than chaos and (ii) deeply understanding the fundamental analytical methods for the behaviors treated in this book leads to an easy comprehension of the theoretical approach to chaos. In addition, partially due to the limitation of pages, there is unfortunately no coverage of dynamical systems in which time is discrete, continuous systems possessing a finite number of degrees of freedom, numerical approaches, and so on. The interested reader in such topics may refer to some advanced books (for example, Nonlinear Oscillations, Dynamical Systems, and Bifurcations of Vector Fields (Guckenheimer and Holmes 1983)), Introduction to Applied Nonlinear Dynamical Systems and Chaos (Wiggins 1980), Applied nonlinear dynamics: analytical, computational and experimental methods (Nayfeh and Balachandran 2008). The road to chaos (Ueda et al. 1992), and so on).

The videos are placed on the web page https://www.wiley.com/go/yabuno/instabilitiesinmechanicalsystems. Corrections or comments sent to yabuno@iit.tsukuba.ac.jp are welcome. Corrections will be placed on my web page https://drive.google.com/drive/folders/19FKdkk7H3yy-eB726R_RNDniPFhxg1qY?usp=sharing.

I am grateful to my present or former graduate and undergraduate students in my laboratory who performed the experiments shown in the videos and drew many of the figures. I owe much to Professors Masatsugu Yoshizawa, Nobuharu Aoshima, Walter Lacarbonara, Balakumar Balachandran, Alexander Seyranian, Alexei Mailybaev, Yukio Ishida, Ken-ichi. Nagai, Shin-ichi Maruyama, and Kiyotaka Yamashita for fruitful discussions. This book was partly written while I was a visiting professor at Sapienza University of Rome. I would like to express my gratitude to professors in nonlinear dynamics research group, in particular Professors Fabrizio Vestroni and Giuseppe Rega, for their many shared insights. Especial thanks must also be extended to Dr. Michela Taló who meticulously proofread the manuscript. I thank Gabby Robles, Associate Managing Editor at Wiley for steering the completion of the book.

Tokyo, 2020 *Hiroshi Yabuno*

References

Guckenheimer, J. and P. J. Holmes (1983). *Nonlinear oscillations, dynamical systems, and bifurcations of vector fields*. Springer.

Lacarbonara, W. (2013). Nonlinear structural mechanics: theory, dynamical phenomena and modeling. Springer Science & Business Media.

Meirovitch, L. (1975). *Elements of vibration analysis*. McGraw-Hill.

Nayfeh, A. H. and B. Balachandran (2008). Applied nonlinear dynamics: analytical, computational and experimental methods. John Wiley & Sons.

Nayfeh, A. H. and D. T. Mook (2008). *Nonlinear oscillations*. John Wiley & Sons.

Shaw, S. W. and B. Balachandran (2008). A review of nonlinear dynamics of mechanical systems in year 2008. *Journal of System Design and Dynamics* 2(3), 611–640.

Thompson, J. and H. Stewart (2002). *Nonlinear Dynamics and Chaos*. Wiley.

Thomsen, J. J. (2003). *Vibrations and Stability*. Springer.

Troger, H. and A. Steindl (2012). *Nonlinear stability and bifurcation theory: an introduction for engineers and applied scientists*. Springer Science & Business Media.

Ueda, Y., R. H. Abraham, and H. B. Stewart (1992). *The road to chaos*. Aerial Press.

Wiggins, S. (1980). *Introduction to Applied Nonlinear Dynamics and Chaos*. 1980.

About the Companion Website

This book is accompanied by a companion website:

www.wiley.com/go/yabuno/instabilitiesinmechanicalsystems

The website includes:
- Videos

1

Equilibrium States and Their Stability

In this chapter, as some of most important concepts to analyze dynamical systems, the definitions of *equilibrium states* and their *stability* are introduced. Considerations on the representative mechanical systems as a spring-mass system, a pendulum, and a magnetically levitated system, will make these concepts accessible. The stability of the equilibrium states is investigated in this chapter without mathematical analysis for the governing equations, but by inspecting the equation forms from a physical point of view. Then, a stabilization control method for the unstable equilibrium state will be briefly mentioned for the magnetically levitated system.

1.1 Equilibrium States

Let us start with the introduction of a *static equilibrium state* or an *equilibrium state* that is one of most important concepts to gain insights on the nonlinear dynamics of mechanical systems.

Definition 1.1 A point $x = x_{st}$ is said to be an equilibrium state, if it has the property of remaining at position x_{st} independently of the time at which the state of the system starts at x_{st}.

The equilibrium state $x = x_{st}$ has zero velocity and zero acceleration. Let us examine the equilibrium states of three fundamental mechanical systems.

1.1.1 Spring-Mass System

Consider a spring-mass system subject to gravity force. Figure 1.1a shows the position of the mass m at a certain instant while it is moving.

First, let us derive the equation of motion using Newton's second law. We introduce a static coordinate system x' whose origin is located at the lower end of the spring without mass as Figure 1.1b.

Figure 1.1c,d shows the free body diagrams at the state where the spring is elongated of Δx. $F_s^{\#}$ is the force acting on the spring from the mass and Δx is the resulting elongation

Linear and Nonlinear Instabilities in Mechanical Systems: Analysis, Control and Application,
First Edition. Hiroshi Yabuno.

Companion website: www.wiley.com/go/yabuno/instabilitiesinmechanicalsystems

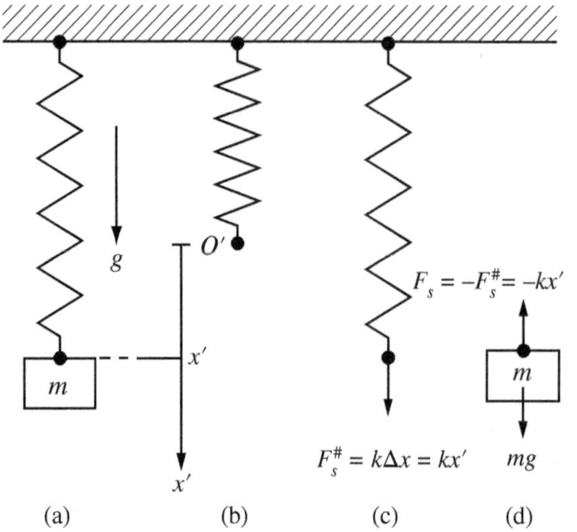

Figure 1.1 Spring-mass system and free body diagram. (a) Position of the vibrating mass m at a certain instant. (b) Coordinate x' whose origin O' is set to the lower end of the non-elongated spring; (c) and (d) Free body diagrams of spring and mass, respectively, where the elongation of the spring under the application of force $F_s^{\#}$ is denoted as Δx.

(see Figure 1.2a). In the ideal linear spring, Δx is proportional to $F_s^{\#}$ through a positive proportional constant k, i.e. a positive slope in the force–displacement characteristic of Figure 1.2b; the positive k means *positive stiffness*. Then, the mass is subject to the spring force $F_s = -F_s^{\#} = -k\Delta x$ representing the reaction force to $F_s^{\#}$ as shown in Figure 1.1d; the relationship between Δx and F_s is shown in Figure 1.2c, and the *negative* slope $(-k)$ indicates that the spring force direction is always opposite to that of Δx, i.e. the spring force acts on the mass as a *restoring* force. In this system, Δx is equal to x'.

Since the external force consists of the spring force F_s and the gravity force mg, the equation of motion is expressed as

$$m\frac{d^2x'}{dt^2} = -kx' + mg. \tag{1.1}$$

Thus, let us seek for the equilibrium state of the spring-mass system. The equilibrium state x_{st} that satisfies the following equation is obtained by assuming as conditions that velocity $\frac{dx'}{dt}$ and acceleration $\frac{d^2x'}{dt^2}$ are zero:

$$0 = -kx_{st} + mg. \tag{1.2}$$

Thus, the equilibrium state is

$$x_{st} = \frac{mg}{k}. \tag{1.3}$$

At this equilibrium state, the gravity force is balanced by the spring force.

Next, we rewrite the equation of motion by introducing a new coordinate x whose origin is shifted to the equilibrium state (see Figure 1.3). By substituting

$$x'(t) = x_{st} + x(t) \tag{1.4}$$

into Eq. (1.1) and taking into account that the equilibrium state x_{st} is time-independent, i.e. $\frac{dx_{st}}{dt} = \frac{d^2x_{st}}{dt^2} = 0$, we obtain

$$m\frac{d^2x}{dt^2} = -k(x + x_{st}) + mg. \tag{1.5}$$

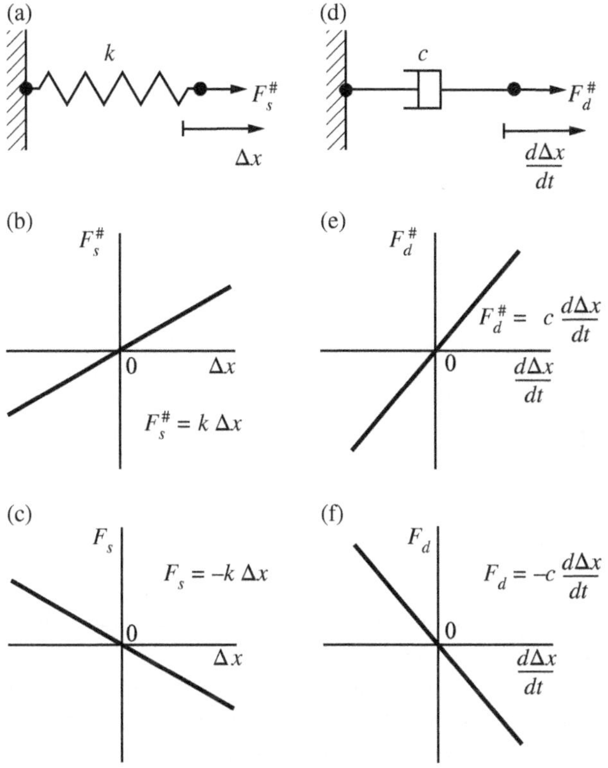

Figure 1.2 Characteristics of linear spring and linear damper. (a) Spring is elongated of Δx by applying force $F_s^\#$. (b) Force–displacement relationship $F_s^\# = k\Delta x$ showing that the ideal spring has a positive k. Such a stiffness effect is called *positive stiffness*. (c) Force $F_s(= -F_s^\# = -k\Delta x)$ produced by the spring with elongation Δx. When a mass is attached to the right end of the spring, the directions of the mass displacement and of the spring force acting on the mass are opposite. Thus, the spring force F_s acts on a mass as a *restoring* force. (d) By applying the force $F_d^\#$, the damper has velocity $\frac{d\Delta x}{dt}$. (e) Force–velocity relationship $F_d = c\frac{d\Delta x}{dt}$ showing that the ideal damper has a positive c. Such a damping effect is called a *positive damping*. (f) Force $F_d\left(= -F_d^\# = -c\frac{d\Delta x}{dt}\right)$ produced by the damper with velocity $\frac{d\Delta x}{dt}$. When a mass is attached to the right end of the damper, the directions of the mass velocity and the damping force acting on the mass are opposite.

By using the equilibrium equation (1.2), we can obtain another equation of motion with a different form as

$$m\frac{d^2x}{dt^2} = -kx. \tag{1.6}$$

By the way, unlike Eq. (1.1), the right-hand side does not include a constant term and Eq. (1.6) has the *trivial solution* ($x = 0$). In general, an equation having the trivial solution is called *homogeneous* equation. An equation not having trivial solution ($x' = 0$) as Eq. (1.1) is called *inhomogeneous* equation. The complete solution of the inhomogeneous equation consists of the homogeneous and particular solutions. Therefore, Eq. (1.6) describes the dynamics of the system in a simpler form from a mathematical point of view. We will solve Eq. (1.6) in Section 2.4.4.

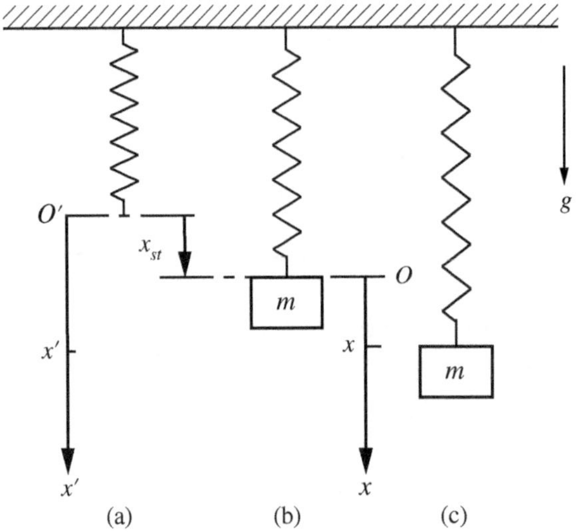

Figure 1.3 Coordinate transformation from x' to x ($x' = x + x_{st}$). (a) Origin O' of coordinate x' is set to the lower end of the non-elongated spring. (b) The origin O of coordinate x is located at the equilibrium state. (c) Position of mass at a certain instant while vibrating.

1.1.2 Magnetically Levitated System

As second example, we consider a simplified model of magnetically levitated vehicle (electro-magnetic suspension, EMS) shown in Figure 1.4. The vehicle is magnetically levitated from the guideway by using electromagnets attached to the vehicle. To investigate the dynamics of the levitated system in the vertical direction, we can focus on the motion of a magnet on the system, i.e. the part within the dotted circle in Figure 1.4a. The static coordinate x' is introduced and used to describe the position of the magnet. Its origin is located at the guideway as shown in Figure 1.4b. The electromagnetic force is assumed to be proportional to the square of the current i^2 and to be inversely proportional to the square of the gap between the guideway and the magnet x'^2 as (Dorf and Bishop 2011)

$$F_m = -k\frac{i^2}{x'^2}, \tag{1.7}$$

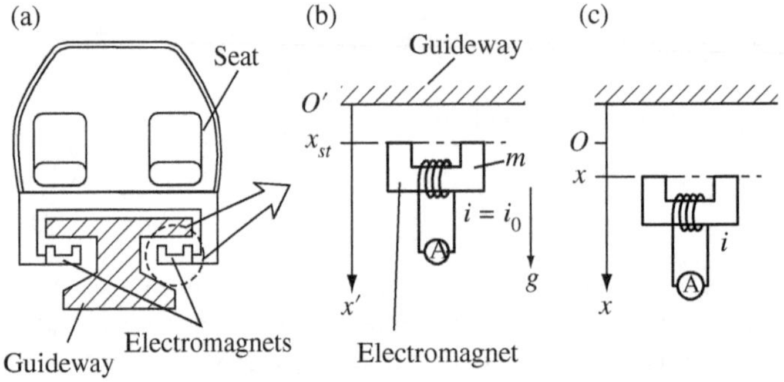

Figure 1.4 Magnetically levitated vehicle (MAGLEV). (a) Cross section of the magnetically levitated vehicle. (b) Static equilibrium state of the magnet expressed by coordinate x' in case of $i = i_0$. (c) Position of the magnet at a certain instant expressed by coordinate x whose origin is located at the static equilibrium state. Reprinted with permission Yabuno (2004).

where k is a positive constant which is determined by the materials of the magnet and the guideway, the shape of the magnet, and so on. Hence, the equation of motion is expressed as

$$m\frac{d^2x'}{dt^2} = -k\frac{i^2}{x'^2} + mg. \tag{1.8}$$

This equation is nonlinear since the first term of the magnetic force is not proportional to x'.

The equilibrium state x_{st} under the constant current of $i = i_0$ is determined by adopting a similar approach to the case of the spring-mass system. By assuming $d^2x'/dt^2 = 0$, Eq. (1.8) leads to the equilibrium equation

$$0 = -k\frac{i_0^2}{x_{st}^2} + mg. \tag{1.9}$$

Then, the equilibrium state $x_{st}(> 0)$ is expressed as

$$x_{st} = \sqrt{\frac{ki_0^2}{mg}}. \tag{1.10}$$

At the gap of $x' = x_{st}$, the gravity force is balanced by the magnetic force. Furthermore, we apply the coordinate transformation in the spring-mass system in Section 1.1.1, i.e.

$$x' = x_{st} + x, \tag{1.11}$$

where the origin of x is located at the equilibrium state O in Figure 1.4c. We obtain

$$\begin{aligned} m\frac{d^2x}{dt^2} &= -k\left(\frac{i_0}{x_{st} + x}\right)^2 + mg \\ &= -k\frac{i_0^2}{x_{st}^2}\left\{\frac{1}{1 + (x/x_{st})}\right\}^2 + mg. \end{aligned} \tag{1.12}$$

At this stage, unlike the spring-mass system, Eq. (1.8) cannot be transformed into the homogeneous equation due to the nonlinearity of the magnetic force.

By focusing on the motion in the neighborhood of the equilibrium state, i.e. in the state of $|x/x_{st}| \ll 1$, we approximate Eq. (1.7) to obtain the linear expression. Using Taylor expansion, we represent the magnetic force in terms of a power series with respect to the small value $|x/x_{st}|$ as (see Section 2.3)

$$F = -k\frac{i_0^2}{x_{st}^2}\left\{1 - 2\left(\frac{x}{x_{st}}\right) + 3\left(\frac{x}{x_{st}}\right)^2 - \cdots\right\}. \tag{1.13}$$

In the region of $|x/x_{st}| \ll 1$, this uniform expansion can be truncated by a finite number of terms depending on the desired accuracy. Neglecting higher order terms than the first order term with respect to x/x_{st}, we obtain the linearized magnetic force as

$$F = -k\frac{i_0^2}{x_{st}^2}\left\{1 - 2\left(\frac{x}{x_{st}}\right)\right\} = -k\frac{i_0^2}{x_{st}^2} + \frac{2ki_0^2}{x_{st}^3}x + O\left(\left(\frac{x}{x_{st}}\right)^2\right), \tag{1.14}$$

where O is the Landau symbol (Nayfeh 1981; Witelski and Bowen 2015); see Appendix D. Hence, the approximated equation of motion is written as

$$m\frac{d^2x}{dt^2} = -k\frac{i_0^2}{x_{st}^2} + \frac{2ki_0^2}{x_{st}^3}x + mg. \tag{1.15}$$

Recalling the equilibrium equation (1.9), we obtain the equation of motion in the form of linear homogeneous equation as

$$m\frac{d^2x}{dt^2} = k_m x, \tag{1.16}$$

where $k_m \overset{\text{def}}{=} \frac{2ki_0^2}{x_{st}^3} > 0$ is constant. It is the same form as Eq. (1.6), but the sign of the coefficient k_m of x in the right-hand side is different from that of x in Eq. (1.6). Figure 1.2c for the spring-mass system corresponds to Fig. 1.5c for the magnetically levitated system because of $F_m = k_m x$. The positive slope indicates that the magnetic force does not act on the levitated system as a restoring force but augments the deviation $|x|$ from the equilibrium state because such force acts in the same direction of the deviation from the equilibrium state. The corresponding force–displacement relationship is reported in Figure 1.5b, and it is characterized by *negative* slope ($F_m^\# = -k_m x$ in which x corresponds to Δx in Figure 1.5b). Such a feature is called *negative stiffness*, which produces a buckling phenomenon as mentioned in Section 5.1. We will solve Eq. (1.16) in Section 2.4.2.

1.1.3 Simple Pendulum

Let us move on to the third example of a simple pendulum as shown in Figure 1.6. The equation of motion is

$$ml\frac{d^2\theta'}{dt^2} = -mg\sin\theta'. \tag{1.17}$$

Letting $d^2\theta'/dt^2 = 0$, we obtain the equilibrium equation

$$0 = -mg\sin\theta_{st}. \tag{1.18}$$

The pendulum has two equilibrium states corresponding to the downward vertical position $\theta_{st} = 2n\pi \overset{\text{def}}{=} \theta_{st1}$ and the upright position $\theta_{st} = (2n+1)\pi \overset{\text{def}}{=} \theta_{st2}$ $(n = \dots, -1, 0, 1, 2, \dots)$.

Similar to the preceding two examples, we transform the coordinate as follows:

$$\theta' = \theta_{st} + \theta. \tag{1.19}$$

Under the assumption of $|\theta| \ll 1$, we can linearize the equation of motion in the neighborhood of each equilibrium state separately. By substituting Eq. (1.19) into Eq. (1.17) and considering Eq. (1.18), the result is expanded with respect to the small term θ:

$$ml\frac{d^2\theta}{dt^2} = -mg\left.\frac{d\sin(\theta_{st}+\theta)}{d\theta}\right|_{\theta=0}\theta + O(\theta^2). \tag{1.20}$$

In the neighborhood of θ_{st1} and θ_{st2}, Eq. (1.17) is expressed respectively as:

$$ml\frac{d^2\theta}{dt^2} = -mg\theta \tag{1.21}$$

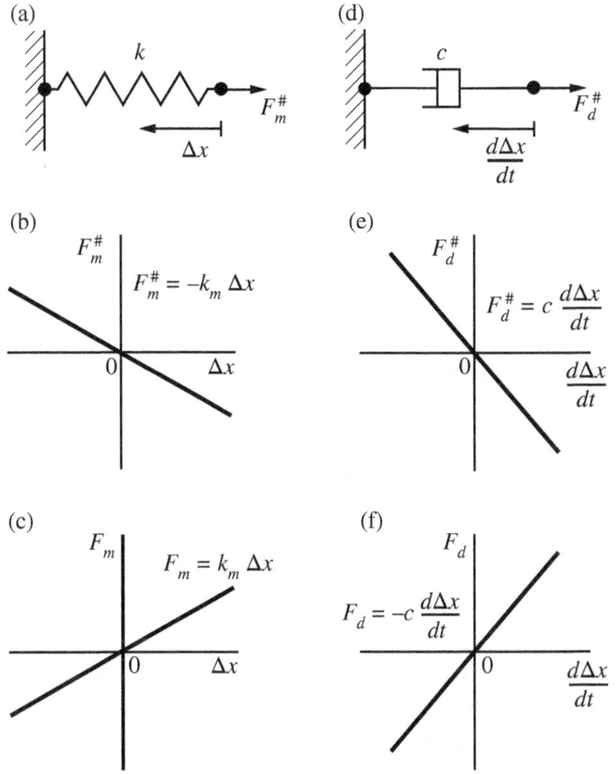

Figure 1.5 Negative stiffness and negative damping. (a) Virtual spring is shrunk with Δx by applying force $F_m^{\#}$. (b) Force–displacement relationship $F_m = -k_m \Delta x$, where $-k_m$ is negative. Such a stiffness effect is called *negative stiffness*. (c) Corresponding force $F_m (= -F_m^{\#} = k_m \Delta x)$ produced by the spring with elongation Δx. If a mass is attached to the right end of the virtual spring, the displacement directions of mass and spring force acting on the mass are the same. Therefore, the virtual spring force F_m does not act on the mass as a restoring force. (d) Virtual damper has velocity $\frac{d\Delta x}{dt}$ by applying the force $F_d^{\#}$. (e) Force–velocity relationship $F_d = -c\frac{d\Delta x}{dt}$, where $-c$ is negative. Such a damping effect is called *negative damping*. (f) Corresponding force $F_d \left(= -F_d^{\#} = c\frac{d\Delta x}{dt}\right)$ produced by the damper with $\frac{d\Delta x}{dt}$. If a mass is attached to the right end of the damper, the directions of the velocity of the mass and the damping force acting on the mass are the same.

Figure 1.6 Pendulum.

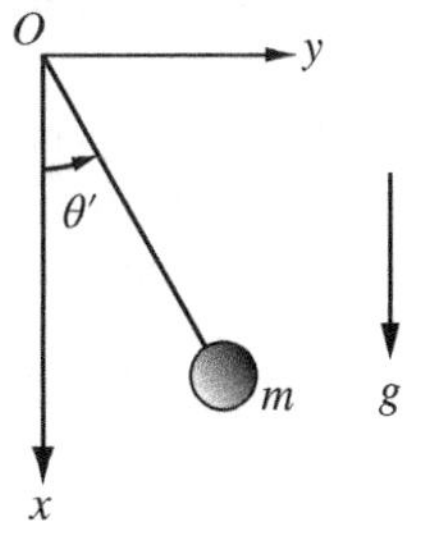

and

$$ml\frac{d^2\theta}{dt^2} = mg\theta. \tag{1.22}$$

The right-hand side of these equations can be regarded as an external force in the systems as previously done in Sections 1.1.1 and 1.1.2.

In the neighborhood of the downward vertical position $\theta = \theta_{st1}$, as for the spring-mass system, the coefficient on the right-hand term in Eq. (1.21) is negative (see Figure 1.2c in which Δx corresponds to θ). The equivalent external force is a restoring force and is characterized by a positive stiffness (see Figure 1.2b). The external force causes a vibration around the equilibrium state $\theta = \theta_{st1}$.

On the other hand, in the neighborhood of the upright position $\theta = \theta_{st2}$, due to the positive sign of the coefficient on the right-hand side (see Figure 1.5c in which Δx corresponds to θ), the equivalent external force is not a restoring force as already observed for the magnetically levitated system and is characterized by a negative stiffness (see Figure 1.5b). The external force monotonically increases the deviation $|\theta|$ from the equilibrium state $\theta = \theta_{st2}$.

We dealt with the four kinds of equilibrium states. The dynamics in the neighborhood of every equilibrium state are governed by the second-order ordinary differential equations. The main difference among the cases is only in the sign of the coefficient placed on their right-hand side. In the case of a negative coefficient, the equivalent external force is a restoring force, whereas it is not in the case of a positive coefficient. Hence, the behaviors of the spring-mass system and the simple pendulum in the neighborhood of the downward vertical position are qualitatively the same as it is for the behaviors of the magnetically levitated system and the simple pendulum in the neighborhood of the upright position. The above feature is one of the many cases in which the mathematical form is directly related to the physical characteristics. It will be noticed throughout this book that there are such various analogies between the dynamics in mechanical systems and the mathematical forms of their governing equations.

However, we intuitively know the qualitative difference between the equilibrium states of the downward vertical position and the upright position in the simple pendulum. The difference cannot be distinguished only through the concept of equilibrium state. It will be physically characterized by introducing the concept of *stability* in Section 1.3.

1.2 Work and Potential Energy

We consider a particle of mass m moving from a position P_1 to another position P_2 under the action of an external force $\boldsymbol{F}$ as shown in Figure 1.7. Let the position vector of m be $\boldsymbol{r}$. The elementary work done by $\boldsymbol{F}$ in an infinitesimal segment $d\boldsymbol{r}$ of the path is

$$\overline{dW} = \boldsymbol{F} \cdot d\boldsymbol{r}, \tag{1.23}$$

where the dot stands for inner product. $\overline{dW}$ does not generally denote the perfect differential of W, but the infinitesimal work associated to the infinitesimal displacement $d\boldsymbol{r}$ (Lanczos 1986).

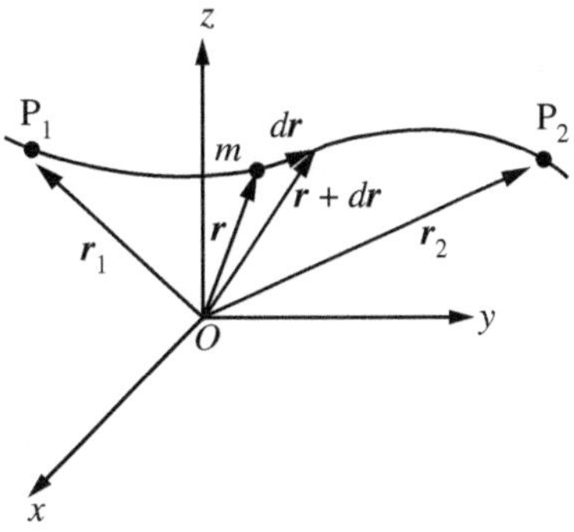

Figure 1.7 Path of particle m and an infinitesimal segment dr of the path.

Definition 1.2 If $\overline{dW}$ can be expressed in the form of the perfect differential of a scalar function V as

$$\overline{dW} = \boldsymbol{F} \cdot d\boldsymbol{r} = dV, \tag{1.24}$$

$\boldsymbol{F}$ is called a *conservative force* and $U \overset{\text{def}}{=} -V$ is called *potential energy*.

Then, $\boldsymbol{F}$ can be expressed as

$$\boldsymbol{F} = -\nabla U, \tag{1.25}$$

where ∇ is a differential operator called *nabla* and written in the terms of the Cartesian components, x, y, and z, in the form

$$\nabla = \frac{\partial}{\partial x}\boldsymbol{e}_x + \frac{\partial}{\partial y}\boldsymbol{e}_y + \frac{\partial}{\partial z}\boldsymbol{e}_z. \tag{1.26}$$

Under the conservative force, the motion of the particle of mass m is governed by the following equation of motion:

$$m\frac{d^2\boldsymbol{r}}{dt^2} = -\nabla U. \tag{1.27}$$

At an equilibrium state, the potential energy satisfies

$$\nabla U = \boldsymbol{0}. \tag{1.28}$$

Problem 1.2 Derive Eq. (1.25) from $\boldsymbol{F} \cdot d\boldsymbol{r} = d(-U)$ in Cartesian coordinate.

Ans: Letting

$$d\boldsymbol{r} = dx\boldsymbol{e}_x + dy\boldsymbol{e}_y + dz\boldsymbol{e}_z, \tag{1.29}$$

we have the following relationship:

$$\begin{aligned} dV = d(-U) &= -\left(\frac{\partial U}{\partial x}dx + \frac{\partial U}{\partial y}dy + \frac{\partial U}{\partial z}dz\right) \\ &= -\left(\frac{\partial U}{\partial x}\boldsymbol{e}_x + \frac{\partial U}{\partial y}\boldsymbol{e}_y + \frac{\partial U}{\partial z}\boldsymbol{e}_z\right) \cdot (dx\boldsymbol{e}_x + dy\boldsymbol{e}_y + dz\boldsymbol{e}_z) \\ &= -\nabla\, U \cdot d\boldsymbol{r}. \end{aligned} \tag{1.30}$$

Comparing with Eq. (1.24) yields Eq. (1.25).

The work done from position $P_1(\boldsymbol{r} = \boldsymbol{r}_1)$ to position $P_2(\boldsymbol{r} = \boldsymbol{r}_2)$ by $\boldsymbol{F}$ is expressed using Eq. (1.23) as

$$\int_{\boldsymbol{r}_1}^{\boldsymbol{r}_2} \boldsymbol{F} \cdot d\boldsymbol{r}. \tag{1.31}$$

In particular, if $\boldsymbol{F}$ is a conservative force, the relationship $\boldsymbol{F} \cdot d\boldsymbol{r} = d(-U)$ holds, and Eq. (1.31) can be written as

$$\int_{\boldsymbol{r}_1}^{\boldsymbol{r}_2} \boldsymbol{F} \cdot d\boldsymbol{r} = -\int_{\boldsymbol{r}_1}^{\boldsymbol{r}_2} dU = U(\boldsymbol{r}_1) - U(\boldsymbol{r}_2). \tag{1.32}$$

The work does not depend on the path between the points $P_1(\boldsymbol{r}_1)$ and $P_2(\boldsymbol{r}_2)$, but only on their positions. Then, the potential energy U is derived from $\boldsymbol{F}$ as

$$U(\boldsymbol{r}_2) - U(\boldsymbol{r}_1) = -\int_{\boldsymbol{r}_1}^{\boldsymbol{r}_2} \boldsymbol{F} \cdot d\boldsymbol{r}. \tag{1.33}$$

The kinetic energy is defined using the velocity $\boldsymbol{v} = d\boldsymbol{r}/dt$ of the mass m as

$$K = \frac{1}{2} m \left(\frac{d\boldsymbol{r}}{dt} \cdot \frac{d\boldsymbol{r}}{dt} \right). \tag{1.34}$$

Here, considering Newton's second law for the case with constant mass

$$m \frac{d^2\boldsymbol{r}}{dt^2} = \boldsymbol{F}, \tag{1.35}$$

the first derivative of the kinetic energy with respect to time t is

$$\frac{dK}{dt} = m \frac{d\boldsymbol{r}}{dt} \cdot \frac{d^2\boldsymbol{r}}{dt^2} = \frac{d\boldsymbol{r}}{dt} \cdot \boldsymbol{F}, \tag{1.36}$$

which is called *power*. We consider the case when $\boldsymbol{F}$ is conservative, i.e. $\boldsymbol{F} = -\nabla U$. Because of

$$\frac{d\boldsymbol{r}}{dt} \cdot \boldsymbol{F} = \left(\frac{dx}{dt}\boldsymbol{e}_x + \frac{dy}{dt}\boldsymbol{e}_y + \frac{dz}{dt}\boldsymbol{e}_z \right) \cdot \left(-\frac{\partial U}{\partial x}\boldsymbol{e}_x - \frac{\partial U}{\partial y}\boldsymbol{e}_y - \frac{\partial U}{\partial z}\boldsymbol{e}_z \right) = -\frac{dU}{dt}, \tag{1.37}$$

from Eq. (1.36), the *total energy* defined as the sum of kinetic energy and potential energy:

$$E = K + U \tag{1.38}$$

is constant, i.e.

$$\frac{dE}{dt} = \frac{d}{dt}(K + U) = 0. \tag{1.39}$$

As a result, if the external force is conservative, the total energy is conserved. Since the external forces in the three systems investigated in Section 1.1 can be expressed as Eq. (1.25), those systems are conservative. In the following sections, the potential energy is used for the determination of *stability*.

1.3 Stability of the Equilibrium State in Conservative Systems

We investigate the dynamics of conservative systems in the neighborhood of an equilibrium state. The variation of the total energy E can be expressed as

$$\Delta E = \Delta K + \Delta U = 0. \tag{1.40}$$

According to the definition of equilibrium state, $\boldsymbol{r}_{st}$ satisfying $\boldsymbol{F}(\boldsymbol{r}_{st}) = \boldsymbol{0}$, i.e. $\nabla U(\boldsymbol{r}_{st}) = \boldsymbol{0}$, is an *equilibrium state*. The potential energy, which is a function of the position vector $\boldsymbol{r}$, can be expanded in the neighborhood of the equilibrium state $\boldsymbol{r} = \boldsymbol{r}_{st}$

$$\begin{aligned}
U(\boldsymbol{r}) &= U(\boldsymbol{r}_{st}) + \left.\frac{\partial U}{\partial x}\right|_{\boldsymbol{r}=\boldsymbol{r}_{st}} \Delta x + \left.\frac{\partial U}{\partial y}\right|_{\boldsymbol{r}=\boldsymbol{r}_{st}} \Delta y + \left.\frac{\partial U}{\partial z}\right|_{\boldsymbol{r}=\boldsymbol{r}_{st}} \Delta z \\
&\quad + \frac{1}{2}\left.\frac{\partial^2 U}{\partial x^2}\right|_{\boldsymbol{r}=\boldsymbol{r}_{st}} \Delta x^2 + \frac{1}{2}\left.\frac{\partial^2 U}{\partial y^2}\right|_{\boldsymbol{r}=\boldsymbol{r}_{st}} \Delta y^2 + \frac{1}{2}\left.\frac{\partial^2 U}{\partial z^2}\right|_{\boldsymbol{r}=\boldsymbol{r}_{st}} \Delta z^2 \\
&\quad + \left.\frac{\partial^2 U}{\partial x \partial y}\right|_{\boldsymbol{r}=\boldsymbol{r}_{st}} \Delta x \Delta y + \left.\frac{\partial^2 U}{\partial y\, \partial z}\right|_{\boldsymbol{r}=\boldsymbol{r}_{st}} \Delta y \Delta z + \left.\frac{\partial^2 U}{\partial z \partial x}\right|_{\boldsymbol{r}=\boldsymbol{r}_{st}} \Delta z \Delta x + O(|\Delta \boldsymbol{r}|^3) \\
&= U(\boldsymbol{r}_{st}) + \left.\begin{bmatrix} \dfrac{\partial U}{\partial x} & \dfrac{\partial U}{\partial y} & \dfrac{\partial U}{\partial z} \end{bmatrix}\right|_{\boldsymbol{r}=\boldsymbol{r}_{st}} \begin{bmatrix} \Delta x \\ \Delta y \\ \Delta z \end{bmatrix} \\
&\quad + \frac{1}{2}[\Delta x \quad \Delta y \quad \Delta z] \left.\begin{bmatrix} \dfrac{\partial^2 U}{\partial x^2} & \dfrac{\partial^2 U}{\partial x\, \partial y} & \dfrac{\partial^2 U}{\partial x\, \partial z} \\ \dfrac{\partial^2 U}{\partial y\, \partial x} & \dfrac{\partial^2 U}{\partial y^2} & \dfrac{\partial^2 U}{\partial y\, \partial z} \\ \dfrac{\partial^2 U}{\partial z\, \partial x} & \dfrac{\partial^2 U}{\partial z\, \partial y} & \dfrac{\partial^2 U}{\partial z^2} \end{bmatrix}\right|_{\boldsymbol{r}=\boldsymbol{r}_{st}} \begin{bmatrix} \Delta x \\ \Delta y \\ \Delta z \end{bmatrix} + O(|\Delta \boldsymbol{r}|^3),
\end{aligned} \tag{1.41}$$

where $\Delta \boldsymbol{r} = \boldsymbol{r} - \boldsymbol{r}_{st} = \Delta x \boldsymbol{e}_x + \Delta y \boldsymbol{e}_y + \Delta z \boldsymbol{e}_z$. By using the *Jacobian matrix* and *Hessian matrix* of the scalar valued function U at $\boldsymbol{r}_{st}$, respectively, as

$$DU(\boldsymbol{r}_{st}) \overset{\text{def}}{=} [\nabla U(\boldsymbol{r})|_{\boldsymbol{r}=\boldsymbol{r}_{st}}]^T = \left.\begin{bmatrix} \dfrac{\partial U}{\partial x} & \dfrac{\partial U}{\partial y} & \dfrac{\partial U}{\partial z} \end{bmatrix}\right|_{\boldsymbol{r}=\boldsymbol{r}_{st}}, \tag{1.42}$$

$$HU(\boldsymbol{r}_{st}) \overset{\text{def}}{=} \left.\begin{bmatrix} \dfrac{\partial^2 U}{\partial x^2} & \dfrac{\partial^2 U}{\partial x\, \partial y} & \dfrac{\partial^2 U}{\partial x\, \partial z} \\ \dfrac{\partial^2 U}{\partial y\, \partial x} & \dfrac{\partial^2 U}{\partial y^2} & \dfrac{\partial^2 U}{\partial y\, \partial z} \\ \dfrac{\partial^2 U}{\partial z\, \partial x} & \dfrac{\partial^2 U}{\partial z\, \partial y} & \dfrac{\partial^2 U}{\partial z^2} \end{bmatrix}\right|_{\boldsymbol{r}=\boldsymbol{r}_{st}}, \tag{1.43}$$

Eq. (1.41) is described as

$$U(\boldsymbol{r}) = U(\boldsymbol{r}_{st}) + DU(\boldsymbol{r}_{st})\Delta \boldsymbol{r} + \frac{1}{2}\Delta \boldsymbol{r}^T HU(\boldsymbol{r}_{st})\Delta \boldsymbol{r} + O(|\Delta \boldsymbol{r}|^3). \tag{1.44}$$

Since $DU(\boldsymbol{r}_{st}) = 0$, the variation of the potential energy $\Delta U = U(\boldsymbol{r}) - U(\boldsymbol{r}_{st})$ can be expressed as

$$\Delta U = \frac{1}{2}[\Delta x \quad \Delta y \quad \Delta z] HU(\boldsymbol{r}_{st}) \begin{bmatrix} \Delta x \\ \Delta y \\ \Delta z \end{bmatrix}. \tag{1.45}$$

In addition, from Eq. (1.40), we have

$$\Delta K = -\Delta U = -\frac{1}{2}[\Delta x \quad \Delta y \quad \Delta z] HU(\boldsymbol{r}_{st}) \begin{bmatrix} \Delta x \\ \Delta y \\ \Delta z \end{bmatrix}. \tag{1.46}$$

Therefore, if the Hessian matrix is positive-definite (e.g. Strang et al. 1993) at the equilibrium state, i.e. $\Delta \boldsymbol{r}^T HU(\boldsymbol{r}_{st})\Delta \boldsymbol{r} > 0$ when $\Delta \boldsymbol{r} \neq 0$, the increase of U by the displacement, i.e. by the deviation from the equilibrium state $\Delta \boldsymbol{r}$, decreases the kinetic energy K. In this state, one says that the equilibrium state $\boldsymbol{r}_{st}$ is stable. In the positive-semidefinite case, i.e. $\Delta \boldsymbol{r}^T HU(\boldsymbol{r}_{st})\Delta \boldsymbol{r} \geq 0$ when $\Delta \boldsymbol{r} \neq 0$, the terms in the order of $O(|\Delta \boldsymbol{r}^3|)$ or much higher order are required to determine the stability. If the Hessian matrix is not positive-definite or positive-semidefinite, one says that the equilibrium state $\boldsymbol{r}_{st}$ is unstable. We will return to the discussion of stability in Section 2.2.

1.4 Stability of Mechanical Systems

We recall the equilibrium states of the three systems investigated in Section 1.1 and discuss their stability using the method introduced in Section 1.3.

1.4.1 Stability of Spring-Mass System

Let us calculate the potential energy of the spring mass system using Eq. (1.6). Knowing the external force $F = -F_s^{\#} = -kx$, the potential energy is

$$U = -\int F\, dx = -\int (-kx)dx = \frac{1}{2}kx^2, \tag{1.47}$$

where the integral constant is set to zero. The Hessian matrix of U at the equilibrium state $x = 0$ is k and is positive definite. The potential energy curve is schematically depicted as the well in Figure 1.8a (see Problem 1.4.1) and has concave feature, which is due to the positive stiffness ($k > 0$), i.e. the restoring characteristic of the spring.

Problem 1.4.1 The potential energy of the spring-mass system under the gravity effect in Figure 1.1 is directly calculated by the definition as

$$U = -\int (-kx' + mg)dx' = \frac{1}{2}kx'^2 - mgx', \tag{1.48}$$

where the integral constant is set to zero. By the coordinate transformation from x' to x, derive Eq. (1.47).

Ans:

$$\begin{aligned} U &= \frac{1}{2}kx'^2 - mgx' \\ &= \frac{1}{2}k(x_{st} + x)^2 - mg(x_{st} + x) \\ &= \frac{1}{2}kx^2 - \frac{1}{2}mgx_{st}. \end{aligned} \tag{1.49}$$

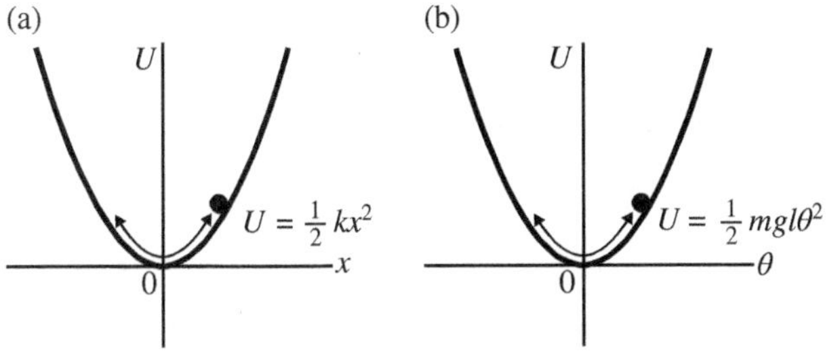

Figure 1.8 Potential energy curves. (a) Potential energy curve of the spring-mass system. (b) Potential energy curve of the simple pendulum in the neighborhood of the downward vertical equilibrium position.

Neglecting the constant second term yields Eq. (1.47).

By considering $\Delta x = x$, $\Delta y = 0$, and $\Delta z = 0$, Eq. (1.46) leads to

$$\Delta K = -\frac{1}{2}kx^2. \tag{1.50}$$

Because of positive k, as the deviation $|x(t)|$ from the equilibrium state $x = 0$ increases, the kinetic energy K decreases. The mass oscillates around the equilibrium state $x = 0$ as the motion of an imaginary particle on the potential energy curve in Figure 1.8a.

1.4.2 Stability of Magnetically Levitated System

Next, we examine Eq. (1.16) related to the motion of the magnetically levitated system. Also in this case, the potential energy U can be derived by the external force $F = -F_m^\# = k_m x$ as

$$U = -\int F\,dx = -\int k_m x\,dx = -\frac{1}{2}k_m x^2, \tag{1.51}$$

where the integral constant is set zero. The total energy

$$E = \frac{1}{2}m\left(\frac{dx}{dt}\right)^2 - \frac{1}{2}k_m x^2 \tag{1.52}$$

is conserved. Unlike the spring-mass system, because of the negative stiffness of magnetic suspension, i.e. $-k_m < 0$ corresponding to the negative slope in Figure 1.5b, the Hessian matrix of U at the equilibrium state $x = 0$ is $-k_m$ and negative definite.

As the deviation $|x(t)| \neq 0$ from the equilibrium state $x = 0$ increases, the kinetic energy increases. The absolute value of velocity increases monotonically as the motion of an imaginary particle on the potential energy curve in Figure 1.9a. Regardless of the magnitude of the initial deviation, the deviation $|\Delta x(t)|$ from the equilibrium state is not bounded. In other words, in the case when the position is changed downward from the equilibrium state due to a disturbance, the system continues going down and finally the levitated system falls to the guide rail. In the case when the position is changed upward from the equilibrium state due to a disturbance, the system continues going up and finally the electromagnet touches to the guide rail. Therefore, the system does not carry out the *stable levitation*. In practical systems, feedback control is equipped to stabilize the unstable equilibrium state as will be mentioned in Section 1.4.4.

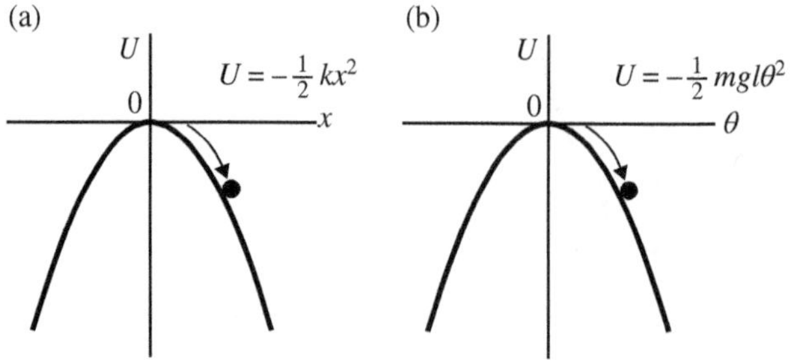

Figure 1.9 Potential energy curves. (a) Potential energy curve of the magnetically levitated system. (b) Potential energy curve of the simple pendulum in the neighborhood of the upright vertical equilibrium position.

1.4.3 Pendulum

The pendulum has two equilibrium states, i.e. the downward vertical and upright positions, $\theta_{st1} = 2n\pi$ and $\theta_{st2} = (2n+1)\pi$, respectively. The equations governing the motion near these equilibrium states are Eqs. (1.21) and (1.22), respectively. Considering the sign of the coefficient on their right-hand side, these equations are mathematically equivalent to Eqs. (1.6) and (1.16), respectively. The analogy with the governing equations of the spring-mass system and the magnetically levitated system leads to the determination of stability of the equilibrium states θ_{st1} and θ_{st2}. In fact, the potential energy curves are calculated in the neighborhood of $\theta_{st1} = 0$ and $\theta_{st2} = \pi$, respectively, as

$$U = -\int (-mg\theta)l\,d\theta = \frac{1}{2}mgl\theta^2, \tag{1.53}$$

$$U = -\int (mg\theta)l\,d\theta = -\frac{1}{2}mgl\theta^2, \tag{1.54}$$

where the integral constants are set to zero. The potential energy curves are shown as Figures 1.8b and 1.9b, respectively.

1.4.4 Stabilization Control of Magnetically Levitated System

As elucidated in Section 1.4.2, the equilibrium state of the magnetically levitated system is unstable. Let us propose a method to change the unstable equilibrium state into a stable one. For stabilization, the negative stiffness in the electromagnetic suspension needs to be turned into a positive one. To this end, we modify the current in the electromagnet depending on x as

$$i = i_0 + cx, \tag{1.55}$$

where c is a position feedback gain. Equation (1.12) is rewritten as

$$m\frac{d^2x}{dt^2} = -k\left(\frac{i_0 + cx}{x_{st} + x}\right)^2 + mg. \tag{1.56}$$

Under the condition of Eq. (1.55), the equilibrium state is $x' = x_{st} = \sqrt{\frac{ki_0^2}{mg}}$, i.e. $x = x' - x_{st} = 0$ is still the equilibrium state and Eq. (1.9) is still the equilibrium equation. Equation

(1.56) is linearized in the neighborhood of the equilibrium state $x = 0$ as

$$m\frac{d^2x}{dt^2} = -k_m\left(\frac{x_{st}c}{i_0} - 1\right)x + O(2), \tag{1.57}$$

where $k_m = 2k\frac{i_0^2}{x_{st}^3} > 0$, and $O(2)$ denotes higher order terms than the second order of $\frac{cx}{i_0}$ and $\frac{x}{x_{st}}$.

Problem 1.4.4 Derive Eq. (1.57).

Ans: Assuming $\left|\frac{x}{x_{st}}\right| \ll 1$ and demanding small cost of the feedback, i.e. $\left|\frac{cx}{i_0}\right| \ll 1$, Eq. (1.56) can be expanded into the power series with respect to $\frac{x}{x_{st}}$ and $\frac{cx}{i_0}$ as

$$\begin{aligned} m\frac{d^2x}{dt^2} &= -k\left(\frac{i_0 + cx}{x_{st} + x}\right)^2 + mg \\ &= -k\frac{i_0^2}{x_{st}^2}\left(\frac{1 + (cx/i_0)}{1 + (x/x_{st})}\right)^2 + mg \\ &= -k\frac{i_0^2}{x_{st}^2}\left(1 + \frac{cx}{i_0}\right)^2\left(1 + \frac{x}{x_{st}}\right)^{-2} + mg \\ &= -k\frac{i_0^2}{x_{st}^2}\left(1 + 2 \times \frac{cx}{i_0}\right)\left(1 - 2 \times \frac{x}{x_{st}}\right) + mg + O(2) \\ &= -k_m\left(\frac{cx_{st}}{i_0} - 1\right)x + O(2). \end{aligned} \tag{1.58}$$

The sign of the right-hand side determines whether the stiffness is negative or positive. If $k_m\left(\frac{cx_{st}}{i_0} - 1\right) > 0$, the stiffness is positive and the magnetic force expressed by the right-hand side in Eq. (1.57) acts as an ideal spring. Then, the equilibrium position $x = 0$ becomes stable. Therefore, by setting the feedback gain as $c > i_0/x_{st}$, we can stabilize the equilibrium state of the magnetically levitated vehicle.

References

Dorf, R. C. and R. H. Bishop (2011). *Modern control systems*. Pearson.

Lanczos, C. (1986). *The Variational Principles of Mechanics*. Dover.

Nayfeh, A. H. (1981). *Introduction to Perturbation Technique*. Wiley.

Strang, G., Strang G., Strang G., and Strang G. (1993). *Introduction to linear algebra*, Vol. 3. Wellesley-Cambridge Press Wellesley, MA.

Witelski, T. and M. Bowen (2015). *Methods of mathematical modelling: Continuous systems and differential equations*. Springer.

Yabuno, H. (2004). *Kougaku no tameno Hisenkei-kaiseki Nyuumon, The Elements of Nonlinear Analysis for Engineering*. Saiensu-sha.

2

Linear Dynamical Systems

In this chapter, we investigate analytical approaches for linear systems. The stability of equilibrium states physically discussed using the potential energy concept in Chapter 1 is defined consistently from a mathematical point of view. Stable and unstable states are better classified using a phase space from a geometrical point of view.

2.1 Vector Field and Phase Space

Equations of motion derived from Newton's second law are coupled equations including the first or second derivatives with respect to time t. They can be often transformed into the following coupled first order differential equations with respect to state vector $\boldsymbol{x} \in \mathbb{R}^n$ as

$$\frac{d\boldsymbol{x}}{dt} = \boldsymbol{f}(\boldsymbol{x}, t), \tag{2.1}$$

where $\boldsymbol{f}$ is a vector valued function of $\mathbb{R}^n$, or

$$\begin{cases} \dfrac{dx_1}{dt} = f_1(x_1, \ldots, x_n, t), \\ \quad\vdots \\ \dfrac{dx_n}{dt} = f_n(x_1, \ldots, x_n, t), \end{cases} \tag{2.2}$$

where $\boldsymbol{x} = [x_1, \ldots, x_n]^T$ and $\boldsymbol{f}(\boldsymbol{x}, t) = [f_1(x_1, \ldots, x_n, t), \ldots, f_n(x_1, \ldots, x_n, t)]^T$.

Example 2.1

1. Transform the equation of motion of the spring-mass system given by Eq. (1.6) into the coupled first order differential equations. By assuming $x_1 = x$ and $x_2 = dx/dt$, since

$$\boldsymbol{f}(\boldsymbol{x}, t) = \begin{bmatrix} 0 & 1 \\ -\dfrac{k}{m} & 0 \end{bmatrix} \begin{bmatrix} x_1 \\ x_2 \end{bmatrix}, \tag{2.3}$$

Eq. (1.6) is rewritten as

$$\frac{d}{dt}\begin{bmatrix} x_1 \\ x_2 \end{bmatrix} = \begin{bmatrix} 0 & 1 \\ -\dfrac{k}{m} & 0 \end{bmatrix} \begin{bmatrix} x_1 \\ x_2 \end{bmatrix}. \tag{2.4}$$

Linear and Nonlinear Instabilities in Mechanical Systems: Analysis, Control and Application, First Edition. Hiroshi Yabuno.

Companion website: www.wiley.com/go/yabuno/instabilitiesinmechanicalsystems

2. Transform the equation of motion of the pendulum given by Eq. (1.17) into the coupled first order differential equations. By assuming $x_1 = \theta'$ and $x_2 = d\theta'/dt$, i.e. $\boldsymbol{x} = [x_1\ x_2]^T = [\theta'\ \ d\theta'/dt]^T$, since

$$\boldsymbol{f}(\boldsymbol{x}, t) = \begin{bmatrix} x_2 \\ -\frac{g}{l}\sin x_1 \end{bmatrix}, \tag{2.5}$$

Eq. (1.17) is rewritten in the above form as

$$\frac{d}{dt}\begin{bmatrix} x_1 \\ x_2 \end{bmatrix} = \begin{bmatrix} x_2 \\ -\frac{g}{l}\sin x_1 \end{bmatrix}. \tag{2.6}$$

In the case in which $\boldsymbol{f}(\boldsymbol{x}, t)$ features the following special form as:

$$\boldsymbol{f}(\boldsymbol{x}, t) = A(t)\boldsymbol{x} + \boldsymbol{b}(t), \tag{2.7}$$

Eq. (2.1) is said to be a *linear differential equation*, and the matrix $A(t)$ is called a *linear operator*. Otherwise, it is defined as a *nonlinear differential equation*. The systems whose dynamics are governed by linear and nonlinear differential equations are defined as *linear systems* and *nonlinear systems*, respectively. Furthermore, if the equation has the *trivial solution*, i.e. $\boldsymbol{x} = \boldsymbol{0}$, it is called a homogeneous equation. Otherwise, it is called an *inhomogeneous equation*. If the equation does not explicitly include the time, it is called an *autonomous equation*. Otherwise, it is called a *nonautonomous equation* (see Problem 2.1).

Problem 2.1

1. How are the following ordinary differential equations classified?

(a) $$m\frac{d^2x}{dt^2} + kx = 0. \tag{2.8}$$

(b) $$m\frac{d^2x}{dt^2} + c\frac{dx}{dt} + kx = 0. \tag{2.9}$$

(c) $$\frac{d^2\theta}{dt^2} + \frac{g}{l}\sin\theta = 0. \tag{2.10}$$

(d) $$m\frac{d^2x}{dt^2} + k(1 + a\cos\nu t)x = 0. \tag{2.11}$$

(e) $$m\frac{d^2x}{dt^2} + kx = f\cos\nu t. \tag{2.12}$$

Ans:

(a) Equation (2.8): linear, homogeneous, and autonomous.
(b) Equation (2.9): linear, homogeneous, and autonomous.
(c) Equation (2.10): nonlinear, homogeneous, and autonomous.
(d) Equation (2.11): linear, homogeneous, and nonautonomous.
(e) Equation (2.12): linear, inhomogeneous, nonautonomous.

2. Compare the dynamics governed by the autonomous Eq. (2.8) and the nonautonomous Eq. (2.12) for the cases in which the same initial conditions, $x = x_0(\neq 0)$ and $dx/dt = 0$, are given at two different times: (i) $t = 0$ and (ii) $t = t_0(\neq 0)$.

Ans: Introducing a new time coordinate t' as

$$t' = t - t_0, \tag{2.13}$$

the initial conditions for case (ii) can be rewritten as $x = x_0(\neq 0)$ and $dx/dt = 0$ at $t' = 0$. The first and second derivatives with respect to the original time t are rewritten using the new time t' as

$$\frac{d}{dt} = \frac{d}{dt'}\frac{dt'}{dt} = \frac{d}{dt'},$$

$$\frac{d^2}{dt^2} = \frac{d}{dt}\left(\frac{d}{dt'}\right) = \frac{d}{dt'}\left(\frac{d}{dt'}\right)\frac{dt'}{dt} = \frac{d^2}{dt'^2},$$

and Eqs. (2.8) and (2.12) are rewritten as

$$m\frac{d^2x}{dt'^2} + kx = 0, \tag{2.14}$$

$$m\frac{d^2x}{dt'^2} + kx = f\cos\nu(t' + t_0). \tag{2.15}$$

Equations (2.8) and (2.14) feature the same form, while Eqs. (2.12) and (2.15) are not the same due to the difference of the term on the right-hand side. Such difference is a direct consequence of the explicit time dependency in Eq. (2.12), i.e. the nonautonomous feature. Therefore, regardless of the initial time, the system governed by the autonomous differential equation exhibits the same behavior under the same initial conditions. On the other hand, in the system governed by the nonautonomous differential equation, the difference in the initial time causes different behaviors, even if the initial conditions are the same.

Let us consider the autonomous system as

$$\frac{d\boldsymbol{x}}{dt} = \boldsymbol{f}(\boldsymbol{x}) \tag{2.16}$$

or

$$\begin{cases} \dfrac{dx_1}{dt} = f_1(x_1, \dots, x_n), \\ \quad\vdots \\ \dfrac{dx_n}{dt} = f_n(x_1, \dots, x_n), \end{cases} \tag{2.17}$$

where $\boldsymbol{x} = [x_1, \dots, x_n]^T$. The solution at a time t can be regarded as a point in the n-dimensional space with coordinates $(x_1, \dots, x_n)$, which is called a *phase space*. At each point, the velocity is expressed by the right-hand side of Eq. (2.16) or (2.17). The point moves along a curve in the space with time evolution. The curve is called a *trajectory* or an *integral curve*. A set of representative trajectories on a phase space described for different initial conditions is called a *phase portrait*. The right-hand side of Eq. (2.16), $\boldsymbol{f}(\boldsymbol{x})$, represents the tangent vector on the trajectory at each point in the phase space; the function $\boldsymbol{f}$ assigns the tangent vector for each point in the phase space and is called a ***vector field***.

In the case of $\boldsymbol{f}(\boldsymbol{x}) = \boldsymbol{0}$, i.e. $d\boldsymbol{x}/dt = \boldsymbol{0}$, the point $\boldsymbol{x}$ does not move. Such a state $\boldsymbol{x} \overset{\text{def}}{=} \boldsymbol{x}_{st}$ is an *equilibrium state* and satisfies the following equilibrium equation:

$$\boldsymbol{0} = \boldsymbol{f}(\boldsymbol{x}_{st}) \tag{2.18}$$

or

$$\begin{cases} 0 = f_1(x_{1st}, \dots, x_{nst}), \\ \quad\vdots \\ 0 = f_n(x_{1st}, \dots, x_{nst}), \end{cases} \tag{2.19}$$

where

$$\boldsymbol{x}_{st} = \begin{bmatrix} x_{1st} \\ \vdots \\ x_{nst} \end{bmatrix}. \tag{2.20}$$

2.2 Stability of Equilibrium States

In Chapter 1, the definition of stability is provided from a physical point of view based on potential energy. Let us consider the stability of the equilibrium points $\boldsymbol{x}_{st}$ from a mathematical point of view (e.g. Wiggins 1980).

Definition 2.1 The equilibrium state $\boldsymbol{x}_{st}$ is said to be *stable (stable in the sense of Liapunov)* if for any arbitrary $\epsilon > 0$, there exists a constant $\delta > 0$ such that the satisfaction of the inequality

$$|\boldsymbol{x}(0) - \boldsymbol{x}_{st}| < \delta \tag{2.21}$$

implies the satisfaction of the inequality

$$|\boldsymbol{x}(t) - \boldsymbol{x}_{st}| < \epsilon, \quad t \geq 0, \tag{2.22}$$

where $|\cdot|$ denotes the Euclidean norm. Otherwise, the equilibrium state which is not stable is called *unstable*. Furthermore, the stable equilibrium state satisfying the following condition is called *asymptotically stable*.

Definition 2.2 The equilibrium state $\boldsymbol{x}_{st}$ is said to be *asymptotically stable* if it is stable and there exists a constant $\delta > 0$ such that the satisfaction of the inequality

$$|\boldsymbol{x}(0) - \boldsymbol{x}_{st}| < \delta \tag{2.23}$$

implies the satisfaction of the equation

$$\lim_{t\to\infty} |\boldsymbol{x}(t) - \boldsymbol{x}_{st}| = 0. \tag{2.24}$$

2.3 Linearization and Local Stability

We have already dealt with the linearization of the equations of motion about a magnetically levitated system and a pendulum near their equilibrium states. In this section, we consider the linearization of the general first order differential equations as Eq. (2.16) or (2.17). We introduce the vector $\Delta\boldsymbol{x}(t)$ expressing the deviation from the equilibrium state $\boldsymbol{x}_{st}$ to the state position $\boldsymbol{x}(t)$ at a time t as

$$\boldsymbol{x}(t) = \boldsymbol{x}_{st} + \Delta\boldsymbol{x}(t) \tag{2.25}$$

or

$$\begin{bmatrix} x_1 \\ \vdots \\ x_n \end{bmatrix} = \begin{bmatrix} x_{st1} + \Delta x_1 \\ \vdots \\ x_{stn} + \Delta x_n \end{bmatrix}, \tag{2.26}$$

where

$$\Delta\boldsymbol{x} = \begin{bmatrix} \Delta x_1 \\ \vdots \\ \Delta x_n \end{bmatrix}. \tag{2.27}$$

Substituting Eq. (2.25) into Eq. (2.16) and expressing the right-hand side of the equation with Taylor series near the equilibrium state, i.e. with respect to $\Delta\boldsymbol{x} = [\Delta x_1, \ldots, \Delta x_n]^T$, yields

$$\frac{d\Delta\boldsymbol{x}}{dt} = D\boldsymbol{f}|_{\boldsymbol{x}=\boldsymbol{x}_{st}} \Delta\boldsymbol{x} + O(|\Delta\boldsymbol{x}|^2), \tag{2.28}$$

where the equilibrium equation (2.18), i.e. $\boldsymbol{f}(\boldsymbol{x}_{st}) = \boldsymbol{0}$, is taken into account and $D\boldsymbol{f}|_{\boldsymbol{x}=\boldsymbol{x}_{st}}$ is the Jacobian matrix of $\boldsymbol{f}$ at $\boldsymbol{x}_{st} = \boldsymbol{0}$.

Problem 2.3 Let us consider the case of two variables, i.e. $\boldsymbol{x} = [x_1, x_2]^T$ and $\boldsymbol{f}(\boldsymbol{x}, t) = [f_1(x_1, x_2), f_2(x_1, x_2)]^T$. The equilibrium state and the deviation from that can be expressed as $\boldsymbol{x}_{st} = [x_{1st}, x_{2st}]^T$ and $\Delta\boldsymbol{x} = [\Delta x_1, \Delta x_2]^T = [x_1 - x_{1st}, x_2 - x_{2st}]^T$, respectively. Express the equation corresponding to Eq. (2.28).

Ans:

$$\begin{aligned} \frac{d}{dt}\begin{bmatrix} \Delta x_1 \\ \Delta x_2 \end{bmatrix} &= \begin{bmatrix} f_1(x_{1st}, x_{2st}) \\ f_2(x_{1st}, x_{2st}) \end{bmatrix} + \begin{bmatrix} \dfrac{\partial f_1}{\partial x_1} & \dfrac{\partial f_1}{\partial x_2} \\ \dfrac{\partial f_2}{\partial x_1} & \dfrac{\partial f_2}{\partial x_2} \end{bmatrix}\Bigg|_{\substack{x_1 = x_{1st} \\ x_2 = x_{2st}}} \begin{bmatrix} \Delta x_1 \\ \Delta x_2 \end{bmatrix} + O(|\Delta\boldsymbol{x}|^2) \\ &= \begin{bmatrix} \dfrac{\partial f_1}{\partial x_1} & \dfrac{\partial f_1}{\partial x_2} \\ \dfrac{\partial f_2}{\partial x_1} & \dfrac{\partial f_2}{\partial x_2} \end{bmatrix}\Bigg|_{\substack{x_1 = x_{1st} \\ x_2 = x_{2st}}} \begin{bmatrix} \Delta x_1 \\ \Delta x_2 \end{bmatrix} + O(|\Delta\boldsymbol{x}|^2). \end{aligned} \tag{2.29}$$

Notice that in the linearization, by using the Taylor expansion, it is important that $|\Delta\boldsymbol{x}|$ must be sufficiently less than 1. Otherwise the higher order term $O(|\Delta\boldsymbol{x}|^2)$ cannot be neglected and the linearized equation (2.28) is nonsense. In fact, Eq. (1.13) was expressed by the power series with the small dimensionless variable x/x_{st}. By focusing on the dynamics near the equilibrium state x_{st}, the nondimensional variable $|x/x_{st}|$ could be assumed sufficiently less than 1 (e.g. Witelski and Bowen 2015).

Example 2.2 We linearize Eq. (2.6) near the equilibrium points, i.e. the downward vertical position and the upright one. The equilibrium equation corresponding to Eq. (2.18) is expressed as

$$0 = \begin{bmatrix} x_{2st} \\ -\frac{g}{l}\sin x_{1st} \end{bmatrix}. \tag{2.30}$$

The equilibrium states for the downward vertical position x_{st-1} and the upright one x_{st-2} are

$$\boldsymbol{x}_{st-1} \overset{\text{def}}{=} \begin{bmatrix} 2n\pi \\ 0 \end{bmatrix} \tag{2.31}$$

and

$$\boldsymbol{x}_{st-2} \overset{\text{def}}{=} \begin{bmatrix} (2n+1)\pi \\ 0 \end{bmatrix}, \tag{2.32}$$

respectively, where n is an integer. Hence, because the Jacobian matrix at the equilibrium points is

$$Df(\boldsymbol{x}_{st}) = \left.\begin{bmatrix} \frac{\partial f_1}{\partial x_1} & \frac{\partial f_1}{\partial x_2} \\ \frac{\partial f_2}{\partial x_1} & \frac{\partial f_2}{\partial x_2} \end{bmatrix}\right|_{x_1=x_{1st},x_2=x_{2st}} = \begin{bmatrix} 0 & 1 \\ -\frac{g}{l}\cos x_{1st} & 0 \end{bmatrix}, \tag{2.33}$$

the Jacobian matrices for each equilibrium state are expressed as:

$$Df|_{x=x_{st-1}} = \begin{bmatrix} 0 & 1 \\ -\frac{g}{l} & 0 \end{bmatrix} \tag{2.34}$$

and

$$Df|_{x=x_{st-2}} = \begin{bmatrix} 0 & 1 \\ \frac{g}{l} & 0 \end{bmatrix}. \tag{2.35}$$

Finally, we obtain the linearized systems near the downward vertical position $\boldsymbol{x}_{st-1}$ and the upright one $\boldsymbol{x}_{st-2}$, respectively, as

$$\frac{d\Delta\boldsymbol{x}}{dt} = \begin{bmatrix} 0 & 1 \\ -\frac{g}{l} & 0 \end{bmatrix} \Delta\boldsymbol{x} \tag{2.36}$$

and

$$\frac{d\Delta\boldsymbol{x}}{dt} = \begin{bmatrix} 0 & 1 \\ \frac{g}{l} & 0 \end{bmatrix} \Delta\boldsymbol{x}. \tag{2.37}$$

Applying the method presented in Section 2.4, the solutions of Eqs. (2.36) and (2.37) are obtained and the local stability of the equilibrium points can be determined.

2.4 Eigenvalues of Linear Operators and Phase Portraits in a Single-Degree-of-Freedom System

In this section, we analyze single-degree-of-freedom systems, i.e. two-dimensional systems in the generalized form as follows:

$$\frac{d\boldsymbol{x}}{dt} = A\boldsymbol{x}, \quad \boldsymbol{x} \in \mathbb{R}^2, \tag{2.38}$$

where $\boldsymbol{x} = [x_1, x_2]^T$ and A is 2×2 matrix. This analysis is very important to qualitatively characterize the local stability of the equilibrium states in two-dimensional nonlinear dynamical systems. Furthermore, the method can be straightforward expanded and adopted for higher-dimensional dynamical systems.

2.4.1 Description of the Solution by Matrix Exponential Function

The solution of Eq. (2.38) for any linear operator A is expressed as (Hirsh and Samle 1974)

$$\boldsymbol{x} = \exp\ tA\ \boldsymbol{x}(0), \tag{2.39}$$

where $\boldsymbol{x}(0)$ is the initial condition of $\boldsymbol{x}$ and the *matrix exponential function* $\exp\ tA$ is defined by the following convergent series:

$$\exp\ tA \overset{\text{def}}{=} I + tA + \frac{t^2}{2!}A^2 + \frac{t^3}{3!}A^3 + \cdots + \frac{t^n}{n!}A^n + \cdots . \tag{2.40}$$

The calculation method for the matrix exponential function depends on the eigenvalue of A.

2.4.2 Case with Distinct Eigenvalues

We consider the case when A has two real distinct eigenvalues, λ_1 and λ_2 ($\lambda_1 > \lambda_2$). First, we try to obtain the solution of Eq. (2.38) without the use of Eq. (2.39). Letting the eigenvectors corresponding to λ_1 and λ_2 be $\boldsymbol{p}_1$ and $\boldsymbol{p}_2$, respectively, we have the following relationship:

$$\begin{cases} A\boldsymbol{p}_1 = \lambda_1 \boldsymbol{p}_1 = [\boldsymbol{p}_1, \boldsymbol{p}_2] \begin{bmatrix} \lambda_1 \\ 0 \end{bmatrix}, & (2.41) \\ A\boldsymbol{p}_2 = \lambda_2 \boldsymbol{p}_2 = [\boldsymbol{p}_1, \boldsymbol{p}_2] \begin{bmatrix} 0 \\ \lambda_2 \end{bmatrix}. & (2.42) \end{cases}$$

These equations are combined as

$$A[\boldsymbol{p}_1, \boldsymbol{p}_2] = [\boldsymbol{p}_1, \boldsymbol{p}_2] \begin{bmatrix} \lambda_1 & 0 \\ 0 & \lambda_2 \end{bmatrix}. \tag{2.43}$$

Therefore, introducing the transformation matrix

$$P = [\boldsymbol{p}_1, \boldsymbol{p}_2], \tag{2.44}$$

the matrix A can be diagonalized as

$$P^{-1}AP = \begin{bmatrix} \lambda_1 & 0 \\ 0 & \lambda_2 \end{bmatrix} \overset{\text{def}}{=} B. \tag{2.45}$$

Under the coordinate transformation:

$$x = Py, \tag{2.46}$$

where $\boldsymbol{y} = [y_1, \ y_2]^T$, Eq. (2.38) can be transformed into

$$\frac{d\boldsymbol{y}}{dt} = B\boldsymbol{y}. \tag{2.47}$$

The equations are decoupled as

$$\begin{cases} \dfrac{dy_1}{dt} = \lambda_1 y_1, & (2.48) \\ \dfrac{dy_2}{dt} = \lambda_2 y_2. & (2.49) \end{cases}$$

This enables us to solve Eq. (2.38) with the same ease with which we solve independently two one dimensional systems. We easily obtain the solutions as follows:

$$\begin{cases} y_1 = y_1(0)e^{\lambda_1 t}, & (2.50) \\ y_2 = y_2(0)e^{\lambda_2 t}, & (2.51) \end{cases}$$

where $y_1(0)$ and $y_2(0)$ are the initial values for y_1 and y_2 at $t = 0$. Of course, because B is diagonal, these solutions are also directly derived from the definition of the matrix exponential function of Eq. (2.40) as follows:

$$\begin{aligned} \begin{bmatrix} y_1 \\ y_2 \end{bmatrix} &= \exp\ tB\, \boldsymbol{y}_0 \\ &= \begin{bmatrix} 1 + \lambda_1 t + \frac{1}{2!}(\lambda_1 t)^2 + \cdots & 0 \\ 0 & 1 + \lambda_2 t + \frac{1}{2!}(\lambda_2 t)^2 + \cdots \end{bmatrix} \begin{bmatrix} y_1(0) \\ y_2(0) \end{bmatrix} \\ &= \begin{bmatrix} e^{\lambda_1 t} & 0 \\ 0 & e^{\lambda_2 t} \end{bmatrix} \begin{bmatrix} y_1(0) \\ y_2(0) \end{bmatrix}, \end{aligned} \tag{2.52}$$

where

$$\boldsymbol{y}_0 = P^{-1}\boldsymbol{x}_0 \tag{2.53}$$

or

$$\begin{bmatrix} y_1(0) \\ y_2(0) \end{bmatrix} = P^{-1} \begin{bmatrix} x_1(0) \\ x_2(0) \end{bmatrix}. \tag{2.54}$$

From Eq. (2.46),

$$\boldsymbol{x} = P\boldsymbol{y} = P \begin{bmatrix} y_1 \\ y_2 \end{bmatrix} = y_1(0)e^{\lambda_1 t}\boldsymbol{p}_1 + y_2(0)e^{\lambda_2 t}\boldsymbol{p}_2. \tag{2.55}$$

If the initial condition is given at $\boldsymbol{x}(0) = c\boldsymbol{p}_1$, i.e. the initial condition is set on y_1 axis parallel to the eigenvector $\boldsymbol{p}_1$, we have $y_1(0) = c$ and $y_2(0) = 0$. The flow approaches (or leaves) the origin remaining on y_1 axis if $\lambda_1 < 0$ (or $\lambda_1 > 0$). Generally, the space spanned by the eigenvectors corresponding to negative (or positive) eigenvalues is referred to as a *stable* (or *unstable*) *subspace*, and the space spanned by the eigenvectors corresponding to zero eigenvalues is referred to as a *center subspace* (see also Section 2.5).

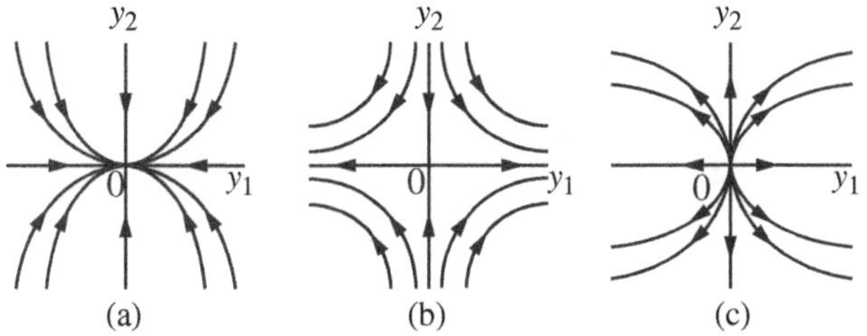

Figure 2.1 Phase portraits (distinct nonzero eigenvalues: (a) $\lambda_2 < \lambda_1 < 0$, (b) $\lambda_2 < 0 < \lambda_1$, (c) $0 < \lambda_2 < \lambda_1$).

We examine the characteristics of the trajectories. First, let us consider the case of $\lambda_1\lambda_2 \neq 0$ and $y_1y_2 \neq 0$. Eliminating dt by Eqs. (2.48) and (2.49) yields

$$\lambda_1 \frac{dy_2}{y_2} = \lambda_2 \frac{dy_1}{y_1}, \tag{2.56}$$

$$|y_2| = c|y_1|^{\frac{\lambda_2}{\lambda_1}}, \tag{2.57}$$

where c is a positive integral constant and we have the following equations:

$$\frac{dy_2}{dy_1} = c\frac{\lambda_2}{\lambda_1} y_1^{\frac{\lambda_2}{\lambda_1}-1} \quad (y_1y_2 > 0), \tag{2.58}$$

$$\frac{dy_2}{dy_1} = -c\frac{\lambda_2}{\lambda_1} y_1^{\frac{\lambda_2}{\lambda_1}-1} \quad (y_1y_2 < 0). \tag{2.59}$$

Representative portraits are described in Figure 2.1; (a), (b), and (c) are called *stable node, saddle, and unstable node,* respectively.

Next, we consider the cases including a zero-eigenvalue, i.e. $0 = \lambda_2 < \lambda_1$ and $\lambda_2 < \lambda_1 = 0$. These cases do not have a flow in y_2-direction and y_1-direction as shown in Figure 2.2a,b, respectively. There is no arrow on the axis parallel to the eigenvector associated to the zero eigenvalue; the axes of y_2 and y_1 are center subspaces in Figure 2.2a,b, respectively.

2.4.3 Case with Repeated Eigenvalues

In the case with repeated eigenvalues, A can be transformed through transformation $B = P^{-1}AP$ into

$$B = \begin{bmatrix} \lambda & 0 \\ 0 & \lambda \end{bmatrix} \tag{2.60}$$

Figure 2.2 Phase portraits (distinct eigenvalues including zero eigenvalue: (a) $0 = \lambda_2 < \lambda_1$, and (b) $\lambda_2 < \lambda_1 = 0$).

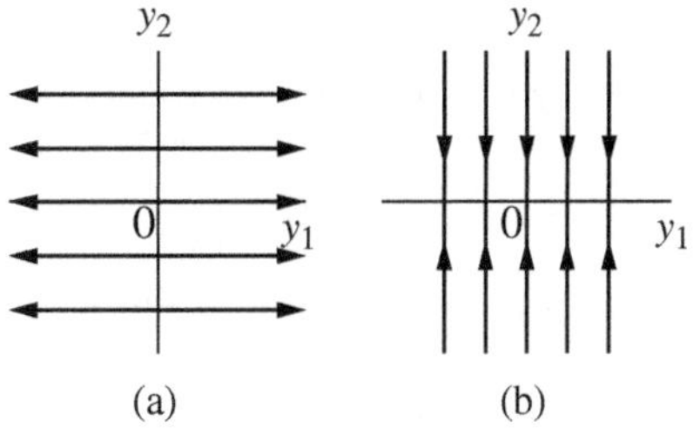

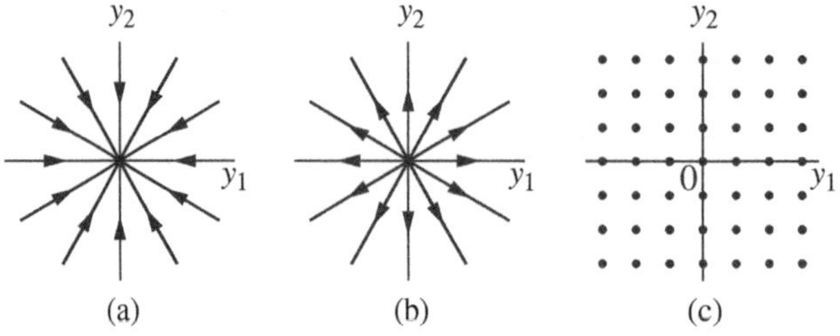

Figure 2.3 Phase portraits (repeated eigenvalues with diagonal A matrix): (a) $\lambda_1 = \lambda_2 < 0$, (b) $0 < \lambda_1 = \lambda_2$, and (c) $\lambda_1 = \lambda_2 = 0$.

or

$$B = \begin{bmatrix} \lambda & 1 \\ 0 & \lambda \end{bmatrix}. \tag{2.61}$$

The original matrix A which can be transformed into Eq. (2.60) is B itself because of $A = PBP^{-1} = P\lambda IP^{-1} = \lambda I = B$. Thus, the phase portraits are reported in Figure 2.3a,b for $\lambda > 0$ and $\lambda < 0$, respectively; these phase portraits are called *star nodes*. In the special case of $\lambda = 0$, there is no tangent vector and no trajectory as shown in Figure 2.3c. The whole plane is filled with equilibrium states.

In the case when A is not diagonal matrix, the transformed matrix is given by Eq. (2.61). Let us seek the transformation matrix P. Matrix A has only one eigenvector $\boldsymbol{p}_1$ satisfying

$$[A - \lambda I]\boldsymbol{p}_1 = 0. \tag{2.62}$$

Also, a vector $\boldsymbol{p}_2$ independent of $\boldsymbol{p}_1$ satisfies

$$[A - \lambda I]\boldsymbol{p}_2 = \boldsymbol{p}_1. \tag{2.63}$$

Equations (2.62) and (2.63) are rewritten respectively as

$$\begin{cases} A\boldsymbol{p}_1 = \lambda \boldsymbol{p}_1 = [\boldsymbol{p}_1, \ \boldsymbol{p}_2] \begin{bmatrix} \lambda \\ 0 \end{bmatrix}, & (2.64) \\ A\boldsymbol{p}_2 = \boldsymbol{p}_1 + \lambda \boldsymbol{p}_2 = [\boldsymbol{p}_1, \ \boldsymbol{p}_2] \begin{bmatrix} 1 \\ \lambda \end{bmatrix}. & (2.65) \end{cases}$$

Furthermore, similar to the derivation of Eq. (2.43), these equations are combined as

$$A[\boldsymbol{p}_1, \quad \boldsymbol{p}_2] = [\boldsymbol{p}_1, \quad \boldsymbol{p}_2] \begin{bmatrix} \lambda & 1 \\ 0 & \lambda \end{bmatrix}. \tag{2.66}$$

Therefore, by using the transformation matrix

$$P = [\boldsymbol{p}_1, \quad \boldsymbol{p}_2], \tag{2.67}$$

the matrix A can be transformed into Jordan canonical form of Eq. (2.61) as

$$B = P^{-1}AP = \begin{bmatrix} \lambda & 1 \\ 0 & \lambda \end{bmatrix}. \tag{2.68}$$

By $\boldsymbol{x} = P\boldsymbol{y}$, Eq. (2.38) is transformed into

$$\frac{d\boldsymbol{y}}{dt} = B\boldsymbol{y}. \tag{2.69}$$

We express the matrix B as the sum of the diagonal matrix and the nilpotent matrix with the remaining elements as

$$B = B_{diag} + B', \tag{2.70}$$

where

$$B_{diag} = \begin{bmatrix} \lambda & 0 \\ 0 & \lambda \end{bmatrix}, \quad B' = \begin{bmatrix} 0 & 1 \\ 0 & 0 \end{bmatrix}. \tag{2.71}$$

Since B_{diag} and B' are commutative, i.e. $B_{diag}B' = B'B_{diag}$, we have

$$e^{t(B_{diag}+B')} = e^{tB_{diag}}e^{tB'}, \tag{2.72}$$

where $e^{tB_{diag}}$ can be directly calculated from the definition of matrix exponential function of Eq. (2.40) as the derivation of Eq. (2.52):

$$e^{tB_{diag}} = \begin{bmatrix} e^{\lambda t} & 0 \\ 0 & e^{\lambda t} \end{bmatrix} \tag{2.73}$$

and $e^{tB'}$ can be also calculated directly from Eq. (2.40) as

$$e^{tB'} = \begin{bmatrix} 1 & t \\ 0 & 1 \end{bmatrix}. \tag{2.74}$$

Problem 2.4.3

1. Show that B_{diag} and B' are commutative.

Ans:

$$B_{diag}B' = \lambda IB' = B'\lambda I = B'B_{diag}, \tag{2.75}$$

where I is the unit matrix.

2. Show that only if the matrices A_1 and A_2 are commutative, i.e. $A_1A_2 = A_2A_1$, then

$$e^{A_1+A_2} = e^{A_1}e^{A_2} \tag{2.76}$$

holds like the general exponential function.

Ans: According to the definition, calculate the left and right hand sides and compare them.

$$e^{A_1+A_2} = I + (A_1 + A_2) + \frac{1}{2!}(A_1^2 + A_1A_2 + A_2A_1 + A_2^2) + \cdots, \tag{2.77}$$

$$\begin{aligned} e^{A_1}e^{A_2} &= \left(I + A_1 + \frac{1}{2!}A_1^2 + \cdots\right)\left(I + A_2 + \frac{1}{2!}A_2^2 + \cdots\right) \\ &= I + (A_1 + A_2) + \frac{1}{2!}(A_1^2 + 2A_1A_2 + A_2^2) + \cdots. \end{aligned} \tag{2.78}$$

Therefore, Eq. (2.76) holds only in the case of $A_1A_2 = A_2A_1$.

Since B_{diag} and B' are commutative, the solution of Eq. (2.69) is expressed as

$$\begin{aligned} \begin{bmatrix} y_1 \\ y_2 \end{bmatrix} &= e^{tB_{diag}}e^{tB'}\boldsymbol{y}_0 \\ &= \begin{bmatrix} e^{\lambda t} & 0 \\ 0 & e^{\lambda t} \end{bmatrix}\begin{bmatrix} 1 & t \\ 0 & 1 \end{bmatrix}\begin{bmatrix} y_1(0) \\ y_2(0) \end{bmatrix} \end{aligned}$$

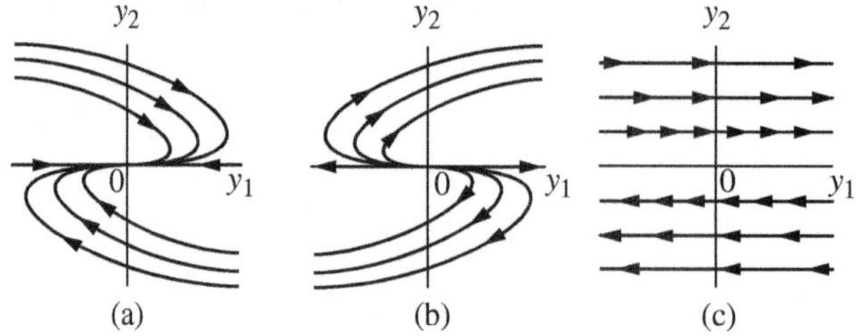

Figure 2.4 Phase portraits (repeated eigenvalues with not diagonal A matrix): (a) $\lambda_1 = \lambda_2 < 0$, (b) $0 < \lambda_1 = \lambda_2$, and (c) $\lambda_1 = \lambda_2 = 0$.

$$= \begin{bmatrix} e^{\lambda t} & te^{\lambda t} \\ 0 & e^{\lambda t} \end{bmatrix} \begin{bmatrix} y_1(0) \\ y_2(0) \end{bmatrix}. \tag{2.79}$$

The phase portraits for the cases of $\lambda_1 = \lambda_2 < 0$ and $0 < \lambda_1 = \lambda_2$ are shown in Figure 2.4a,b, respectively. When $\lambda_1 = \lambda_2 < 0$, y_1 axis is a stable subspace, but y_2 axis is not a subspace because any flows under the nonzero initial condition on y_2 axis do not remain on y_2 axis in the time revolution. On the other hand, when $0 < \lambda_1 = \lambda_2$, y_1 axis is an unstable subspace, while y_2 axis is not a subspace because any flows under the nonzero initial condition on y_2 axis do not remain on y_2 axis in the time revolution. The phase portraits of Figure 2.4a,b are called *stable inflected node* and *unstable inflected node*, respectively. In the special case $\lambda_1 = \lambda_2 = 0$, $y_1 = y_1(0) + y_2(0)t$ and $y_2 = y_2(0)$ is directly obtained from Eq. (2.69). The tangent vector is parallel to the y_1-axis and the magnitude is proportional to the initial condition $y_2(0)$. The phase portrait is shown in Figure 2.4c.

2.4.4 Case with Complex Eigenvalues

We consider the case when the eigenvalues of A are complex, $\lambda = \lambda_r \pm i\lambda_i$. We first transform matrix A into a *real Jordan canonical form* as

$$B = \begin{bmatrix} \lambda_r & \lambda_i \\ -\lambda_i & \lambda_r \end{bmatrix}. \tag{2.80}$$

The eigenvalue of $\lambda_r + i\lambda_i$ and the associated eigenvector $\boldsymbol{p} = \boldsymbol{p}_r + i\boldsymbol{p}_i$ have the following relationship:

$$A(\boldsymbol{p}_r + i\boldsymbol{p}_i) = (\lambda_r + i\lambda_i)(\boldsymbol{p}_r + i\boldsymbol{p}_i). \tag{2.81}$$

Separating this into the real and imaginary parts yields

$$\begin{cases} A\boldsymbol{p}_r = \boldsymbol{p}_r\lambda_r - \boldsymbol{p}_i\lambda_i = [\boldsymbol{p}_r, \ \boldsymbol{p}_i] \begin{bmatrix} \lambda_r \\ -\lambda_i \end{bmatrix}, & (2.82) \\ A\boldsymbol{p}_i = \boldsymbol{p}_r\lambda_i + \boldsymbol{p}_i\lambda_r = [\boldsymbol{p}_r, \ \boldsymbol{p}_i] \begin{bmatrix} \lambda_i \\ \lambda_r \end{bmatrix}. & (2.83) \end{cases}$$

Similar to the derivation of Eq. (2.43), these equations are combined as follows:

$$A\,[\boldsymbol{p}_r, \ \boldsymbol{p}_i] = [\boldsymbol{p}_r, \ \boldsymbol{p}_i] \begin{bmatrix} \lambda_r & \lambda_i \\ -\lambda_i & \lambda_r \end{bmatrix}. \tag{2.84}$$

Therefore, using the transformation matrix

$$P = [\boldsymbol{p}_r, \quad \boldsymbol{p}_i], \tag{2.85}$$

matrix A can be transformed into a real Jordan canonical form as

$$B = P^{-1}AP = \begin{bmatrix} \lambda_r & \lambda_i \\ -\lambda_i & \lambda_r \end{bmatrix} \tag{2.86}$$

and by exploiting the relationship $\boldsymbol{x} = P\boldsymbol{y}$, Eq. (2.38) is transformed into

$$\frac{d\boldsymbol{y}}{dt} = B\boldsymbol{y}. \tag{2.87}$$

Similar to Eq. (2.70), matrix B is expressed as

$$B = B_{diag} + B_{skew}, \tag{2.88}$$

where

$$B_{diag} = \begin{bmatrix} \lambda_r & 0 \\ 0 & \lambda_r \end{bmatrix}, \quad B_{skew} = \begin{bmatrix} 0 & \lambda_i \\ -\lambda_i & 0 \end{bmatrix}. \tag{2.89}$$

Since B_{diag} and B_{skew} matrices are commutative (see Problem 2.4.4),

$$\mathrm{e}^{t(B_{diag}+B_{skew})} = \mathrm{e}^{tB_{diag}}\mathrm{e}^{tB_{skew}}. \tag{2.90}$$

Thus, while $\mathrm{e}^{tB_{diag}}$ is given in Eq. (2.73) as

$$\mathrm{e}^{tB_{diag}} = \begin{bmatrix} \mathrm{e}^{\lambda_r t} & 0 \\ 0 & \mathrm{e}^{\lambda_r t} \end{bmatrix} \overset{\text{def}}{=} T_{diag}(t), \tag{2.91}$$

$\mathrm{e}^{tB_{skew}}$ can be directly computed from the definition of matrix exponential function of Eq. (2.40) as

$$\mathrm{e}^{tB_{skew}} = \begin{bmatrix} \cos\lambda_i t & \sin\lambda_i t \\ -\sin\lambda_i t & \cos\lambda_i t \end{bmatrix} \overset{\text{def}}{=} T_{skew}(t). \tag{2.92}$$

For the derivation of Eq. (2.92), see Problem 2.4.4.

Problem 2.4.4

1. Show that B_{diag} and B_{skew} are commutative.

Ans:

$$B_{diag}B_{skew} = \lambda_r I B_{skew} = B_{skew}\lambda_r I = B_{skew}B_{diag}, \tag{2.93}$$

where I is the unit matrix.

2. Derive Eq. (2.92).

Ans:

Using the following results:

$$(tB_{skew})^{2n} = (-1)^n(\lambda_i t)^{2n}\begin{bmatrix} 1 & 0 \\ 0 & 1 \end{bmatrix}, \tag{2.94}$$

$$(tB_{skew})^{2n+1} = (-1)^n(\lambda_i t)^{2n+1}\begin{bmatrix} 0 & 1 \\ -1 & 0 \end{bmatrix}, \tag{2.95}$$

we obtain Eq. (2.92) as follows:

$$
\begin{aligned}
e^{tB_{skew}} &= \sum_{n=0}^{\infty}\left\{\frac{1}{(2n)!}(tB_{skew})^{2n}+\frac{1}{(2n+1)!}(tB_{skew})^{2n+1}\right\}\\
&= \sum_{n=0}^{\infty}\frac{(-1)^n(\lambda_i t)^{2n}}{(2n)!}\begin{bmatrix}1 & 0\\ 0 & 1\end{bmatrix}+\sum_{n=0}^{\infty}\frac{(-1)^n(\lambda_i t)^{2n+1}}{(2n+1)!}\begin{bmatrix}0 & 1\\ -1 & 0\end{bmatrix}\\
&= \cos\lambda_i t\begin{bmatrix}1 & 0\\ 0 & 1\end{bmatrix}+\sin\lambda_i t\begin{bmatrix}0 & 1\\ -1 & 0\end{bmatrix}=\begin{bmatrix}\cos\lambda_i t & \sin\lambda_i t\\ -\sin\lambda_i t & \cos\lambda_i t\end{bmatrix}.
\end{aligned}
\tag{2.96}
$$

As a result, the solution of Eq. (2.87) is expressed as

$$
\begin{aligned}
\begin{bmatrix}y_1\\ y_2\end{bmatrix} &= e^{tB_{diag}}e^{tB_{skew}}\boldsymbol{y}_0\\
&= \begin{bmatrix}e^{\lambda_r t} & 0\\ 0 & e^{\lambda_r t}\end{bmatrix}\begin{bmatrix}\cos\lambda_i t & \sin\lambda_i t\\ -\sin\lambda_i t & \cos\lambda_i t\end{bmatrix}\begin{bmatrix}y_1(0)\\ y_2(0)\end{bmatrix}\\
&= \begin{bmatrix}e^{\lambda_r t}\cos\lambda_i t & e^{\lambda_r t}\sin\lambda_i t\\ -e^{\lambda_r t}\sin\lambda_i t & e^{\lambda_r t}\cos\lambda_i t\end{bmatrix}\begin{bmatrix}y_1(0)\\ y_2(0)\end{bmatrix}.
\end{aligned}
\tag{2.97}
$$

Then, $\boldsymbol{x}$ is expressed as

$$\boldsymbol{x} = P\boldsymbol{y} = y_1\boldsymbol{p}_r + y_2\boldsymbol{p}_i. \tag{2.98}$$

The distance $a(t)$ from the origin O to a point on the trajectory on $y_1 - y_2$ plane is expressed from Eq. (2.97) as

$$a(t) = \sqrt{y_1(0)^2 + y_2(0)^2}e^{\lambda_r t}. \tag{2.99}$$

When $\lambda_r > 0$ ($\lambda_r < 0$), the distance, i.e. the amplitude of the oscillation grows (decays) with time. When $\lambda_r = 0$, the oscillation has a constant amplitude which depends on the initial condition. Furthermore, because y_1 and y_2 are given by Eqs. (2.97) and (2.99) as

$$
\begin{cases}
y_1 = a(t)\cos(-\lambda_i t + \phi), & (2.100)\\
y_2 = a(t)\sin(-\lambda_i t + \phi), & (2.101)
\end{cases}
$$

where the initial phase ϕ satisfies

$$
\begin{cases}
\cos\phi = \dfrac{y_1(0)}{\sqrt{y_1(0)^2 + y_2(0)^2}}, & (2.102)\\[2ex]
\sin\phi = \dfrac{y_2(0)}{\sqrt{y_1(0)^2 + y_2(0)^2}}, & (2.103)
\end{cases}
$$

the locus rotates in the clockwise direction with time. The phase portraits for $\lambda_r < 0$, $\lambda_r = 0$, and $\lambda_r > 0$ are shown, respectively, in Figure 2.5a–c, which are called stable focus, center, and unstable focus.

The matrix in Eq. (2.97) expresses the liner map from the initial position to the position at a time t:

$$T(t) = T_{diag}(t)T_{skew}(t) = e^{\lambda_r t}\begin{bmatrix}\cos\lambda_i t & \sin\lambda_i t\\ -\sin\lambda_i t & \cos\lambda_i t\end{bmatrix}. \tag{2.104}$$

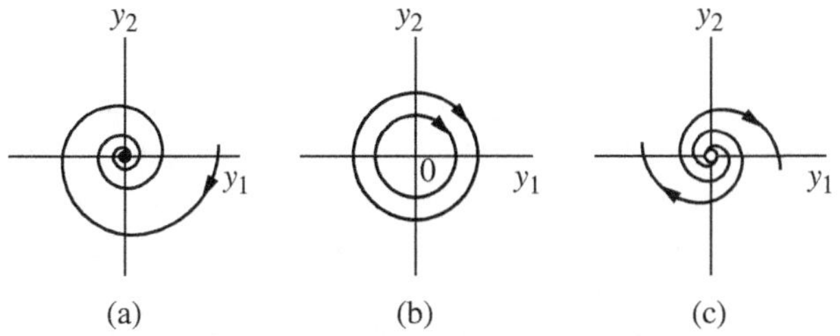

Figure 2.5 Phase portraits (complex eigenvalues): (a) $\lambda_r < 0$, (b) $\lambda_r = 0$, and (c) $\lambda_r > 0$. The behavior (a) oscillatory decays, (b) is neutral, and (c) oscillatory grows.

In general, when the matrix expressing a linear transformation is a function of t, the derivative at $t = 0$ is called an infinitesimal transformation (e.g. Goldstein et al. 1980).

The infinitesimal transformations of $T_{diag}(t)$ and $T_{skew}(t)$ are

$$\left.\frac{dT_{diag}}{dt}\right|_{t=0} = \begin{bmatrix} \lambda_r & 0 \\ 0 & \lambda_r \end{bmatrix} \tag{2.105}$$

and

$$\left.\frac{dT_{skew}}{dt}\right|_{t=0} = \begin{bmatrix} 0 & \lambda_i \\ -\lambda_i & 0 \end{bmatrix}, \tag{2.106}$$

which are respectively the vector fields of dynamical systems governed by

$$\frac{d}{dt}\begin{bmatrix} y_1 \\ y_2 \end{bmatrix} = \begin{bmatrix} \lambda_r & 0 \\ 0 & \lambda_r \end{bmatrix}\begin{bmatrix} y_1 \\ y_2 \end{bmatrix} \tag{2.107}$$

and

$$\frac{d}{dt}\begin{bmatrix} y_1 \\ y_2 \end{bmatrix} = \begin{bmatrix} 0 & \lambda_i \\ -\lambda_i & 0 \end{bmatrix}\begin{bmatrix} y_1 \\ y_2 \end{bmatrix}. \tag{2.108}$$

The phase space described by Eq. (2.107) is featured as in Figure 2.3a,b for $\lambda_r < 0$ and $\lambda_r > 0$, respectively. On the other hand, the phase space described by Eq. (2.108) is featured as in Figure 2.5b.

Since $T_{diag}(0) = T_{skew}(0) = I$, the infinitesimal transformation of the liner map of $T(t) = T_{diag}(t)T_{skew}(t)$ is

$$\left.\frac{dT}{dt}\right|_{t=0} = \left.\frac{dT_{diag}}{dt}\right|_{t=0} T_{skew}(0) + T_{diag}(0)\left.\frac{dT_{skew}}{dt}\right|_{t=0} = \left.\frac{dT_{diag}}{dt}\right|_{t=0} + \left.\frac{dT_{skew}}{dt}\right|_{t=0}, \tag{2.109}$$

which consists of the sum of the infinitesimal transformation $\left.\frac{dT_{diag}}{dt}\right|_{t=0}$ and $\left.\frac{dT_{skew}}{dt}\right|_{t=0}$. Therefore, the phase space of the linear transformation of $T(t)$ is established from the sum of these infinitesimal transformations. In other words, the vector field expressed by Eq. (2.80) is the sum of the vector fields of Eqs. (2.105) and (2.106). It is observed that the phase portraits of Figure 2.5a,c are the geometrical combination of Figures 2.3a and 2.5b and that of Figures 2.3b and 2.5b, respectively.

2.5 Invariant Subspaces

Generally, phase spaces consist of the invariant subspaces which are spanned by the eigenvectors corresponding to eigenvalues of A (or B) with positive real part, eigenvectors corresponding to eigenvalues of A (or B) with negative real part, and eigenvectors corresponding to eigenvalues of A (or B) with zero real part. They are called a unstable subspace, a stable subspace, and a center subspace, respectively, and denoted as E^u, E^s, and E^c, respectively. Because the flow whose initial condition is on a subspace entirely remains in the subspace, thus, such subspace is invariant. For example, y_1 and y_2 in the phase of Figure 2.1b are an unstable subspace and a stable one, respectively. The center subspace can be seen as y_2 axis in Figure 2.2a and y_1 axis in Figure 2.2b. Also, the $y_1 - y_2$ planes in Figure 2.5a–c are stable subspace, center subspace, and unstable subspace, respectively.

Problem 2.5 Show the subspace in the following six vector fields:

$$B_1 = \begin{bmatrix} -0.1 & 1 & 0 \\ -1 & -0.1 & 0 \\ 0 & 0 & -0.1 \end{bmatrix}, \quad B_2 = \begin{bmatrix} 0.1 & 1 & 0 \\ -1 & 0.1 & 0 \\ 0 & 0 & -0.1 \end{bmatrix},$$

$$B_3 = \begin{bmatrix} -0.1 & 1 & 0 \\ -1 & -0.1 & 0 \\ 0 & 0 & 0 \end{bmatrix}, \quad B_4 = \begin{bmatrix} -1 & 0 & 0 \\ 0 & -2 & 0 \\ 0 & 0 & 0 \end{bmatrix}, \quad B_5 = \begin{bmatrix} 0 & 1 & 0 \\ -1 & 0 & 0 \\ 0 & 0 & -0.1 \end{bmatrix},$$

$$B_6 = \begin{bmatrix} -1 & 0 & 0 \\ 0 & 1 & 0 \\ 0 & 0 & -1 \end{bmatrix}.$$

Ans:
The phase portraits for $B_1, \ldots, B_6$ are Figure 2.6a–f, respectively. In (a), $y_1 - y_2$ plane is a stable subspace and y_3 axis is also a stable subspace. In (b), $y_1 - y_2$ plane is an unstable subspace and y_3 axis is a stable subspace. In (c), $y_1 - y_2$ plane is a stable subspace and y_3 axis is a center subspace. In (d), $y_1 - y_2$ plane is a stable subspace and y_3 axis is a center subspace. In (e), $y_1 - y_2$ plane is a center subspace and y_3 axis is a stable subspace. In (f), y_2 plane is an unstable subspace and $y_3 - y_1$ plane is a stable subspace.

2.6 Change of Stability Due to the Variation of System Parameters

In Sections 2.4 and 2.5, it is clarified that the eigenvalues of the linear operator

$$A = \begin{bmatrix} a_{11} & a_{12} \\ a_{21} & a_{22} \end{bmatrix} \tag{2.110}$$

qualitatively characterize the phase portrait and determine the stability of the equilibrium state in local sense. The elements, a_{11}, a_{12}, a_{21}, and a_{22}, of the linear matrix A consist of the system parameters. Therefore, the variation of the parameters can change not only the quantitative features of the dynamics but also the qualitative ones, i.e. the phase portrait

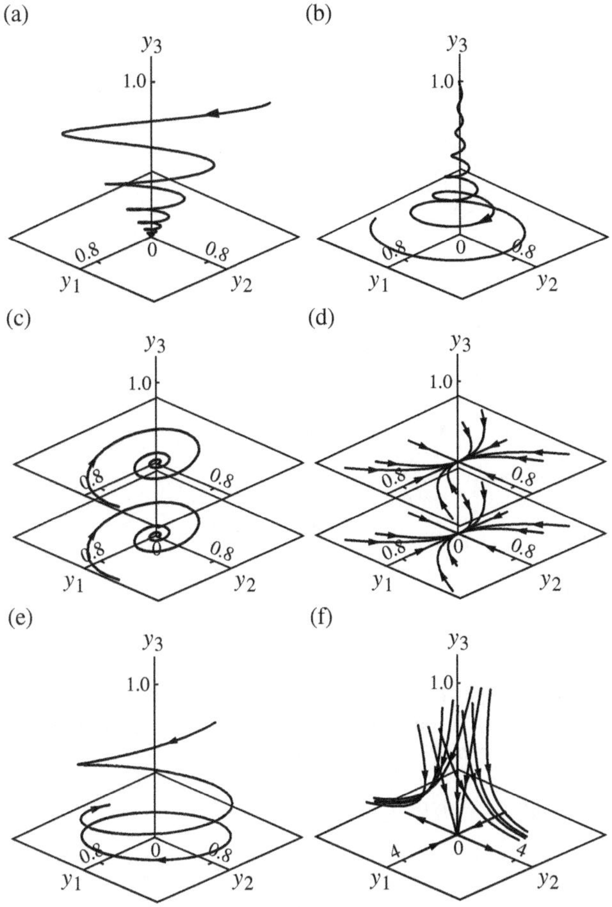

Figure 2.6 Subspaces of three-dimensional systems given in problem 2.6.

and the stability of the system. The eigenvalues λ of A are the solutions of the following characteristic equation:

$$|A - \lambda I| = \lambda^2 - (a_{11} + a_{22})\lambda + a_{11}a_{22} - a_{12}a_{21} = 0 \tag{2.111}$$

or equivalently

$$|A - \lambda I| = \lambda^2 - \mathrm{Tr}(A)\lambda + |A| = 0, \tag{2.112}$$

where $\mathrm{Tr}(A)$ stands for the trace of A. Letting $c' = -(a_{11} + a_{22})$ and $k' = a_{11}a_{22} - a_{12}a_{21}$, we rewrite the characteristic equation as follows:

$$\lambda^2 + c'\lambda + k' = 0. \tag{2.113}$$

We can show the qualitative dynamics depending on the parameters k' and c' on the parameter plane of $k' - c'$ as in Figure 2.7 (Thompson and Stewart 2002); the coordinates in the phase spaces are chosen according to the transformations corresponding to the characteristics of eigenvalue as mentioned in Section 2.4.

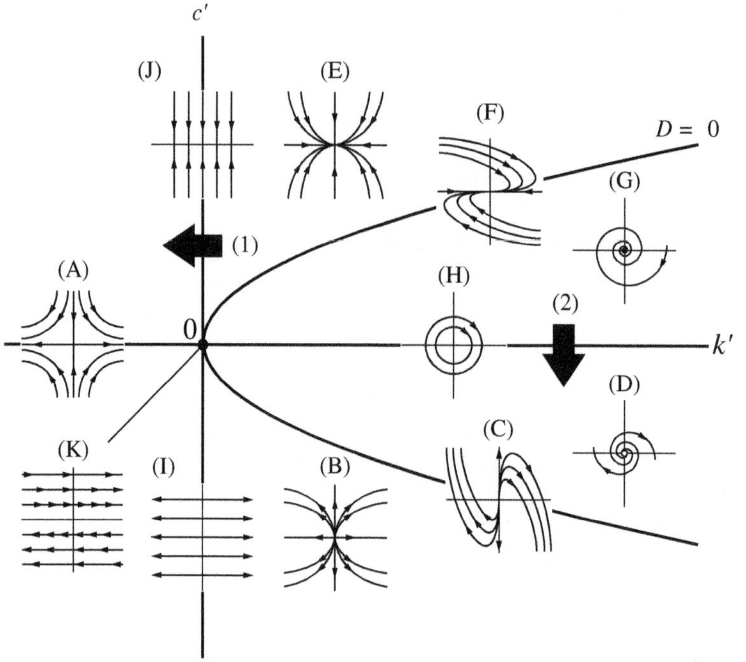

Figure 2.7 Variations of phase portraits on parameter space $k' - c'(D = c'^2 - 4k')$. At the origin of $k' = c' = 0$, there are two cases of phase portrait shown in Figures 2.3c and 2.4c depending on A. Arrow (1) and (2) denote static and dynamic destabilization, respectively.

By the way, the matrix A of the spring-mass-damper system

$$m\frac{d^2x}{dt^2} + c\frac{dx}{dt} + kx = 0 \tag{2.114}$$

or equivalently

$$\frac{d^2x}{dt^2} + c'\frac{dx}{dt} + k'x = 0 \tag{2.115}$$

is

$$A = \begin{bmatrix} 0 & 1 \\ -k' & -c' \end{bmatrix}, \tag{2.116}$$

where $c' = \frac{c}{m}, k' = \frac{k}{m}$. Then, the characteristic equation is Eq. (2.113).

In particular, under the following conditions, the system is in the critical state of stability.

1. Case of $k' > 0$ and $c' = 0$, i.e.

$$B = \begin{bmatrix} 0 & \sqrt{k'} \\ -\sqrt{k'} & 0 \end{bmatrix}, \tag{2.117}$$

and the phase plane is shown in Figure 2.5b.

2. Case of $k' = 0$ and $c' > 0$, i.e.

$$B = \begin{bmatrix} 0 & 0 \\ 0 & -c' \end{bmatrix}, \tag{2.118}$$

and the phase plane is shown in Figure 2.2b.

In the system which is characterized by a real spring with positive stiffness $k > 0$ as shown in Figure 1.2a–c and a damper with positive damping $c > 0$ as shown in Figure 1.2d–f, the system has the dynamics, (E), (F), or (G), described on the first quadrant in Figure 2.7. In these cases, the trivial equilibrium state is stable. (E) is the so-called over-damping case and the initial displacement is monotonically decayed, i.e. without oscillation. There are two one-dimensional subspaces. (G) is the so-called under-damping case and the initial displacement is oscillatory decayed. There is no one-dimensional subspace. (F) is the so called critical damping case which is the boundary between the over- and under-damping cases and has one one-dimensional subspace.

When the combination of $k' = \frac{k}{m}$ and $c' = \frac{c}{m}$ exists in the second, third, and fourth quadrants, i.e. the system has a spring with negative stiffness as shown in Figure 1.5a–c, or a damper with negative damping as shown in Figure 1.5d–f, the trivial equilibrium point is unstable. The half lines of $k' > 0, c' = 0$, and $k' = 0, c' > 0$ are the boundaries of stability, which correspond to the cases of only a spring with positive stiffness and only a damper with positive damping, respectively. When the parameter k' varies along the arrow (1), i.e. the positive stiffness $k > 0$ is changed to the negative stiffness $k < 0$ while keeping the positive damping $c > 0$, the trivial equilibrium state is statically destabilized and the displacement monotonically grows. The phenomenon is called *buckling* and will be investigated in Chapter 5. On the other hand, when the parameter c' varies along the arrow (2), i.e. the positive damping $c > 0$ is changed to the negative damping $c < 0$, the trivial equilibrium state is dynamically destabilized. Then, the displacement oscillatory increases with the growth of amplitude. The produced oscillation is called *self-excited oscillation* (see Chapter 6). In the situation where the parameter c' varies along the arrow (2), the complex conjugate eigenvalues with a negative real part is changed to the complex conjugate eigenvalues with a positive real part through purely imaginary complex conjugate eigenvalues. The bifurcation caused by such a change of eigenvalues is *Hopf bifurcation* (see Section 7.4.3).

Problem 2.6 In addition to the position feedback of Eq. (1.55), we apply the velocity feedback $c_2\frac{dx}{dt}$ as

$$i = i_0 + c_1 x + c_2 \frac{dx}{dt}. \tag{2.119}$$

Select the position and velocity feedback gains so that the stability of the equilibrium state becomes asymptotically stable.

Ans: In the linearized equation corresponding to Eq. (1.58), k' and c' in Eq. (2.115) are

$$k' = k_m\left(\frac{c_1 x_{st}}{i_0} - 1\right), \quad c' = \frac{k_m c_2 x_{st}}{i_0}. \tag{2.120}$$

Therefore, by setting the feedback gains c_1 and c_2 so that the combination of k' and c' lies in the first quadrant, the system becomes asymptotically stable.

The destabilization by the change of the parameter value along arrows (1) and (2) causes the growth of the displacement and the amplitude. In such cases, the linear analysis cannot determine if the growth continues infinitely or not. Further discussion on this subject will be postponed until Chapters 7 and 10 in which the nonlinearity neglected in this chapter is taken into account.

References

Goldstein, H., C. Poole, and J. Safko (2002). *Classical mechanics.*

Hirsh, M. W. and S. Samle (1974). *Differential Equations, Dynamical Systems, and Linear Algebra*. Academic Press.

Thompson, J. and H. Stewart (2002). *Nonlinear Dynamics and Chaos.* Wiley.

Wiggins, S. (1980). *Introduction to Applied Nonlinear Dynamics and Chaos.* 1980.

Witelski, T. and M. Bowen (2015). *Methods of mathematical modelling: Continuous systems and differential equations.* Springer.

3

Dynamic Instability of Two-Degree-of-Freedom-Systems

In Chapter 2, the stability of second-order systems corresponding to single-degree of freedom systems was examined together with its dependence on the system parameters. In this chapter, the dynamics specifically related to multiple degree-of-freedom systems is introduced. We will consider only two-degree-of-freedom systems because these suffice for clarifying the characteristics. In particular, the specific mechanism of dynamic instability, that a single degree-of-freedom model fails to describe, is studied through a two-degree-of-freedom model.

3.1 Positional Forces and Velocity-Dependent Forces

We consider the following two-degree-of-freedom linear system:

$$M\frac{d^2\boldsymbol{x}}{dt^2} = -K'\boldsymbol{x} - D'\frac{d\boldsymbol{x}}{dt}, \tag{3.1}$$

where M, K', and D' are 2×2 matrices. $-K'\boldsymbol{x}$ and $-D'd\boldsymbol{x}/dt$ are positional and velocity-dependent forces, respectively.

The mass matrix M directly defined from the kinetic energy is symmetric, i.e. $M^T = M$, and positive definite (Meirovitch 1980). K' and D' are not generally symmetric, but can be decomposed into symmetric and skew-symmetric matrices as follows:

$$K' = \frac{1}{2}(K' + K'^T) + \frac{1}{2}(K' - K'^T) \overset{\text{def}}{=} K + N, \tag{3.2}$$

$$D' = \frac{1}{2}(D' + D'^T) + \frac{1}{2}(D' - D'^T) \overset{\text{def}}{=} D + G, \tag{3.3}$$

where K and D are symmetric as

$$K = \frac{1}{2}(K' + K'^T), \quad D = \frac{1}{2}(D' + D'^T) \tag{3.4}$$

while N and G are skew-symmetric

$$N = \frac{1}{2}(K' - K'^T), \quad G = \frac{1}{2}(D' - D'^T). \tag{3.5}$$

Linear and Nonlinear Instabilities in Mechanical Systems: Analysis, Control and Application,
First Edition. Hiroshi Yabuno.

Companion website: www.wiley.com/go/yabuno/instabilitiesinmechanicalsystems

The matrices, K, N, D, and G, are called the stiffness, circulatory, viscous damping, and gyroscopic matrices, respectively. Then, Eq. (3.1) is rewritten as

$$M\frac{d^2\boldsymbol{x}}{dt^2} = \boldsymbol{F}_K + \boldsymbol{F}_N + \boldsymbol{F}_D + \boldsymbol{F}_G, \tag{3.6}$$

where

$$\begin{aligned} \boldsymbol{F}_K &= -K\boldsymbol{x}, \quad \boldsymbol{F}_N = -N\boldsymbol{x}, \\ \boldsymbol{F}_D &= -D\frac{d\boldsymbol{x}}{dt}, \quad \boldsymbol{F}_G = -G\frac{d\boldsymbol{x}}{dt}. \end{aligned} \tag{3.7}$$

The forces $\boldsymbol{F}_K$, $\boldsymbol{F}_N$, $\boldsymbol{F}_D$, and $\boldsymbol{F}_G$ are called *conservative* or *potential*, *nonconservative positional* or *circulatory*, *velocity dependent damping* or *dissipative*, and *gyroscopic*, respectively (Kirillov 2013; Seyranian et al. 1994). For the energy-based discussion in the subsequent sections, taking the inner product of both sides of Eq. (3.6) with the velocity vector $\frac{d\boldsymbol{x}}{dt}$, we have

$$M\frac{d^2\boldsymbol{x}}{dt^2}\cdot\frac{d\boldsymbol{x}}{dt} = (\boldsymbol{F}_K + \boldsymbol{F}_N + \boldsymbol{F}_D + \boldsymbol{F}_G)\cdot\frac{d\boldsymbol{x}}{dt}. \tag{3.8}$$

3.2 Total Energy and Its Time Variation

3.2.1 Kinetic Energy

In general, the inner product of vectors $\boldsymbol{a}$ and $\boldsymbol{b}$, $\boldsymbol{a}\cdot\boldsymbol{b}$, is expressed in terms of the product of the corresponding column matrices, $\boldsymbol{a}$ and $\boldsymbol{b}$, as

$$\boldsymbol{a}\cdot\boldsymbol{b} = \overline{\boldsymbol{b}}^T\boldsymbol{a}, \tag{3.9}$$

where the overbar denotes the complex conjugate. The left-hand side of Eq. (3.8) can be calculated considering that M is symmetric and positive definite as follows:

$$\begin{aligned} M\frac{d^2\boldsymbol{x}}{dt^2}\cdot\frac{d\boldsymbol{x}}{dt} &= \left(\frac{d\boldsymbol{x}}{dt}\right)^T M\frac{d^2\boldsymbol{x}}{dt^2} = \frac{1}{2}\left\{\left(\frac{d\boldsymbol{x}}{dt}\right)^T M\frac{d^2\boldsymbol{x}}{dt^2} + \left(\frac{d^2\boldsymbol{x}}{dt^2}\right)^T M\frac{d\boldsymbol{x}}{dt}\right\} \\ &= \frac{d}{dt}\left\{\frac{1}{2}\left(\frac{d\boldsymbol{x}}{dt}\right)^T M\frac{d\boldsymbol{x}}{dt}\right\} = \frac{dT}{dt}, \end{aligned} \tag{3.10}$$

where $T \overset{\text{def}}{=} \frac{1}{2}\left(\frac{d\boldsymbol{x}}{dt}\right)^T M\frac{d\boldsymbol{x}}{dt}$ is the kinetic energy.

3.2.2 Potential Energy Due to Conservative Force $\boldsymbol{F}_K$

The first term on the right-hand side of Eq. (3.8) is calculated considering the symmetric characteristic of K as follows:

$$\begin{aligned} \boldsymbol{F}_K\cdot\frac{d\boldsymbol{x}}{dt} &= \left(\frac{d\boldsymbol{x}}{dt}\right)^T \boldsymbol{F}_K = \left(\frac{d\boldsymbol{x}}{dt}\right)^T(-K\boldsymbol{x}) \\ &= -\frac{1}{2}\left\{\left(\frac{d\boldsymbol{x}}{dt}\right)^T K\boldsymbol{x} + \boldsymbol{x}^T K\frac{d\boldsymbol{x}}{dt}\right\} \end{aligned}$$

Figure 3.1 Path on which the work is computed.

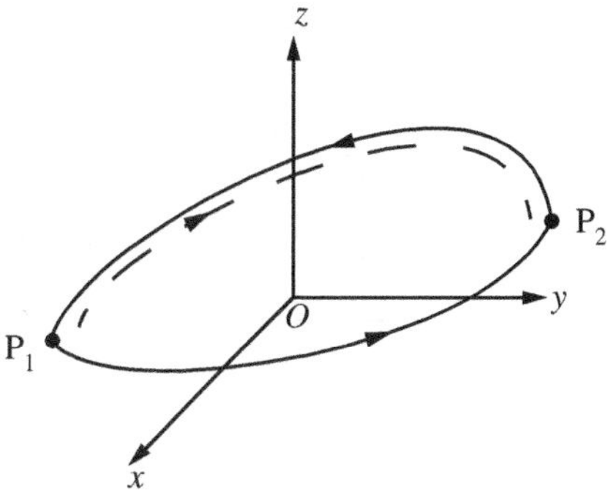

$$= \frac{d}{dt}\left(-\frac{1}{2}\boldsymbol{x}^T K\boldsymbol{x}\right) = \frac{d(-U)}{dt}, \tag{3.11}$$

where $U \overset{\text{def}}{=} \frac{1}{2}\boldsymbol{x}^T K\boldsymbol{x}$. Furthermore, we investigate the characteristic of $\boldsymbol{F}_k$ using the theory of vector analysis (e.g. Marsden and Tromba 2003). Because K is symmetric, rot $\boldsymbol{F}_K = \boldsymbol{0}$. From the Stokes' theorem,

$$\oint \boldsymbol{F}_K \cdot d\boldsymbol{x} = \iint \text{rot } \boldsymbol{F}_K \cdot d\boldsymbol{S} = 0. \tag{3.12}$$

Therefore, under the action of $\boldsymbol{F}_K$, when a particle of mass m is moved from a position P_1 to another position P_2 along a path and is returned from P_2 to P_1 along another path as shown in Figure 3.1, the work done by $\boldsymbol{F}_K$ is 0. In other words, the work done by $\boldsymbol{F}_K$ to move a particle from a point to another is independent of the path between these points.

In addition, by exploiting the theory of vector calculus, if rot $\boldsymbol{F}_K = \boldsymbol{0}$, there is a function U satisfying $\boldsymbol{F}_K = -\text{grad } U$. The function U corresponds to the potential energy. As a simple example, consider $\boldsymbol{F}_K$ in two-degree-of-freedom system as

$$\boldsymbol{F}_K = -\begin{bmatrix} k_{11} & k_{12} \\ k_{12} & k_{22} \end{bmatrix} \begin{bmatrix} x_1 \\ x_2 \end{bmatrix}. \tag{3.13}$$

Due to the symmetric characteristic of K, we can easily derive rot $\boldsymbol{F}_K = \boldsymbol{0}$.

The infinitesimal work done by $\boldsymbol{F}_K$ can be also expressed as

$$\boldsymbol{F}_K \cdot d\boldsymbol{x} = -K\boldsymbol{x} \cdot d\boldsymbol{x} = -d\boldsymbol{x}^T K\boldsymbol{x} = -d\left(\frac{1}{2}\boldsymbol{x}^T K\boldsymbol{x}\right) = -dU, \tag{3.14}$$

where the symmetric characteristic of K is used. Therefore, the infinitesimal work done by $\boldsymbol{F}_K$ can be written as the total derivative of $-U$; this is equivalent to the before-mentioned fact that the work due to $\boldsymbol{F}_K$ is independent of the path. Thus, $\boldsymbol{F}_K$ has the potential $U = \frac{1}{2}\boldsymbol{x}^T K\boldsymbol{x}$. Let us consider a system that does not include any external forces except $\boldsymbol{F}_K$. The matrix K corresponds to the Hessian matrix $\boldsymbol{H}U(\boldsymbol{r}_{st})$ in Eq. (1.45). If K is positive definite, i.e.

$$U = \frac{1}{2}\boldsymbol{x}^T K\boldsymbol{x} > 0, \tag{3.15}$$

the trivial equilibrium state is stable in the sense of Liapunov.

Problem 3.2.2 Show the equations of motion for the two-degree-of-freedom spring-mass system in Figure 3.2 and check if the external force by the springs has potential. Also discuss the stability of the trivial equilibrium state, $x_1 = x_2 = 0$.

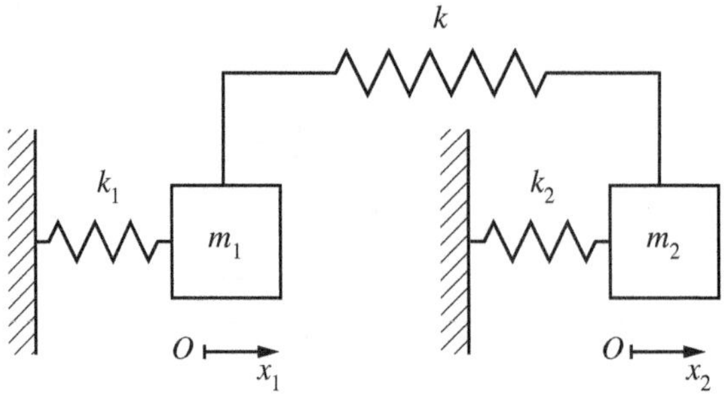

Figure 3.2 Two-degree-of-freedom spring-mass system.

Ans: The equations of motion for m_1 and m_2 are

$$\begin{cases} m_1 \dfrac{d^2x_1}{dt^2} = -k_1x_1 - k(x_1 - x_2), & (3.16) \\ m_2 \dfrac{d^2x_2}{dt^2} = -k_2x_2 - k(x_2 - x_1). & (3.17) \end{cases}$$

Therefore, K' is symmetric

$$K' = \begin{bmatrix} k_1 + k & -k \\ -k & k_2 + k \end{bmatrix} = K \tag{3.18}$$

and the external force has the following potential

$$U = \frac{1}{2}[x_1\ x_2]K\begin{bmatrix} x_1 \\ x_2 \end{bmatrix} = \frac{1}{2}\{k_1x_1^2 + k(x_1 - x_2)^2 + k_2x_2^2\}. \tag{3.19}$$

If all the springs are characterized by positive stiffness, i.e. $k_1 > 0, k_2 > 0$, and $k > 0$, K is positive definite. Then, since $U > 0$ except for $x_1 = x_2 = 0$, the trivial equilibrium state is stable in the sense of Liapunov.

As a simple example, we consider

$$\begin{cases} m_1 \dfrac{d^2x_1}{dt^2} + k_{11}x_1 + k_{12}x_2 = 0, \\ m_2 \dfrac{d^2x_2}{dt^2} + k_{21}x_1 + k_{22}x_2 = 0. \end{cases} \tag{3.20}$$

We show the dynamics in case of $m_1 = m_2 = 1, k_{11} = 9, k_{22} = 16, k_{12} = -1, k_{21} = -1$. Under the initial conditions $x_1 = x_2 = 1$, and $\frac{dx_1}{dt} = \frac{dx_2}{dt} = 0$, the time histories of x_1 and x_2 are shown in Figure 3.3a,b. We show the time variation of the total energy $E = T + U$ in Figure 3.3c. The amplitudes of x_1 and x_2 are finite, and the total energy $E = T + U$ is constant.

From Eqs. (3.10) and (3.11), we rewrite Eq. (3.8) as follows:

$$\begin{aligned} \frac{d}{dt}(T + U) &= \left(\frac{d\boldsymbol{x}}{dt}\right)^T F_N + \left(\frac{d\boldsymbol{x}}{dt}\right)^T F_D + \left(\frac{d\boldsymbol{x}}{dt}\right)^T F_G \\ &= -\left(\frac{d\boldsymbol{x}}{dt}\right)^T N\boldsymbol{x} - \left(\frac{d\boldsymbol{x}}{dt}\right)^T D\frac{d\boldsymbol{x}}{dt} - \left(\frac{d\boldsymbol{x}}{dt}\right)^T G\frac{d\boldsymbol{x}}{dt}. \end{aligned} \tag{3.21}$$

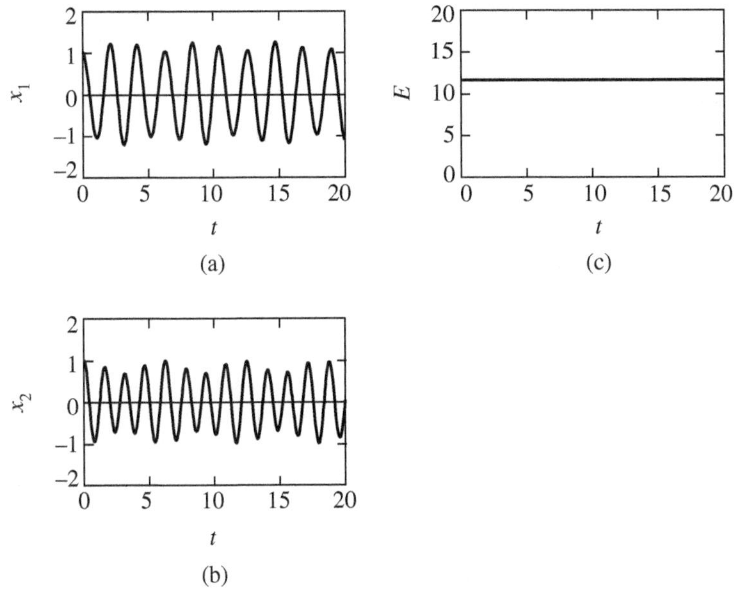

Figure 3.3 Conservative system. (a) and (b) are the time histories of x_1 and x_2, respectively; (c) shows the variation of total energy E.

The time variation of the total energy $T + U$ can depend on the remaining positional force F_N and the velocity dependent forces, F_D and F_G. In the succeeding sections, we discuss the effects of these forces on the variation of the total energy.

3.2.3 Effect of Velocity-Dependent Damping Force F_D

We consider the variation of the total energy by the velocity depending damping force $\boldsymbol{F}_D$. From Eq. (3.21), the time variation of the total energy is governed by

$$\frac{d}{dt}(T+U) = \boldsymbol{F}_D \cdot \frac{d\boldsymbol{x}}{dt} = \left(\frac{d\boldsymbol{x}}{dt}\right)^T \boldsymbol{F}_D = -\left(\frac{d\boldsymbol{x}}{dt}\right)^T D\frac{d\boldsymbol{x}}{dt}. \tag{3.22}$$

The total energy decays with time if and only if the matrix D is positive definite, because

$$\frac{d}{dt}(T+U) = \left(\frac{d\boldsymbol{x}}{dt}\right)^T \boldsymbol{F}_D = -\left(\frac{d\boldsymbol{x}}{dt}\right)^T D\frac{d\boldsymbol{x}}{dt} < 0. \tag{3.23}$$

Problem 3.2.3 Show the matrix D for the two-degree-of-freedom spring-mass-damper system of Figure 3.4 and discuss the stability of the trivial equilibrium state, $x_1 = x_2 = 0$.

Ans: The equations of motion for m_1 and m_2 are

$$\begin{cases} m_1\dfrac{d^2x_1}{dt^2} = -k_1x_1 - k(x_1 - x_2) - c_1\dfrac{dx_1}{dt} - c\left(\dfrac{dx_1}{dt} - \dfrac{dx_2}{dt}\right), & (3.24) \\ m_2\dfrac{d^2x_2}{dt^2} = -k_2x_2 - k(x_2 - x_1) - c_2\dfrac{dx_2}{dt} - c\left(\dfrac{dx_2}{dt} - \dfrac{dx_1}{dt}\right). & (3.25) \end{cases}$$

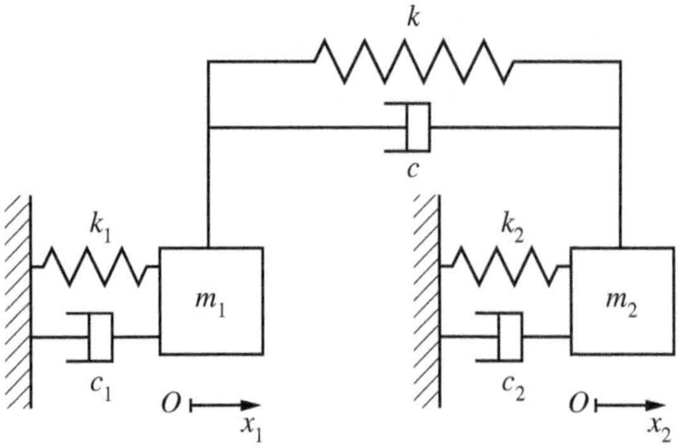

Figure 3.4 Two-degrees-of-freedom spring-mass-damper system.

Therefore, D is symmetric

$$D = \begin{bmatrix} c_1 + c & -c \\ -c & c_2 + c \end{bmatrix} \tag{3.26}$$

and if all dampers have positive damping coefficient, i.e. $c_1 > 0, c_2 > 0,$ and $c > 0$, D is positive definite. Then, since the term on the right-hand side in Eq. (3.23) is

$$\frac{d}{dt}(T + U) = -\left(\frac{d\boldsymbol{x}}{dt}\right)^T D\frac{d\boldsymbol{x}}{dt} < 0 \tag{3.27}$$

except for $\frac{dx_1}{dt} = \frac{dx_2}{dt} = 0$, the total energy decays with time and the trivial equilibrium state is asymptotically stable.

3.2.4 Effect of Circulatory Force F_N

Since N is skew-symmetric, the rotation of the circulatory force $\boldsymbol{F}_N = -N\boldsymbol{x}$ is not equal to zero, i.e.

$$\text{rot } \boldsymbol{F}_N \neq \boldsymbol{0}. \tag{3.28}$$

Therefore, different from rot $\boldsymbol{F}_K$, the circulatory force $\boldsymbol{F}_N$ cannot be expressed by the gradient of any scalar functions and $\boldsymbol{F}_N$ does not have potential energy; $\boldsymbol{F}_N$ is called *nonconservative* force. From the Stokes' theorem,

$$\oint \boldsymbol{F}_N \cdot d\boldsymbol{x} = \iint \text{rot } \boldsymbol{F}_N \cdot d\boldsymbol{S} \neq 0. \tag{3.29}$$

Therefore, under the action of $\boldsymbol{F}_N$, when a particle of mass m is moved from a position P_1 to another position P_2 along a path and is returned from P_2 to P_1 along a different path as shown in Figure 3.1, the work done by $\boldsymbol{F}_N$ is not equal to 0. In other words, the work done by $\boldsymbol{F}_N$ to move a particle of mass m from a point to another depends on the path between these points.

Consider a simple example in two-degree-of-freedom system, i.e.

$$\boldsymbol{F}_N = -N\boldsymbol{x} = \begin{bmatrix} 0 & -n_{12} \\ n_{12} & 0 \end{bmatrix} \begin{bmatrix} x_1 \\ x_2 \end{bmatrix}. \tag{3.30}$$

We can easily obtain rot $\boldsymbol{F}_N = 2n_{12} \neq 0$ due to the skew-symmetric characteristic of $\boldsymbol{F}_N$. Let us examine the effect of the circulatory force on the total energy. Because of

$$\left(\frac{d\boldsymbol{x}}{dt}\right)^T \boldsymbol{F}_N = -\left(\frac{d\boldsymbol{x}}{dt}\right)^T N\frac{d\boldsymbol{x}}{dt} = \begin{bmatrix} \frac{dx_1}{dt} & \frac{dx_2}{dt} \end{bmatrix} \begin{bmatrix} 0 & -n_{12} \\ n_{12} & 0 \end{bmatrix} \begin{bmatrix} x_1 \\ x_2 \end{bmatrix}$$
$$= n_{12}\left(x_1\frac{dx_2}{dt} - \frac{dx_1}{dt}x_2\right) \neq 0, \tag{3.31}$$

the circulatory force expends a non-zero work along closed loop in Figure 3.1 and the total energy is not conserved as follows:

$$\frac{d}{dt}(T+U) = \left(\frac{d\boldsymbol{x}}{dt}\right)^T F_N \neq 0. \tag{3.32}$$

As a simple example, we investigate the following system with the conservative force $\boldsymbol{F}_K$ and the circulatory force $\boldsymbol{F}_N$:

$$\begin{cases} m_1\dfrac{d^2x_1}{dt^2} + k_{11}x_1 + k_{12}x_2 = 0, \\ m_2\dfrac{d^2x_2}{dt^2} + k_{21}x_1 + k_{22}x_2 = 0. \end{cases} \tag{3.33}$$

We show the dynamics $m_1 = m_2 = 1, k_{11} = 9, k_{22} = 16, k_{12} = 1, k_{21} = -1$. Under the initial conditions $x_1 = x_2 = 1$, and $\frac{dx_1}{dt} = \frac{dx_2}{dt} = 0$, the time histories of x_1 and x_2 are shown in Figure 3.5a,b. We show the time variation of the total energy $E = T + U$ in Figure 3.5c.

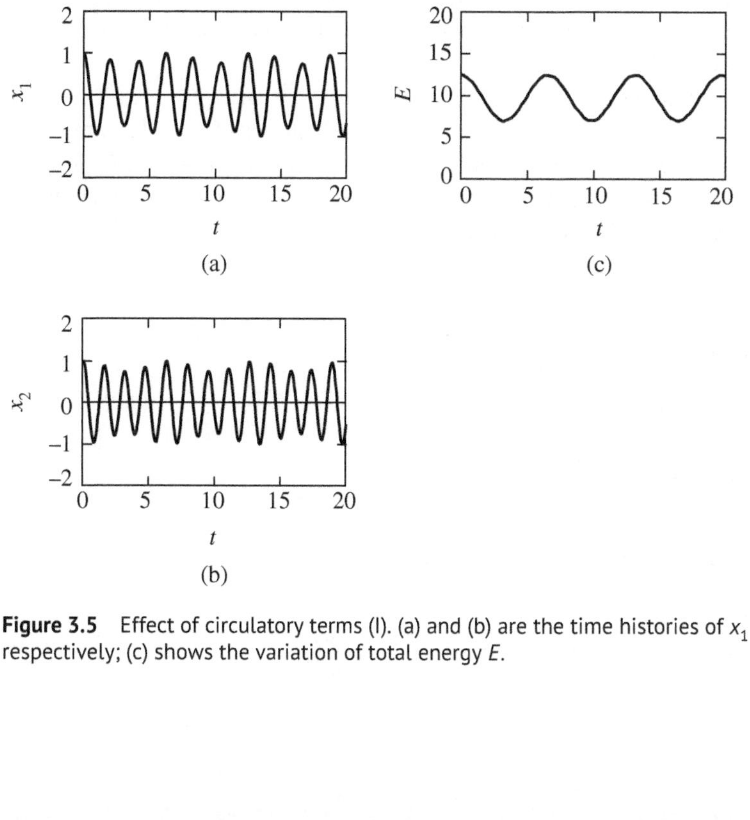

Figure 3.5 Effect of circulatory terms (I). (a) and (b) are the time histories of x_1 and x_2, respectively; (c) shows the variation of total energy E.

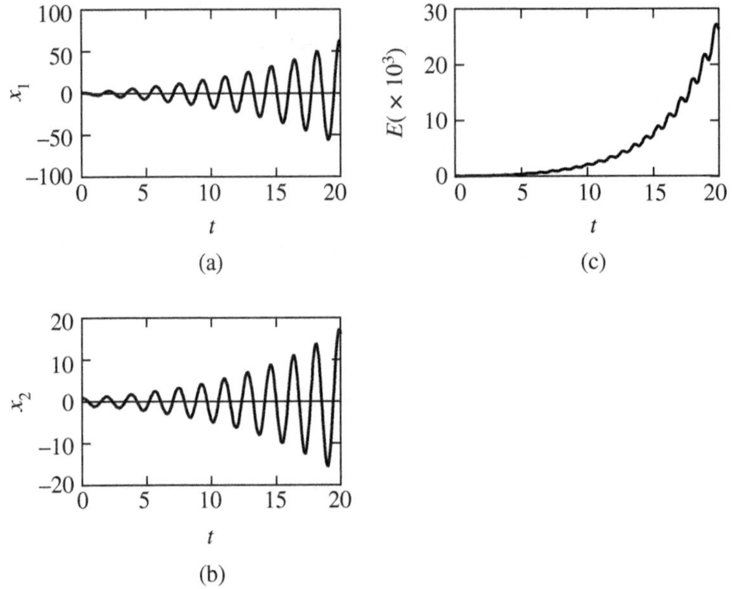

Figure 3.6 Effect of circulatory terms (II). (a) and (b) are the time histories of x_1 and x_2, respectively; (c) shows the variation of total energy E.

As mentioned in Section 1.2.2, if $N = D = G = 0$, the amplitudes of x_1 and x_2 are finite, and E is constant (see the example with respect to Eq. (3.20)). However, in case with circulatory force, i.e. $N \neq 0$, the amplitudes of x_1 and x_2 are finite, but E is changed as in Figure 3.5c. This case corresponds to the case in which the system has real nonorthogonal modal vectors as will be discussed in Section 4.4.1.

Next, we show the dynamics in the case in which only k_{12} is changed to 13. Under the initial conditions $x_1 = x_2 = 1$, and $\frac{dx_1}{dt} = \frac{dx_2}{dt} = 0$, the time histories of x_1 and x_2 are shown in Figure 3.6a,b. We show the time variation of the total energy $E = T + U$ in Figure 3.6c. The amplitudes x_1 and x_2 grow with time, and the total energy is changed. This case corresponds to the case in which the system has complex modal vectors as will be discussed in Section 4.4.2.

3.2.5 Effect of Gyroscopic Force F_G

Since G is skew-symmetric, we obtain

$$\left(\frac{d\boldsymbol{x}}{dt}\right)^T \boldsymbol{F}_G = -\left(\frac{d\boldsymbol{x}}{dt}\right)^T G\frac{d\boldsymbol{x}}{dt} = 0, \tag{3.34}$$

where $\left\{\left(\frac{dx}{dt}\right)^T G\frac{dx}{dt}\right\}^T = -\left(\frac{dx}{dt}\right)^T G\frac{dx}{dt}$ is used, and

$$\frac{d}{dt}(T + U) = \boldsymbol{F}_G \cdot \frac{d\boldsymbol{x}}{dt} = \left(\frac{d\boldsymbol{x}}{dt}\right)^T \boldsymbol{F}_G = -\left(\frac{d\boldsymbol{x}}{dt}\right)^T G\frac{d\boldsymbol{x}}{dt} = 0. \tag{3.35}$$

Therefore, the gyroscopic force does not change the total energy.

As a simple example, we investigate the following system with the conservative force $\boldsymbol{F}_K$ and the gyroscopic force $\boldsymbol{F}_G$:

$$\begin{cases} m_1 \dfrac{d^2x_1}{dt^2} - \gamma \dfrac{dx_2}{dt} + k_{11}x_1 = 0, \\ m_2 \dfrac{d^2x_2}{dt^2} + \gamma \dfrac{dx_1}{dt} + k_{22}x_2 = 0. \end{cases} \tag{3.36}$$

Then,

$$\begin{cases} \dfrac{d}{dt}\left\{ \dfrac{1}{2}m_1 \left(\dfrac{dx_1}{dt} \right)^2 + \dfrac{1}{2}k_{11}x_1^2 \right\} = \gamma \dfrac{dx_1}{dt}\dfrac{dx_2}{dt}, \\ \dfrac{d}{dt}\left\{ \dfrac{1}{2}m_2 \left(\dfrac{dx_2}{dt} \right)^2 + \dfrac{1}{2}k_{22}x_2^2 \right\} = -\gamma \dfrac{dx_1}{dt}\dfrac{dx_2}{dt}. \end{cases} \tag{3.37}$$

Since the total energies of m_1 and m_2 are

$$\begin{cases} E_1 = \dfrac{1}{2}m_1 \left(\dfrac{dx_1}{dt} \right)^2 + \dfrac{1}{2}k_{11}x_1^2, \\ E_2 = \dfrac{1}{2}m_2 \left(\dfrac{dx_2}{dt} \right)^2 + \dfrac{1}{2}k_{22}x_2^2, \end{cases} \tag{3.38}$$

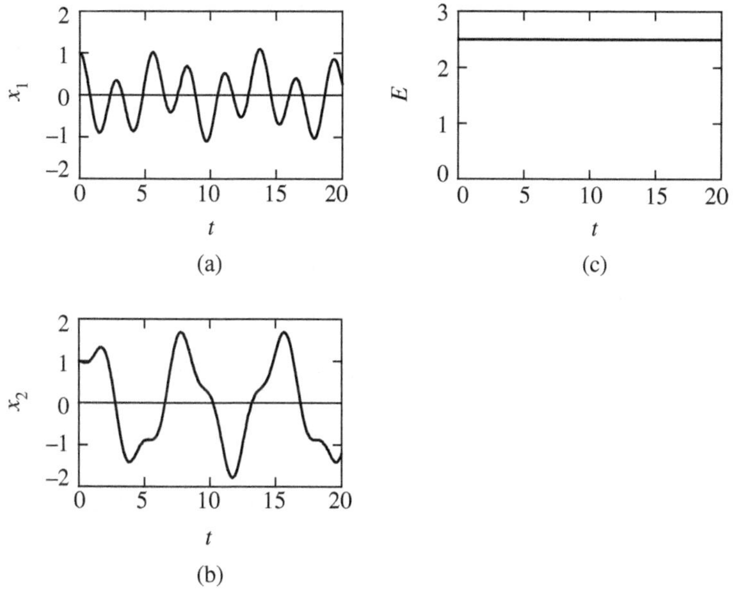

Figure 3.7 Effect of gyroscopic terms. (a) and (b) are the time histories of x_1 and x_2, respectively; (c) shows the total energy E.

the total energy of m_1 and m_2, $E = E_1 + E_2$, is constant.

We show the dynamics in case of $m_1 = m_2 = 1, k_{11} = 4, k_{22} = 1, \gamma = 1$. Under the initial conditions $x_1 = x_2 = 1$, and $\frac{dx_1}{dt} = \frac{dx_2}{dt} = 0$, the time histories of x_1 and x_2 are shown in Figure 3.7a,b. We show the time variation of $E = E_1 + E_2$ in Figure 3.7c. The total energies E is kept constant and the gyroscopic force does not work.

References

Kirillov, O. N. (2013). *Nonconservative stability problems of modern physics*, vol. 14. Walter de Gruyter.

Marsden, J. and A. Tromba (2003). *Vector Calculus*. W. H. Freeman.

Meirovitch, L. (1980). *Computational Methods in Structural Dynamics*. Mechanics: Dynamical Systems. Springer Netherlands.

Seyranian, A. P., E. Lund, and N. Olhoff (1994). Multiple eigenvalues in structural optimization problems. *Structural and Multidisciplinary Optimization* 8(4), 207–227.

4

Modal Analysis of Systems Subject to Conservative and Circulatory Forces

In Chapter 3, we clarified that the positional force expressed by a skew symmetric matrix, which is the circulatory force, makes systems nonconservative. In this chapter, we shift our attention to the solution method for the nonconservative systems. After a review on the orthogonality of modal vectors in the case without circulatory force, we discuss the nonorthogonality of each modal vector due to the circulatory force. Then, the inertial and positional forces in the equations of motion are decoupled by the modal analysis with adjoint vectors; the modal vectors associated with the circulatory force can be imaginary. The system exhibits a different dynamics with respect to that of single-degree-of-freedom systems mentioned in Section 2.4.

4.1 Decomposition of the Matrix *M*

We consider a two-degree-of-freedom system without velocity-dependent force, i.e. $D'd\boldsymbol{x}/dt = \boldsymbol{0}$ in Eq. (3.1)

$$M\frac{d^2\boldsymbol{x}}{dt^2} + K'\boldsymbol{x} = \boldsymbol{0}, \tag{4.1}$$

where the 2×1 matrix $\boldsymbol{x}$ is the displacement vector, the 2×2 mass matrix M is generally symmetric, and the 2×2 matrix K' expressing positional forces is the sum of the symmetric and skew-symmetric matrices (see Eqs. (3.4) and (3.5)). In this chapter, we investigate the orthogonality of the modal vectors spanning the solution vector $\boldsymbol{x}$ and clarify that the circulatory force expressed by the skew-symmetric circulatory matrix $N(= -N^T)$ makes each mode nonorthogonal.

First, we decompose the mass matrix M. Because M is symmetric and positive definite, it is diagonalized by the orthogonal matrix P $(P^T = P^{-1})$ as

$$PMP^T = M', \tag{4.2}$$

where

$$M' = \begin{bmatrix} m_{11} & 0 \\ 0 & m_{22} \end{bmatrix}. \tag{4.3}$$

Linear and Nonlinear Instabilities in Mechanical Systems: Analysis, Control and Application, First Edition. Hiroshi Yabuno.

Companion website: www.wiley.com/go/yabuno/instabilitiesinmechanicalsystems

The m_{11} and m_{22} diagonal elements are the positive eigenvalues of M. Then, M' is decomposed as

$$M' = Q^T Q, \tag{4.4}$$

where

$$Q = \begin{bmatrix} \sqrt{m_{11}} & 0 \\ 0 & \sqrt{m_{22}} \end{bmatrix}. \tag{4.5}$$

Therefore, M is expressed by using Q as

$$M = P^T M' P = P^T Q^T Q P = (QP)^T (QP) = (M^{1/2})^T M^{1/2}, \tag{4.6}$$

where $M^{1/2} = QP$. Then, Eq. (4.1) is rewritten as

$$(M^{1/2})^T M^{1/2} \frac{d^2\boldsymbol{x}}{dt^2} + K'\boldsymbol{x} = 0. \tag{4.7}$$

We introduce $\boldsymbol{u}$ satisfying

$$\boldsymbol{u} = (M^{1/2})\boldsymbol{x} \quad \text{or} \quad \boldsymbol{x} = (M^{1/2})^{-1}\boldsymbol{u}. \tag{4.8}$$

Substituting this into Eq. (4.7) and premultiplying the result by $\{(M^{1/2})^{-1}\}^T (= \{(M^{1/2})^T\}^{-1})$ yields

$$\frac{d^2\boldsymbol{u}}{dt^2} + ((M^{1/2})^{-1})^T K' (M^{1/2})^{-1}\boldsymbol{u} = 0. \tag{4.9}$$

The matrix K' in Eq. (4.1) is the sum of the symmetric matrix K and the skew-symmetric matrix N from Eq. (3.2) as

$$K' = K + N. \tag{4.10}$$

The coefficient of the second term of Eq. (4.9) is rewritten as

$$\begin{aligned} ((M^{1/2})^{-1})^T K' (M^{1/2})^{-1} &= ((M^{1/2})^{-1})^T (K + N)(M^{1/2})^{-1} \\ &= ((M^{1/2})^{-1})^T K (M^{1/2})^{-1} + ((M^{1/2})^{-1})^T N (M^{1/2})^{-1}. \end{aligned} \tag{4.11}$$

Because the first and second terms in the right-hand side are symmetric and skew-symmetric (see Problem 4.1), K and N are transformed into symmetric and skew-symmetric matrices again.

Problem 4.1 Assuming that K and N are symmetric and skew-symmetric, respectively, shows that $((M^{1/2})^{-1})^T K (M^{1/2})^{-1}$ and $((M^{1/2})^{-1})^T N (M^{1/2})^{-1}$ are symmetric and skew-symmetric, respectively.

Ans:

$$\begin{aligned} \{((M^{1/2})^{-1})^T K (M^{1/2})^{-1}\}^T &= ((M^{1/2})^{-1})^T K^T (M^{1/2})^{-1} \\ &= ((M^{1/2})^{-1})^T K (M^{1/2})^{-1}, \\ \{((M^{1/2})^{-1})^T N (M^{1/2})^{-1}\}^T &= ((M^{1/2})^{-1})^T N^T (M^{1/2})^{-1} \\ &= -((M^{1/2})^{-1})^T N (M^{1/2})^{-1}. \end{aligned}$$

Therefore, without loss of generality, we consider that the mass matrix M in Eq. (4.1) is replaced by the identity matrix in the following equation:

$$\frac{d^2\boldsymbol{x}}{dt^2} + K'\boldsymbol{x} = 0 \quad \text{or} \quad \frac{d^2\boldsymbol{x}}{dt^2} + (K + N)\boldsymbol{x} = 0. \tag{4.12}$$

4.2 Characteristic Equation and Modal Vector

Letting the solution be

$$\boldsymbol{x} = \boldsymbol{\Phi}\mathrm{e}^{\lambda t} \tag{4.13}$$

yields

$$L\boldsymbol{\Phi} = \boldsymbol{0}. \tag{4.14}$$

The linear operator L is

$$L = \lambda^2 I + K' \quad \text{or} \quad L = \Lambda I + K', \tag{4.15}$$

where $\Lambda = \lambda^2$. For nontrivial $\boldsymbol{\Phi}$, we have the following characteristic equation:

$$|\lambda^2 I + K'| = 0 \quad \text{or} \quad |\Lambda I + K'| = 0. \tag{4.16}$$

This is a fourth-order polynomial in λ and a quadratic polynomial in $\Lambda = \lambda^2$. We consider the case in which Eq. (4.16) has two distinct solutions Λ_i $(i = 1, 2)$. Let the corresponding ith vectors be $\boldsymbol{\Phi}_i$, with $i = 1, 2$. $\boldsymbol{\Phi}_i$ is called a *modal vector*. Then, we have

$$(\lambda_i^2 I + K')\boldsymbol{\Phi}_i = \boldsymbol{0}(i = 1, 2) \quad \text{or} \quad (\Lambda_i I + K')\boldsymbol{\Phi}_i = \boldsymbol{0}(i = 1, 2). \tag{4.17}$$

In the case in which a circulatory force exists, i.e. K' is not symmetric, the solutions Λ_i of Eq. (4.16) can be complex numbers. On the other hand, if K' does not include the circular matrix N, i.e. $K' = K = K^T$, is symmetric, Λ_i and $\boldsymbol{\Phi}_i$ are real for $i = 1, 2$ (see Problem 4.2). Then, $\boldsymbol{\Phi}_i$ can be expressed by real values χ_i and ψ_i as

$$\boldsymbol{\Phi}_i = \frac{1}{\sqrt{1 + \chi_i^2}} \begin{bmatrix} 1 \\ \chi_i \mathrm{e}^{\mathrm{i}\psi_i} \end{bmatrix}, \tag{4.18}$$

where ψ_i is 0 or π since the elements of $\boldsymbol{\Phi}_i$ are real.

Problem 4.2 In the case where K' is symmetric, i.e. $K' = K$, show that Λ_i and $\boldsymbol{\Phi}_i$ are real for $i = 1, 2$.

Ans: Taking the inner product of Eq. (4.17) and $\boldsymbol{\Phi}_i$ yields

$$\overline{\boldsymbol{\Phi}}_i^T (\Lambda_i I + K)\boldsymbol{\Phi}_i = 0, \quad \Lambda_i = -\frac{\overline{\boldsymbol{\Phi}}_i^T K \boldsymbol{\Phi}_i}{\overline{\boldsymbol{\Phi}}_i^T \boldsymbol{\Phi}_i}. \tag{4.19}$$

Because of

$$\overline{\overline{\boldsymbol{\Phi}}_i^T K \boldsymbol{\Phi}_i} = \overline{(\overline{\boldsymbol{\Phi}}_i^T K \boldsymbol{\Phi}_i)^T} = \overline{\boldsymbol{\Phi}}_i^T K \boldsymbol{\Phi}_i, \quad \overline{\overline{\boldsymbol{\Phi}}_i^T \boldsymbol{\Phi}_i} = \overline{(\overline{\boldsymbol{\Phi}}_i^T \boldsymbol{\Phi}_i)^T} = \overline{\boldsymbol{\Phi}}_i^T \boldsymbol{\Phi}_i, \tag{4.20}$$

Λ_i is real from Eq. (4.19). Then, $\boldsymbol{\Phi}_i$ is real from Eq. (4.17).

4.3 Modal Analysis in Case Without Circulatory Force

In general, if the inner product of a pair of vectors $\boldsymbol{x}_i$ and $\boldsymbol{x}_j$, where $i \neq j$, is zero, i.e.

$$\boldsymbol{x}_i \cdot \boldsymbol{x}_j = \overline{\boldsymbol{x}}_j^T \boldsymbol{x}_i = 0, \quad i \neq j, \tag{4.21}$$

the vectors $\boldsymbol{x}_i$ and $\boldsymbol{x}_j$ are said to be *orthogonal*. If, in addition, the vectors have unit norm, i.e.

$$\boldsymbol{x}_i \cdot \boldsymbol{x}_i = \overline{\boldsymbol{x}}_i^T \boldsymbol{x}_i = 1, \quad i = 1, 2, \tag{4.22}$$

the vectors are said to be *orthonormal*. Let us consider

$$\begin{cases} \Lambda_i \boldsymbol{\Phi}_i + K\boldsymbol{\Phi}_i = \boldsymbol{0}, & (4.23) \\ \Lambda_j \boldsymbol{\Phi}_j + K\boldsymbol{\Phi}_j = \boldsymbol{0}, & (4.24) \end{cases}$$

where $i \neq j$. Taking the inner product of Eq. (4.23) and $\boldsymbol{\Phi}_j$, and that of Eq. (4.24) and $\boldsymbol{\Phi}_i$ yields respectively

$$\begin{cases} \Lambda_i \boldsymbol{\Phi}_j^T \boldsymbol{\Phi}_i + \boldsymbol{\Phi}_j^T K\boldsymbol{\Phi}_i = 0, & (4.25) \\ \Lambda_j \boldsymbol{\Phi}_i^T \boldsymbol{\Phi}_j + \boldsymbol{\Phi}_i^T K\boldsymbol{\Phi}_j = 0. & (4.26) \end{cases}$$

The difference of Eq. (4.25) and the transpose of Eq. (4.26) is

$$(\Lambda_i - \Lambda_j)\boldsymbol{\Phi}_j^T \boldsymbol{\Phi}_i = -\boldsymbol{\Phi}_j^T (K - K^T)\boldsymbol{\Phi}_i. \tag{4.27}$$

Because of $K = K^T$, Eq. (4.27) leads to

$$(\Lambda_i - \Lambda_j)\boldsymbol{\Phi}_j^T \boldsymbol{\Phi}_i = 0. \tag{4.28}$$

Therefore, in the case of $i \neq j$, $\boldsymbol{\Phi}_j^T \boldsymbol{\Phi}_i$ must vanish identically. Hence, $\boldsymbol{\Phi}_i$ and $\boldsymbol{\Phi}_j$ are orthogonal. We determine $\boldsymbol{\Phi}_i (i = 1, 2)$ so as to satisfy

$$\boldsymbol{\Phi}_i^T \boldsymbol{\Phi}_i = 1, \quad i = 1, 2. \tag{4.29}$$

Then, $\boldsymbol{\Phi}_i$ $(i = 1, 2)$ are orthonormal and satisfy

$$\boldsymbol{\Phi}_j^T \boldsymbol{\Phi}_i = \delta_{ji}, \tag{4.30}$$

where δ_{ji} is Kronecker's delta:

$$\delta_{ji} = \begin{cases} 1 & (i = j), \\ 0 & (i \neq j). \end{cases} \tag{4.31}$$

From Eqs. (4.25) and (4.26), we have

$$\begin{cases} \boldsymbol{\Phi}_j^T \boldsymbol{\Phi}_i + \frac{1}{\Lambda_i} \boldsymbol{\Phi}_j^T K\boldsymbol{\Phi}_i = 0, & (4.32) \\ \boldsymbol{\Phi}_i^T \boldsymbol{\Phi}_j + \frac{1}{\Lambda_j} \boldsymbol{\Phi}_i^T K\boldsymbol{\Phi}_j = 0. & (4.33) \end{cases}$$

Because of $K = K^T$, the difference of Eq. (4.32) and the transpose of Eq. (4.33) leads to

$$\left(\frac{1}{\Lambda_i} - \frac{1}{\Lambda_j} \right) \boldsymbol{\Phi}_j^T K\boldsymbol{\Phi}_i = 0. \tag{4.34}$$

In the case of $i \neq j$, $\boldsymbol{\Phi}_j^T K \boldsymbol{\Phi}_i$ must vanish identically. Equations (4.30) and (4.32) lead to

$$\boldsymbol{\Phi}_j^T K \boldsymbol{\Phi}_i = -\Lambda_i \delta_{ji}. \tag{4.35}$$

Thus, the eigenvectors $\boldsymbol{\Phi}_i (i = 1, 2)$ are said to be *K-orthogonal.*

Let $\boldsymbol{x}$ be

$$\boldsymbol{x} = \xi_1 \boldsymbol{\Phi}_1 + \xi_2 \boldsymbol{\Phi}_2 = T\xi, \tag{4.36}$$

where

$$T = \left[\boldsymbol{\Phi}_1, \; \boldsymbol{\Phi}_2\right], \quad \xi = \begin{bmatrix} \xi_1 \\ \xi_2 \end{bmatrix}. \tag{4.37}$$

The matrix T is referred to as a *modal matrix.* The coordinates ξ_1 and ξ_2 are referred to as *modal coordinates.*

Substituting Eq. (4.36) into Eq. (4.12) and premultiplying the result by T^T yields

$$T^T T \ddot{\xi} + T^T K T \xi = \mathbf{0}. \tag{4.38}$$

From Eq. (4.30), $T^T T$ is expressed as

$$T^T T = \begin{bmatrix} \boldsymbol{\Phi}_1^T \boldsymbol{\Phi}_1 & \boldsymbol{\Phi}_1^T \boldsymbol{\Phi}_2 \\ \boldsymbol{\Phi}_2^T \boldsymbol{\Phi}_1 & \boldsymbol{\Phi}_2^T \boldsymbol{\Phi}_2 \end{bmatrix} = I, \tag{4.39}$$

where I is the identity matrix. In addition, exploiting the K-orthogonality expressed in Eq. (4.35),

$$T^T K T = \begin{bmatrix} \boldsymbol{\Phi}_1^T K \boldsymbol{\Phi}_1 & \boldsymbol{\Phi}_1^T K \boldsymbol{\Phi}_2 \\ \boldsymbol{\Phi}_2^T K \boldsymbol{\Phi}_1 & \boldsymbol{\Phi}_2^T K \boldsymbol{\Phi}_2 \end{bmatrix} \tag{4.40}$$

is diagonalized as

$$T^T K T = \begin{bmatrix} -\Lambda_1 & 0 \\ 0 & -\Lambda_2 \end{bmatrix}. \tag{4.41}$$

Hence, Eq. (4.38) can be expressed as

$$\frac{d^2}{dt^2} \begin{bmatrix} \xi_1 \\ \xi_2 \end{bmatrix} + \begin{bmatrix} -\Lambda_1 & 0 \\ 0 & -\Lambda_2 \end{bmatrix} \begin{bmatrix} \xi_1 \\ \xi_2 \end{bmatrix} = \mathbf{0}. \tag{4.42}$$

This equation consists of the following decoupled systems:

$$\frac{d^2 \xi_i}{dt^2} - \Lambda_i \xi_i = 0, \; i = 1, 2. \tag{4.43}$$

Since Λ_i is real (see Problem 4.2), λ in Eq. (4.13) may be real or pure imaginary as follows:

$$\lambda_i = \pm\sqrt{\Lambda_i} \; (\Lambda_i \geq 0) \quad \text{or} \quad \lambda_i = \pm\sqrt{-\Lambda_i} i \; (\Lambda_i < 0). \tag{4.44}$$

In cases of $\Lambda_i > 0$ and $\Lambda_i = 0$, ξ is expressed respectively as

$$\xi_i = a_{i1} e^{\sqrt{\Lambda_i} t} + a_{i2} e^{-\sqrt{\Lambda_i} t} \tag{4.45}$$

and

$$\xi_i = a_{i1} + a_{i2} t, \tag{4.46}$$

where a_{i1} and a_{i2} are real and determined by the initial condition. The phase portrait of the dynamics is described in Figures 2.1b and 2.4c, respectively.

In the case of $\Lambda_i < 0$, using a complex amplitude A_i, we obtain

$$\begin{aligned}\xi_i &= A_i e^{i\sqrt{-\Lambda_i}t} + \overline{A}_i e^{-i\sqrt{-\Lambda_i}t} \\ &= a_i \cos(\sqrt{-\Lambda_i}t + \phi_i),\end{aligned} \tag{4.47}$$

where $A_i = a_i e^{i\phi_i}/2$, and the real numbers a_i and ϕ_i are determined by the initial conditions. The phase portrait of this dynamics is described in Figure 2.5b. From Eqs. (4.18), (4.36), and (4.47), the behavior of the system Eq. (4.1) is expressed as

$$\begin{aligned}\boldsymbol{x} &= \xi_1 \boldsymbol{\Phi}_1 + \xi_2 \boldsymbol{\Phi}_2 \\ &= \frac{a_1}{\sqrt{1+\chi_1^2}} \cos(\sqrt{-\Lambda_1}t + \phi_1) \begin{bmatrix} 1 \\ \chi_1 e^{i\psi_1} \end{bmatrix} \\ &+ \frac{a_2}{\sqrt{1+\chi_2^2}} \cos(\sqrt{-\Lambda_2}t + \phi_2) \begin{bmatrix} 1 \\ \chi_2 e^{i\psi_2} \end{bmatrix},\end{aligned} \tag{4.48}$$

where $\sqrt{-\Lambda_1} < \sqrt{-\Lambda_2}$. The first and second terms are called first and second modes, respectively, and $\sqrt{-\Lambda_1}$ and $\sqrt{-\Lambda_2}$ are called the natural frequencies of first mode and of second mode, respectively. ψ_i is 0 or π which is the synchronous oscillation of in-phase or out-of-phase, respectively.

For example let us consider the system Eq. (3.20) again. For the parameter values and initial condition which are the same as those in Figure 3.3, we obtain $a_1 = 1.12$, $a_2 = 0.85$,

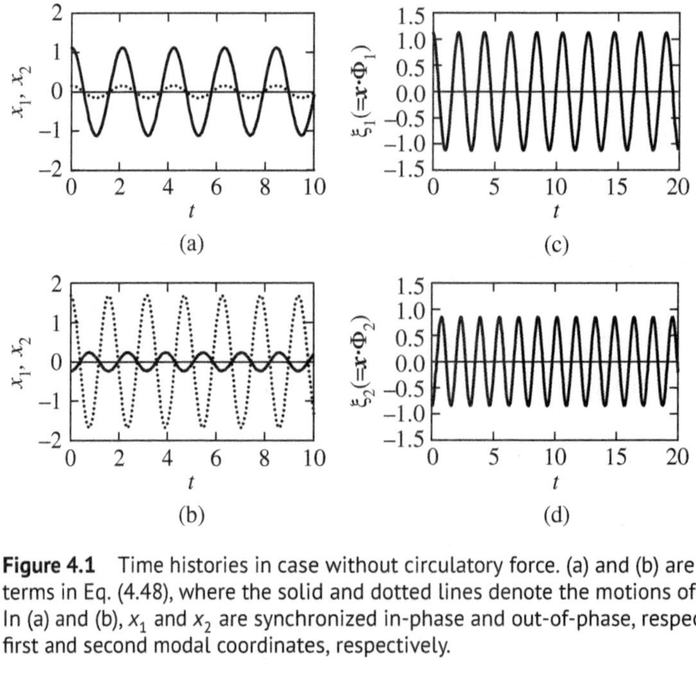

Figure 4.1 Time histories in case without circulatory force. (a) and (b) are the first and second terms in Eq. (4.48), where the solid and dotted lines denote the motions of x_1 and x_2, respectively. In (a) and (b), x_1 and x_2 are synchronized in-phase and out-of-phase, respectively. (c) and (d) are the first and second modal coordinates, respectively.

$\phi_1 = 0$, and $\phi_2 = \pi$ and from the parameter values, we calculate $\psi_1 = 0$, $\psi_2 = \pi$, $\chi_1 = \frac{\sqrt{53}-7}{2}$, and $\chi_2 = \frac{\sqrt{53}+7}{2}$. The time histories of the first and second terms in Eq. (4.48) are Figure 4.1a,b, which are the synchronous oscillation of in-phase and out-of-phase, respectively. Since $\boldsymbol{\Phi}_i$ is orthonormal, ξ_i is obtained by the inner product between Eq. (4.36) and $\boldsymbol{\Phi}_i$ as follows:

$$\xi_i = \boldsymbol{\Phi}_i^T \boldsymbol{x}. \tag{4.49}$$

Therefore, by taking the inner product between the time history $\boldsymbol{x}$ and $\boldsymbol{\Phi}_i$, we can extract the modal coordinate ξ_i shown in Figure 4.1c,d.

4.4 Modal Analysis in Case with Circulatory Force

We consider the case with nonsymmetric positional force $K'\boldsymbol{x}$ ($K' \neq K'^T$) that includes the circulatory force $N\boldsymbol{x}$. In this case, Λ_i can be imaginary because the first equation of Eq. (4.20) does not hold. Because the equation corresponding to Eq. (4.27) is

$$(\Lambda_i - \Lambda_j)\boldsymbol{\Phi}_j^T\boldsymbol{\Phi}_i = -\boldsymbol{\Phi}_j^T(K' - K'^T)\boldsymbol{\Phi}_i \neq 0, \tag{4.50}$$

the eigenvectors are not orthogonal each other. In particular, it should be noted that K'-orthogonality does not hold. In order to carry out the modal analysis for such systems, we begin with the introduction of the concepts of *self-adjoint* and *nonself-adjoint*.

When the linear operator L satisfies the following relationship for any two vectors, $\boldsymbol{a}$ and $\boldsymbol{b}$:

$$(L\boldsymbol{a} \cdot \boldsymbol{b}) - (\boldsymbol{a} \cdot L\boldsymbol{b}) = 0, \tag{4.51}$$

the linear operator L is called *self-adjoint*. If Eq. (4.51) does not hold, it is called *nonself-adjoint*.

Problem 4.4 When the linear operator L is a matrix, seek the condition that L is self-adjoint.

Ans:

$$(L\boldsymbol{a} \cdot \boldsymbol{b}) - (\boldsymbol{a} \cdot L\boldsymbol{b}) = \overline{\boldsymbol{b}}^T L\boldsymbol{a} - \overline{(L\,\boldsymbol{b})}^T \boldsymbol{a} = \overline{\boldsymbol{b}}^T(L - \overline{L}^T)\boldsymbol{a}. \tag{4.52}$$

Therefore, if and only if $L = \overline{L}^T$, L is self-adjoint. Otherwise, L is nonself-adjoint.

When a circulatory force exists ($\overline{K'}^T \neq K'$), the linear operator L in Eq. (4.15) is nonself-adjoint. Thus, $\boldsymbol{\Phi}_i$ is not orthogonal. We define an *adjoint operator* of L as $L^* \overset{\text{def}}{=} \overline{L}^T$; hereafter, the superscript * to a matrix denotes the conjugate transposed matrix. Therefore, using $L^*(\Lambda_i)$, we define the vector $\boldsymbol{\Psi}_i$ that satisfies

$$L^*(\Lambda_i)\boldsymbol{\Psi}_i = \boldsymbol{0}, \tag{4.53}$$

where $\boldsymbol{\Psi}_i$ is called an *adjoint vector* of $\boldsymbol{\Phi}_i$. For matrix

$$L(\Lambda_i) = \Lambda_i I + K', \tag{4.54}$$

the adjoint matrix is defined as

$$L^*(\Lambda_i) = \overline{\Lambda}_i I + K'^T, \tag{4.55}$$

where every element of K' is real. We consider

$$\begin{cases} L(\Lambda_i)\mathbf{\Phi}_i = 0 \Rightarrow \Lambda_i \mathbf{\Phi}_i + K'\mathbf{\Phi}_i = \mathbf{0}, & (4.56) \\ L^*(\Lambda_j)\mathbf{\Psi}_j = 0 \Rightarrow \overline{\Lambda}_j \mathbf{\Psi}_j + K'^T \mathbf{\Psi}_j = \mathbf{0}. & (4.57) \end{cases}$$

Taking the inner product of Eqs. (4.56) and (4.57) with $\mathbf{\Psi}_j$ and $\mathbf{\Phi}_i$, respectively, yields

$$\begin{cases} \Lambda_i \mathbf{\Psi}_j^* \mathbf{\Phi}_i + \mathbf{\Psi}_j^* K' \mathbf{\Phi}_i = 0, & (4.58) \\ \overline{\Lambda}_j \mathbf{\Phi}_i^* \mathbf{\Psi}_j + \mathbf{\Phi}_i^* K'^T \mathbf{\Psi}_j = 0. & (4.59) \end{cases}$$

The transpose and conjugate of Eq. (4.59) is

$$\Lambda_j \mathbf{\Psi}_j^* \mathbf{\Phi}_i + \mathbf{\Psi}_j^* K' \mathbf{\Phi}_i = 0. \tag{4.60}$$

The difference between Eq. (4.58) and Eq. (4.60) is

$$(\Lambda_i - \Lambda_j)\mathbf{\Psi}_j^* \mathbf{\Phi}_i = 0. \tag{4.61}$$

It is concluded that $\mathbf{\Phi}_i$ is orthogonal to $\mathbf{\Psi}_j$ $(i \neq j)$. We determine $\mathbf{\Phi}_i$ and $\mathbf{\Psi}_i$ for $i = 1, 2$ so as to satisfy

$$\mathbf{\Psi}_i^* \mathbf{\Phi}_i = 1, \quad i = 1, 2. \tag{4.62}$$

Then, we obtain the equation corresponding to Eq. (4.30) as

$$\mathbf{\Psi}_j^* \mathbf{\Phi}_i = \delta_{ji}. \tag{4.63}$$

The geometrical relationship between $\mathbf{\Phi}_j$ and $\mathbf{\Psi}_j$ is described in Appendix F. On the other hand, Eqs. (4.58) and (4.60) are respectively rewritten as

$$\mathbf{\Psi}_j^* \mathbf{\Phi}_i + \frac{1}{\Lambda_i} \mathbf{\Psi}_j^* K' \mathbf{\Phi}_i = 0 \tag{4.64}$$

and

$$\mathbf{\Psi}_j^* \mathbf{\Phi}_i + \frac{1}{\Lambda_j} \mathbf{\Psi}_j^* K' \mathbf{\Phi}_i = 0. \tag{4.65}$$

The difference between Eq. (4.64) and Eq. (4.65) is

$$\left(\frac{1}{\Lambda_i} - \frac{1}{\Lambda_j} \right) \mathbf{\Psi}_j^* K' \mathbf{\Phi}_i = 0. \tag{4.66}$$

In the case of $i \neq j$, $\mathbf{\Psi}_j^* K' \mathbf{\Phi}_i$ must vanish identically. Taking into account Eqs. (4.58) and (4.63) yields the equation corresponding to Eq. (4.35) as

$$\mathbf{\Psi}_j^* K' \mathbf{\Phi}_i = -\Lambda_i \delta_{ji}. \tag{4.67}$$

Substituting Eq. (4.36) into Eq. (4.12) and taking the inner product of the result and $\tilde{T}$, where

$$\tilde{T} = [\mathbf{\Psi}_1, \mathbf{\Psi}_2] \tag{4.68}$$

yields

$$\tilde{T}^* T \ddot{\xi} + \tilde{T}^* K' T \xi = \mathbf{0}. \tag{4.69}$$

From Eq. (4.63), we have

$$\tilde{T}^*T = \begin{bmatrix} \boldsymbol{\Psi}_1^*\boldsymbol{\Phi}_1 & \boldsymbol{\Psi}_1^*\boldsymbol{\Phi}_2 \\ \boldsymbol{\Psi}_2^*\boldsymbol{\Phi}_1 & \boldsymbol{\Psi}_2^*\boldsymbol{\Phi}_2 \end{bmatrix} = I. \tag{4.70}$$

Moreover, because of the orthogonality of K' in Eq. (4.67), we have

$$\tilde{T}^*K'T = \begin{bmatrix} \boldsymbol{\Psi}_1^*K'\boldsymbol{\Phi}_1 & \boldsymbol{\Psi}_1^*K_1'\boldsymbol{\Phi}_2 \\ \boldsymbol{\Psi}_2^*K'\boldsymbol{\Phi}_1 & \boldsymbol{\Psi}_2^*K'\boldsymbol{\Phi}_2 \end{bmatrix} = \begin{bmatrix} -\Lambda_1 & 0 \\ 0 & -\Lambda_2 \end{bmatrix}. \tag{4.71}$$

By Eqs (4.70) and (4.71), Eq. (4.69) can be expressed in the decoupled form as

$$\frac{d^2}{dt^2}\begin{bmatrix} \xi_1 \\ \xi_2 \end{bmatrix} - \begin{bmatrix} \Lambda_1 & 0 \\ 0 & \Lambda_2 \end{bmatrix}\begin{bmatrix} \xi_1 \\ \xi_2 \end{bmatrix} = \mathbf{0}. \tag{4.72}$$

This equation governs the time variations of ξ_1 and ξ_2 in Eq. (4.36).

4.4.1 Case Study 1: Λ_i are Real

Since Λ_i is real, λ_i may be real or pure imaginary as follows:

$$\lambda_i = \pm\sqrt{\Lambda_i}\ (\Lambda_i \geq 0) \quad \text{or} \quad \lambda_i = \pm\sqrt{-\Lambda_i}i\ (\Lambda_i < 0). \tag{4.73}$$

Therefore, as in the case without circulatory force, i.e. $K' = K$, the behavior of the system is expressed by the combination of the dynamics described by the phase portraits of Figures 2.1b, 2.4c, or 2.5b, for $\Lambda_i > 0$, $\Lambda_i = 0$, or $\Lambda_i < 0$, respectively.

Let us consider the case of $\Lambda_i < 0 (i = 1, 2)$. Like the case without circulatory force, the behavior of the system Eq. (4.1) is expressed as

$$\begin{aligned} \boldsymbol{x} &= \boldsymbol{\Phi}_1\xi_1 + \boldsymbol{\Phi}_2\xi_2 \\ &= \frac{a_1}{\sqrt{1+\chi_1^2}}\cos(\sqrt{-\Lambda_1}t + \phi_1)\begin{bmatrix} 1 \\ \chi_1 e^{i\psi_1} \end{bmatrix} \\ &\quad + \frac{a_2}{\sqrt{1+\chi_2^2}}\cos(\sqrt{-\Lambda_2}t + \phi_2)\begin{bmatrix} 1 \\ \chi_2 e^{i\psi_2} \end{bmatrix}, \end{aligned} \tag{4.74}$$

where ψ_i is 0 or π which is the synchronous oscillation of in-phase or out-of-phase, respectively.

For example, let us consider the system Eq. (3.33). We show in Figure 4.2a,b, Eq. (4.74) for the system with $m_1 = m_2 = 1, k_{11} = 9, k_{22} = 16, k_{12} = 1$, and $k_{21} = -1$ under the initial conditions $x_1 = x_2 = 1$ and $\frac{dx_1}{dt} = \frac{dx_2}{dt} = 0$. From the initial condition, we obtain $a_1 = 0.88$, $a_2 = 0.88$, $\phi_1 = 0$, and $\phi_2 = 0$, and from the parameter values, we calculate $\psi_1 = 0$, $\psi_2 = 0$, $\chi_1 = \frac{7-3\sqrt{5}}{2}$, and $\chi_2 = \frac{7+3\sqrt{5}}{2}$.

Unlike the case without circulatory force in Section 4.3, $\boldsymbol{\Phi}_i$ is not orthogonal to $\boldsymbol{\Phi}_j (i \neq j)$, but, from Eq. (4.63), it is orthogonal to $\boldsymbol{\Psi}_j$, where $i \neq j$. Thus, the modal coordinates ξ_i in Eq. (4.36) are obtained by the inner product between Eq. (4.36) and $\boldsymbol{\Psi}_i$ as

$$\xi_i = \boldsymbol{\Psi}_i^*\boldsymbol{x}. \tag{4.75}$$

By taking the inner product between the time history $\boldsymbol{x}$ and $\boldsymbol{\Psi}_i$, we can extract the modal coordinate ξ_i for the above example as shown in Figure 4.2c,d. On the other hand, the inner

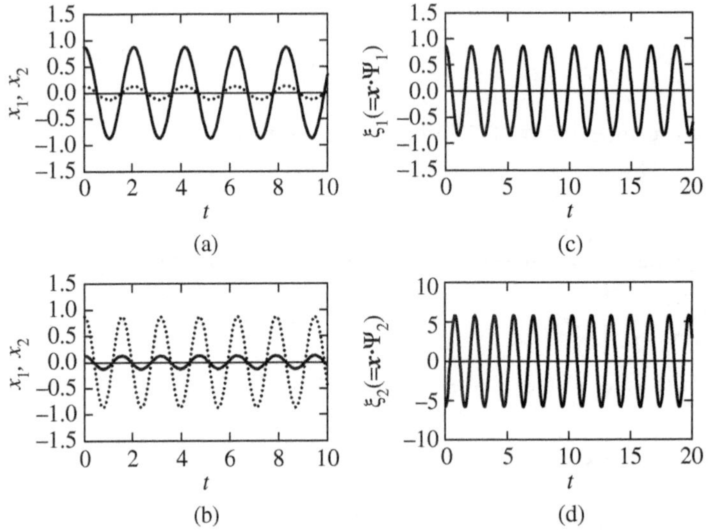

Figure 4.2 Time histories in case with circulatory force (eigenvalues are pure imaginary). (a) and (b) are the first and second terms in Eq. (4.74), where the solid and dashed lines denote the motions of x_1 and x_2, respectively. In (a) and (b), x_1 and x_2 are synchronized in-phase. (c) and (d) are the first and second modal coordinates, respectively.

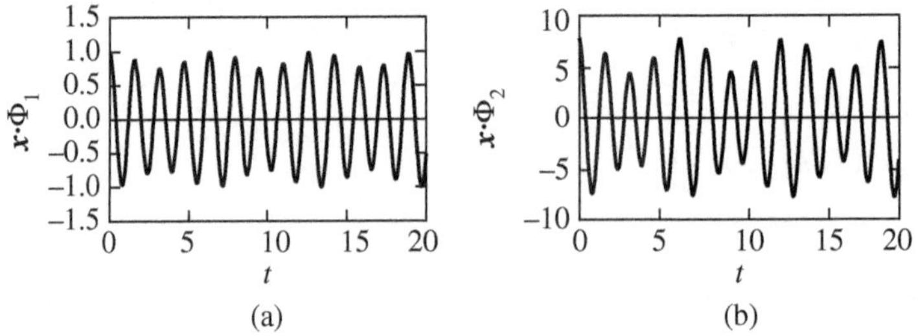

Figure 4.3 Non-orthogonarity of $\boldsymbol{\Phi}_i (i = 1, 2)$. (a) and (b) show $\xi_1 \neq \boldsymbol{\Phi}_1^* \boldsymbol{x}$ and $\xi_2 \neq \boldsymbol{\Phi}_2^* \boldsymbol{x}$, respectively.

product between the time history $\boldsymbol{x}$ and $\boldsymbol{\Phi}_i$ does not lead to the modal coordinates ξ_i; see Figure 4.3a,b.

4.4.2 Case Study 2: Λ_i are Complex

When Eq. (4.16) has a pair of complex conjugate solutions, $\Lambda_1 = \Lambda$ and $\Lambda_2 = \overline{\Lambda}$, the equations corresponding to Eq. (4.17) are expressed by complex mode $\boldsymbol{\Phi}$ as

$$(\Lambda I + K')\boldsymbol{\Phi} = \boldsymbol{0}, \tag{4.76}$$

and

$$(\overline{\Lambda} I + K')\overline{\boldsymbol{\Phi}} = \boldsymbol{0}, \tag{4.77}$$

where $\boldsymbol{\Phi}$ and $\overline{\boldsymbol{\Phi}}$ can be regarded as $\boldsymbol{\Phi}_1$ and $\boldsymbol{\Phi}_2$, respectively. Also, since $\boldsymbol{x}$ is real, using $\xi_1 = \xi$ and $\xi_2 = \bar{\xi}$, we can express $\boldsymbol{x}$ as

$$\boldsymbol{x} = \xi_1\boldsymbol{\Phi}_1 + \xi_2\boldsymbol{\Phi}_2 = \xi\boldsymbol{\Phi} + \overline{\xi\boldsymbol{\Phi}} = T\xi, \tag{4.78}$$

where

$$T = \left[\boldsymbol{\Phi},\ \overline{\boldsymbol{\Phi}}\right], \quad \xi = \begin{bmatrix} \xi \\ \bar{\xi} \end{bmatrix}. \tag{4.79}$$

Substituting Eq. (4.78) into Eq. (4.12) and premultiplying the result by $\tilde{T}^*$, where

$$\tilde{T} = \left[\boldsymbol{\Psi},\ \overline{\boldsymbol{\Psi}}\right], \tag{4.80}$$

we have the equation corresponding to Eq. (4.72) as

$$\frac{d^2}{dt^2}\begin{bmatrix} \xi \\ \bar{\xi} \end{bmatrix} - \begin{bmatrix} \Lambda & 0 \\ 0 & \bar{\Lambda} \end{bmatrix}\begin{bmatrix} \xi \\ \bar{\xi} \end{bmatrix} = \boldsymbol{0} \tag{4.81}$$

thus, resulting in

$$\frac{d^2\xi}{dt^2} - \Lambda\xi = 0, \tag{4.82}$$

and

$$\frac{d^2\bar{\xi}}{dt^2} - \bar{\Lambda}\bar{\xi} = 0. \tag{4.83}$$

These equations exhibit new dynamics that are not included in Chapter 2.

Then, the solutions of Eqs. (4.82) and (4.83) are

$$\begin{cases} \xi = A_u e^{(\mu+i\omega)t} + A_s e^{(-\mu-i\omega)t}, & (4.84) \\ \bar{\xi} = \bar{A}_u e^{(\mu-i\omega)t} + \bar{A}_s e^{(-\mu+i\omega)t}, & (4.85) \end{cases}$$

where A_u and A_s are complex amplitudes. These can be expressed in the polar form by

$$A_u = \frac{a_u}{2}e^{i\phi_u}, \quad A_s = \frac{a_s}{2}e^{-i\phi_s}, \tag{4.86}$$

where a_u, a_s, ϕ_u, and ϕ_s are real and determined from the initial conditions. Thus, the complex modal vector $\boldsymbol{\Phi}$ can be expressed without loss of generality as

$$\boldsymbol{\Phi} = \frac{1}{\sqrt{1+\chi^2}}\begin{bmatrix} 1 \\ \chi e^{i\psi} \end{bmatrix}, \tag{4.87}$$

where χ and ψ are real numbers obtained from Eq. (4.76), and ψ is neither 0 nor π. As a result, $\boldsymbol{x}$ is expressed as

$$\begin{aligned} \boldsymbol{x} &= \xi\boldsymbol{\Phi} + \bar{\xi}\overline{\boldsymbol{\Phi}} \\ &= \{A_u e^{(\mu+i\omega)t} + A_s e^{(-\mu-i\omega)t}\}\boldsymbol{\Phi} + \{\bar{A}_u e^{(\mu-i\omega)t} + \bar{A}_s e^{(-\mu+i\omega)t}\}\overline{\boldsymbol{\Phi}} \\ &= \frac{a_u}{\sqrt{1+\chi^2}}e^{\mu t}\begin{bmatrix} \cos(\omega t + \phi_u) \\ \chi\cos(\omega t + \phi_u + \psi) \end{bmatrix} + \frac{a_s}{\sqrt{1+\chi^2}}e^{-\mu t}\begin{bmatrix} \cos(\omega t + \phi_s) \\ \chi\cos(\omega t + \phi_s - \psi) \end{bmatrix}. \end{aligned} \tag{4.88}$$

The motion consists of two components; the first term tends to infinity and the second ones to zero exponentially. Unlike the case in which $\Lambda_i (i = 1, 2)$ are real numbers, due to the existence of ψ not equal to 0 or π based on the complex normal vector Φ, there is not any synchronous motion. For example, let us consider the system Eq. (3.33). We show the time histories of the first and second terms in Eq. (4.88) for the system with $m_1 = m_2 = 1$, $k_{11} = 9, k_{22} = 16, k_{12} = 13, k_{21} = -1$ under the initial conditions $x_1 = x_2 = 1$, and $\frac{dx_1}{dt} = \frac{dx_2}{dt} = 0$, which correspond to those for the time histories in Figure 3.6a,b. From the initial condition, we obtain $a_u = 5.71$, $a_s = 5.71$, $\phi_u = -1.47$, and $\phi_s = 1.47$, and from the parameter values, we calculate $\psi = 0.24$. Then, the time histories are described in Figure 4.4. Since the time for $x_1 = 0$ and $x_2 = 0$ are different, the motions are not synchronous. The system produces the self-excited oscillation due to the first term of Eq. (4.88).

Furthermore, let us investigate the characteristics of the dynamics in the case with complex eigenmodes governed by Eq. (4.82). Substituting $\xi = \xi_r + i\xi_i$ into Eq. (4.82) yields

$$\frac{d^2}{dt^2}\begin{bmatrix}\xi_r\\ \xi_i\end{bmatrix} - \begin{bmatrix}\Lambda_r & -\Lambda_i\\ \Lambda_i & \Lambda_r\end{bmatrix}\begin{bmatrix}\xi_r\\ \xi_i\end{bmatrix} = \mathbf{0}, \tag{4.89}$$

where $\Lambda = \Lambda_r + i\Lambda_i$, and ξ_r, ξ_i, Λ_r, and Λ_i are real. The characteristic equation of $A' = \begin{bmatrix}\Lambda_r & -\Lambda_i\\ \Lambda_i & \Lambda_r\end{bmatrix}$ is

$$\Lambda'^2 - 2\Lambda_r\Lambda' + \Lambda_r^2 + \Lambda_i^2 = 0. \tag{4.90}$$

The solutions are

$$\Lambda'_+ = \Lambda_r + i\Lambda_i, \quad \Lambda'_- = \Lambda_r - i\Lambda_i. \tag{4.91}$$

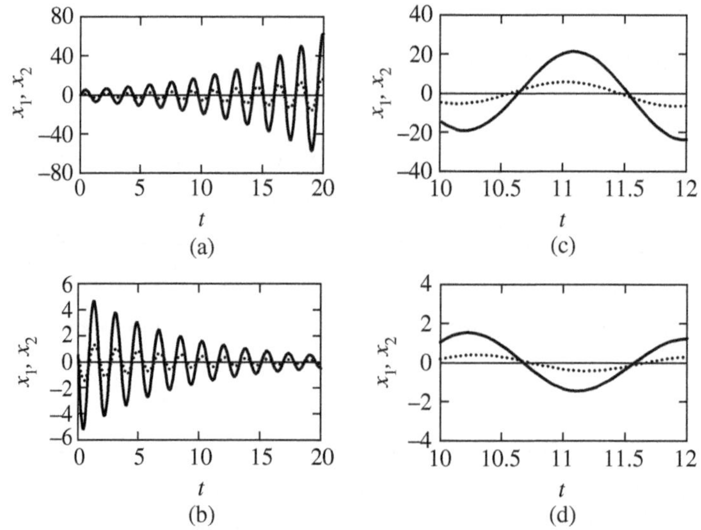

Figure 4.4 Time histories in case with circulatory force (eigenvalues are complex). (a) and (b) are the first and second terms in Eq. (4.88), where the solid and dashed lines denote the motions of x_1 and x_2, respectively. In (a) and (b), x_1 and x_2 are not synchronized in-phase or out-of-phase because ψ is not 0 or π as shown in (c) and (d).

Introducing $x_1 = \xi_r, x_2 = \frac{d\xi_r}{dt}, x_3 = \xi_i$, and $x_4 = \frac{d\xi_i}{dt}$, we consider the state equation

$$\frac{d}{dt}\begin{bmatrix} x_1 \\ x_2 \\ x_3 \\ x_4 \end{bmatrix} - \begin{bmatrix} 0 & 1 & 0 & 0 \\ \Lambda_r & 0 & -\Lambda_i & 0 \\ 0 & 0 & 0 & 1 \\ \Lambda_i & 0 & \Lambda_r & 0 \end{bmatrix}\begin{bmatrix} x_1 \\ x_2 \\ x_3 \\ x_4 \end{bmatrix} = \mathbf{0}, \tag{4.92}$$

where the characteristic equation of $A = \begin{bmatrix} 0 & 1 & 0 & 0 \\ \Lambda_r & 0 & -\Lambda_i & 0 \\ 0 & 0 & 0 & 1 \\ \Lambda_i & 0 & \Lambda_r & 0 \end{bmatrix}$ is

$$\lambda^4 - 2\Lambda_r\lambda^2 + \Lambda_r^2 + \Lambda_i^2 = 0. \tag{4.93}$$

Because of $\Lambda' = \lambda^2$ from Eqs. (4.90) and (4.93), the four eigenvalues of Eq. (4.93) are expressed as follows. Let the two eigenvalues derived from $\sqrt{\Lambda'_+}$ be $\lambda_{\mathrm{I}} = \mu + i\omega$ ($\mu > 0$, $\omega > 0$) and $\lambda_{\mathrm{II}} = \lambda_{\mathrm{I}} e^{i\pi} = -\mu - i\omega$, while the other two eigenvalues derived from $\sqrt{\Lambda'_-}$ can be expressed as $\lambda_{\mathrm{III}} = \overline{\lambda_{\mathrm{I}}} = \mu - i\omega$ and $\lambda_{\mathrm{IV}} = \lambda_{\mathrm{III}} e^{i\pi} = -\mu + i\omega$ (see Figures 4.5 and 4.6a). Focusing on the eigenvectors of A, $\boldsymbol{x}(\mathrm{I})$, and $\boldsymbol{x}(\mathrm{IV})$, corresponding to λ_{I} and λ_{IV}, respectively, we transform A into a *block diagonal form* as shown in Section 2.4.4. By using

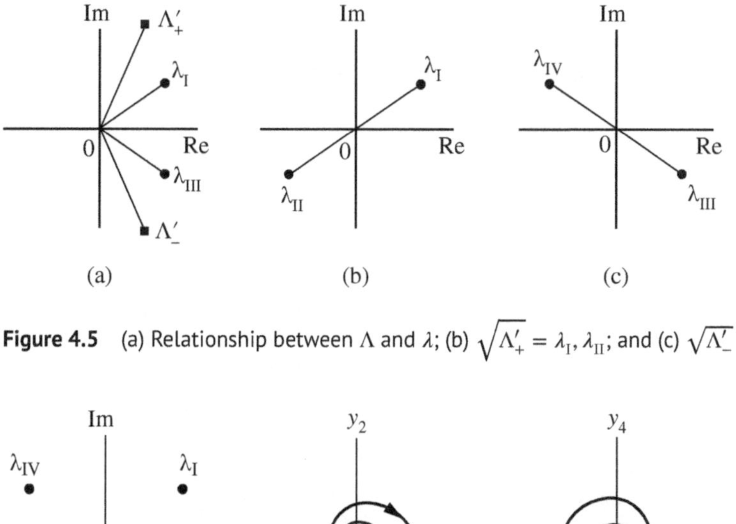

Figure 4.5 (a) Relationship between Λ and λ; (b) $\sqrt{\Lambda'_+} = \lambda_{\mathrm{I}}, \lambda_{\mathrm{II}}$; and (c) $\sqrt{\Lambda'_-} = \lambda_{\mathrm{III}}, \lambda_{\mathrm{IV}}$.

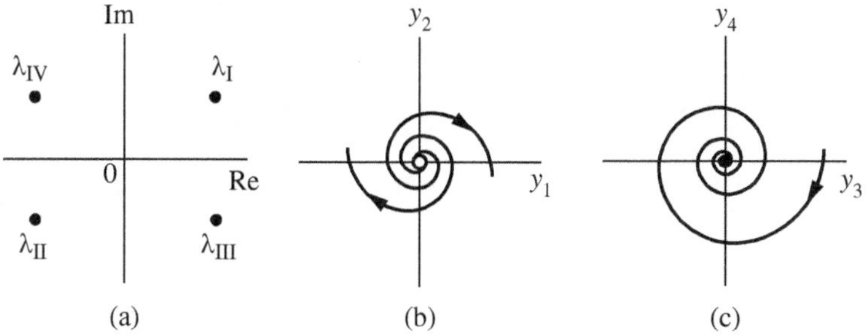

Figure 4.6 The four eigenvalues of λ (a) and phase portraits. λ_{I} and λ_{III} are a pair of complex conjugate numbers with positive real part and the dynamics are expressed by the phase portrait (b). λ_{II} and λ_{IV} are a pair of conjugate complex numbers with negative real part and the dynamics are expressed by the phase portrait (c).

the real vectors,

$$\boldsymbol{x}(\mathrm{I})_r = \begin{bmatrix} x_1(\mathrm{I})_r \\ x_2(\mathrm{I})_r \\ x_3(\mathrm{I})_r \\ x_4(\mathrm{I})_r \end{bmatrix}, \quad \boldsymbol{x}(\mathrm{I})_i = \begin{bmatrix} x_1(\mathrm{I})_i \\ x_2(\mathrm{I})_i \\ x_3(\mathrm{I})_i \\ x_4(\mathrm{I})_i \end{bmatrix}, \tag{4.94}$$

$$\boldsymbol{x}(\mathrm{IV})_r = \begin{bmatrix} x_1(\mathrm{IV})_r \\ x_2(\mathrm{IV})_r \\ x_3(\mathrm{IV})_r \\ x_4(\mathrm{IV})_r \end{bmatrix}, \quad \boldsymbol{x}(\mathrm{IV})_i = \begin{bmatrix} x_1(\mathrm{IV})_i \\ x_2(\mathrm{IV})_i \\ x_3(\mathrm{IV})_i \\ x_4(\mathrm{IV})_i \end{bmatrix}, \tag{4.95}$$

the complex eigenvectors can be expressed as

$$\boldsymbol{x}(\mathrm{I}) = \boldsymbol{x}(\mathrm{I})_r + i\boldsymbol{x}(\mathrm{I})_i, \quad \boldsymbol{x}(\mathrm{IV}) = \boldsymbol{x}(\mathrm{IV})_r + i\boldsymbol{x}(\mathrm{IV})_i. \tag{4.96}$$

Then, from the definitions of eigenvector and eigenvalue, we have

$$\begin{cases} A[\boldsymbol{x}(\mathrm{I})_r + i\boldsymbol{x}(\mathrm{I})_i] = (\mu + i\omega)[\boldsymbol{x}(\mathrm{I})_r + i\boldsymbol{x}(\mathrm{I})_i]. & (4.97) \\ A[\boldsymbol{x}(\mathrm{IV})_r + i\boldsymbol{x}(\mathrm{IV})_i] = (-\mu + i\omega)[\boldsymbol{x}(\mathrm{IV})_r + i\boldsymbol{x}(\mathrm{IV})_i]. & (4.98) \end{cases}$$

Separating each equation into the real and imaginary parts, we have the following four equations.

Real part of Eq. (4.97):

$$A\boldsymbol{x}(\mathrm{I})_r = \mu\boldsymbol{x}(\mathrm{I})_r - \omega\boldsymbol{x}(\mathrm{I})_i = \left[\boldsymbol{x}(\mathrm{I})_r, \;\; \boldsymbol{x}(\mathrm{I})_i, \;\; \boldsymbol{x}(\mathrm{IV})_r, \;\; \boldsymbol{x}(\mathrm{IV})_i\right] \begin{bmatrix} \mu \\ -\omega \\ 0 \\ 0 \end{bmatrix}. \tag{4.99}$$

Imaginary part of Eq.(4.97):

$$A\boldsymbol{x}(\mathrm{I})_i = \omega\boldsymbol{x}(\mathrm{I})_r + \mu\boldsymbol{x}(\mathrm{I})_i = \left[\boldsymbol{x}(\mathrm{I})_r, \boldsymbol{x}(\mathrm{I})_i, \boldsymbol{x}(\mathrm{IV})_r, \boldsymbol{x}(\mathrm{IV})_i\right] \begin{bmatrix} \omega \\ \mu \\ 0 \\ 0 \end{bmatrix}. \tag{4.100}$$

Real part of Eq. (4.98):

$$A\boldsymbol{x}(\mathrm{IV})_r = -\mu\boldsymbol{x}(\mathrm{IV})_r - \omega\boldsymbol{x}(\mathrm{VI})_i = \left[\boldsymbol{x}(\mathrm{I})_r, \;\; \boldsymbol{x}(\mathrm{I})_i, \;\; \boldsymbol{x}(\mathrm{IV})_r, \;\; \boldsymbol{x}(\mathrm{IV})_i\right] \begin{bmatrix} 0 \\ 0 \\ -\mu \\ -\omega \end{bmatrix}. \tag{4.101}$$

Imaginary part of Eq. (4.98):

$$A\boldsymbol{x}(\mathrm{IV})_i = \omega\boldsymbol{x}(\mathrm{IV})_r - \mu\boldsymbol{x}(\mathrm{IV})_i = \left[\boldsymbol{x}(\mathrm{I})_r, \;\; \boldsymbol{x}(\mathrm{I})_i, \;\; \boldsymbol{x}(\mathrm{IV})_r, \;\; \boldsymbol{x}(\mathrm{IV})_i\right] \begin{bmatrix} 0 \\ 0 \\ \omega \\ -\mu \end{bmatrix}. \tag{4.102}$$

According to the method in Section 2.4.4, we can combine the above equations as

$$A\left[\ \boldsymbol{x}(\mathrm{I})_r,\ \ \boldsymbol{x}(\mathrm{I})_i,\ \ \boldsymbol{x}(\mathrm{IV})_r,\ \ \boldsymbol{x}(\mathrm{IV})_i\right]$$

$$=\left[\ \boldsymbol{x}(\mathrm{I})_r,\ \ \boldsymbol{x}(\mathrm{I})_i,\ \ \boldsymbol{x}(\mathrm{IV})_r,\ \ \boldsymbol{x}(\mathrm{IV})_i\right]\begin{bmatrix} \mu & \omega & 0 & 0 \\ -\omega & \mu & 0 & 0 \\ 0 & 0 & -\mu & \omega \\ 0 & 0 & -\omega & -\mu \end{bmatrix}. \tag{4.103}$$

We introduce the transformation matrix $P = \left[\boldsymbol{x}(\mathrm{I})_r,\ \boldsymbol{x}(\mathrm{I})_i,\ \boldsymbol{x}(\mathrm{IV})_r,\ \boldsymbol{x}(\mathrm{IV})_i\right]$. Substituting

$$\begin{bmatrix} x_1 \\ x_2 \\ x_3 \\ x_4 \end{bmatrix} = P \begin{bmatrix} y_1 \\ y_2 \\ y_3 \\ y_4 \end{bmatrix}, \tag{4.104}$$

into Eq. (4.92) and premultiplying the result by P^{-1}, we have the following block diagonal form:

$$\frac{d}{dt}\begin{bmatrix} y_1 \\ y_2 \\ y_3 \\ y_4 \end{bmatrix} = \begin{bmatrix} \mu & \omega & 0 & 0 \\ -\omega & \mu & 0 & 0 \\ 0 & 0 & -\mu & \omega \\ 0 & 0 & -\omega & -\mu \end{bmatrix}\begin{bmatrix} y_1 \\ y_2 \\ y_3 \\ y_4 \end{bmatrix}. \tag{4.105}$$

As a result, the dynamics consists of the decoupled two second order systems, whose linear operators are real Jordan canonical form:

$$\begin{cases} \dfrac{d}{dt}\begin{bmatrix} y_1 \\ y_2 \end{bmatrix} = \begin{bmatrix} \mu & \omega \\ -\omega & \mu \end{bmatrix}\begin{bmatrix} y_1 \\ y_2 \end{bmatrix}, & (4.106) \\[2ex] \dfrac{d}{dt}\begin{bmatrix} y_3 \\ y_4 \end{bmatrix} = \begin{bmatrix} -\mu & \omega \\ -\omega & -\mu \end{bmatrix}\begin{bmatrix} y_3 \\ y_4 \end{bmatrix}. & (4.107) \end{cases}$$

Each solution was already examined in Section 2.4.4 and the phase portraits are described as Figure 4.6b,c, respectively.

Furthermore, by a similarity transformation, Eqs. (4.106) and (4.107) can be expressed as

$$\begin{cases} \dfrac{d}{dt}\begin{bmatrix} y'_1 \\ y'_2 \end{bmatrix} = \begin{bmatrix} 0 & 1 \\ -(\mu^2+\omega^2) & 2\mu \end{bmatrix}\begin{bmatrix} y'_1 \\ y'_2 \end{bmatrix}, & (4.108) \\[2ex] \dfrac{d}{dt}\begin{bmatrix} y'_3 \\ y'_4 \end{bmatrix} = \begin{bmatrix} 0 & 1 \\ -(\mu^2+\omega^2) & -2\mu \end{bmatrix}\begin{bmatrix} y'_3 \\ y'_4 \end{bmatrix}, & (4.109) \end{cases}$$

that is,

$$\begin{cases} \dfrac{d^2y'_1}{dt^2} - 2\mu\dfrac{dy'_1}{dt} + (\mu^2+\omega^2)y'_1 = 0, & (4.110) \\[2ex] \dfrac{d^2y'_3}{dt^2} + 2\mu\dfrac{dy'_3}{dt} + (\mu^2+\omega^2)y'_3 = 0, & (4.111) \end{cases}$$

which are equivalent to the equations of motion for spring-mass-damper systems with negative and positive damping, respectively. The 2×2 matrices in Eqs. (4.108) and (109) are called *controllable canonical form* (Ogata 1997).

When Λ_i are complex, the system of Eq. (4.12) consists of the dynamics expressed by the phase portraits in Figure 4.6b,c, which are the same as those in Figure 2.5c,a. Regardless of the lack of the dissipative force, i.e. $\boldsymbol{F}_D = -D d\boldsymbol{x}/dt$, as Eq. (4.1), the total energy is not kept constant due to the existence of the circulatory force $\boldsymbol{F}_N = N\boldsymbol{x}$.

Similar to the self-excited oscillation due to the negative damping in Section 2.6, further discussions to determine if the growth continues infinitely or not will be postponed until Chapter 10 in which the nonlinearity neglected in this Chapter is taken into account.

Problem 4.4.2 Derive Eq. (4.108) from Eq. (4.106) and Eq. (4.109) from Eq. (4.107).

Ans: Let the matrixes in Eqs. (4.106) and (4.107)

$$\begin{bmatrix} \mu & \omega \\ -\omega & \mu \end{bmatrix} \quad \text{and} \quad \begin{bmatrix} -\mu & \omega \\ -\omega & -\mu \end{bmatrix}, \tag{4.112}$$

be defined as B. Their characteristics equations are in the form of

$$\lambda^2 + \alpha_1 \lambda + \alpha_0 = 0. \tag{4.113}$$

For Eq. (4.106), $\alpha_1 = -2\mu$ and $\alpha_0 = \mu^2 + \omega^2$; thus, the eigenvalues are

$$\lambda = \mu \pm i\omega. \tag{4.114}$$

For Eq. (4.107), $\alpha_1 = 2\mu$ and $\alpha_0 = \mu^2 + \omega^2$; thus, the eigenvalues are

$$\lambda = -\mu \pm i\omega. \tag{4.115}$$

We introduce the vectors $\boldsymbol{t}_1$, $\boldsymbol{t}_2$, and $\boldsymbol{b}$ as follows:

$$\begin{cases} \boldsymbol{t}_1 = (B + \alpha_1 I)\boldsymbol{b}, & (4.116) \\ \boldsymbol{t}_2 = \boldsymbol{b}. & (4.117) \end{cases}$$

From the Hamilton–Cayley theorem, we have

$$B^2 + \alpha_1 B + \alpha_0 I = 0, \tag{4.118}$$

where I is the unit matrix. Then, we have

$$\begin{cases} B\boldsymbol{t}_1 = -\alpha_0 \boldsymbol{t}_2 = [\boldsymbol{t}_1,\ \boldsymbol{t}_2] \begin{bmatrix} 0 \\ -\alpha_0 \end{bmatrix}, & (4.119) \\ B\boldsymbol{t}_2 = \boldsymbol{t}_1 - \alpha_1 \boldsymbol{t}_2 = [\boldsymbol{t}_1,\ \boldsymbol{t}_2] \begin{bmatrix} 1 \\ -\alpha_1 \end{bmatrix}. & (4.120) \end{cases}$$

Combing these equations yields

$$B[\boldsymbol{t}_1,\ \boldsymbol{t}_2] = [\boldsymbol{t}_1,\ \boldsymbol{t}_2] \begin{bmatrix} 0 & 1 \\ -\alpha_0 & -\alpha_1 \end{bmatrix}. \tag{4.121}$$

Setting the transformation matrix $P = [\boldsymbol{t}_1,\ \boldsymbol{t}_2]$, substituting

$$\begin{bmatrix} y_1 \\ y_2 \end{bmatrix} = P \begin{bmatrix} y_1' \\ y_2' \end{bmatrix}, \tag{4.122}$$

into Eq. (4.106), and premultiplying the result by P^{-1} yields Eq. (4.108). Similarly, substituting

$$\begin{bmatrix} y_3 \\ y_4 \end{bmatrix} = P \begin{bmatrix} y'_3 \\ y'_4 \end{bmatrix}, \tag{4.123}$$

into Eq. (4.107) and premultiplying the result by P^{-1} yields Eq. (4.109).

4.5 Synchronous and Nonsynchronous Motions in a Fluid-Conveying Pipe (Video)

Let us see videos on the motions of a pipe conveying fluid whose theoretical investigation is in Section 6.5. The investigated synchronous and nonsynchronous motions in a practical system are observed in the videos. We employ a vertically hung pipe subject to a damping force. Such a practical system is different from the systems discussed in this chapter, indeed we can observe the synchronous and nonsynchronous characteristics depending on the existence of a circulatory force. As shown in Section 6.5, if the velocity of the fluid in the pipe is zero, the system is not subject to the circulatory force and the synchronous motion is expected to occur. On the other hand, if the velocity of the fluid in the pipe is not zero, the system is subject to the circulatory force and the nonsynchronous motion is predicted to occur. Videoclip 16.5(1) shows the case without flow velocity and Videoclip 16.5(2) is its slow motion. From these videos, the motions at each point on the pipe is synchronized. Videoclip 16.5(3) shows the case with flow velocity and Videoclip 16.5(4) is its slow motion. From these videos, the motions at each point on the pipe is nonsynchronized. The nonsynchronous motion is based on the circulatory force caused by the flow, that is on one of follower forces (more in detail, see Sections 6.5 and 16.5).

References

Ogata, K. (1997). *Modern Control Engineering*. Prentice Hall.

5

Static Instability and Practical Examples

When the parameter k' is varied along arrow (1) in Figure 2.7, i.e. through c'-axis ($k' = 0$) from the first quadrant to the second one, the trajectory is changed from monotonic convergence to monotonic divergence. Such a destabilization phenomenon is called a *static destabilization*. In this chapter, we investigate the dynamics of three typical systems producing static destabilization.

5.1 Two-Link Model for a Slender Straight Elastic Rod Subject to Compressive Forces

5.1.1 Static Instability Due to Compressive Forces

We consider the stability of the slender beam in Figure 5.1 subject to the compressive force. Above a critical load, the equilibrium state of the straight position (a) is destabilized and the beam can exhibit the buckling shapes of the first mode (b), second one, (c) and so on.

We here focus on the buckling with the first mode (b) and introduce an equivalent discrete model of two-link system as shown in Figure 5.2 (Stoker 1992). The mass can be freely moved in x-direction. The points A and B, and B and C are connected by a massless rigid link with length l. The points A and C can be freely moved in y-direction. The mass is supported from both sides by two identical springs with spring constant $k/2$ in the state without initial compression or extension; the total spring force is expressed as

$$F_s^{\#} = kx, \tag{5.1}$$

where x is equivalent to Δx in the force-displacement relationship shown in Figure 1.2a and the force acting on the mass is $F_s = -F_s^{\#} = -kx$ (see Appendix A).

In the static equilibrium state of the straight position shown in Figure 5.2a, the springs have their natural length; the springs provide in the lateral direction the restoring force equivalent to that due to the bending of the beam. Furthermore, the compressive forces $\frac{P}{2}$ at points A and C are applied as in Figure 5.2b. The compressive forces act on the mass in x-direction through the links as the following external force F_p (see Problem 5.1.1):

$$F_p = P\frac{x}{\sqrt{l^2 - x^2}}. \tag{5.2}$$

Linear and Nonlinear Instabilities in Mechanical Systems: Analysis, Control and Application, First Edition. Hiroshi Yabuno.

Companion website: www.wiley.com/go/yabuno/instabilitiesinmechanicalsystems

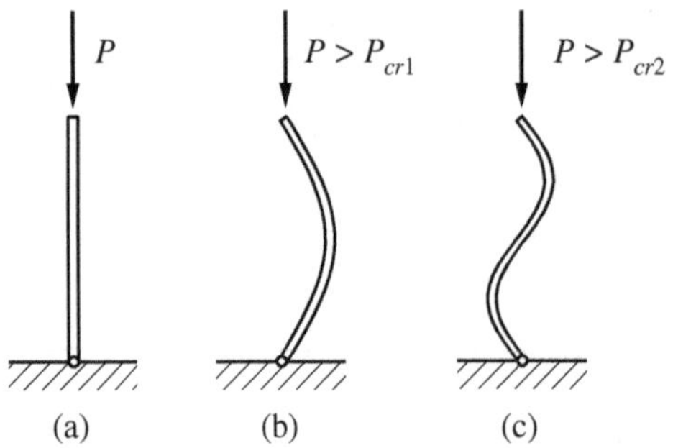

Figure 5.1 Slender beam subject to compressive force P. (a) Below a critical force. (b) Buckling with the first mode. (c) Buckling with the second mode.

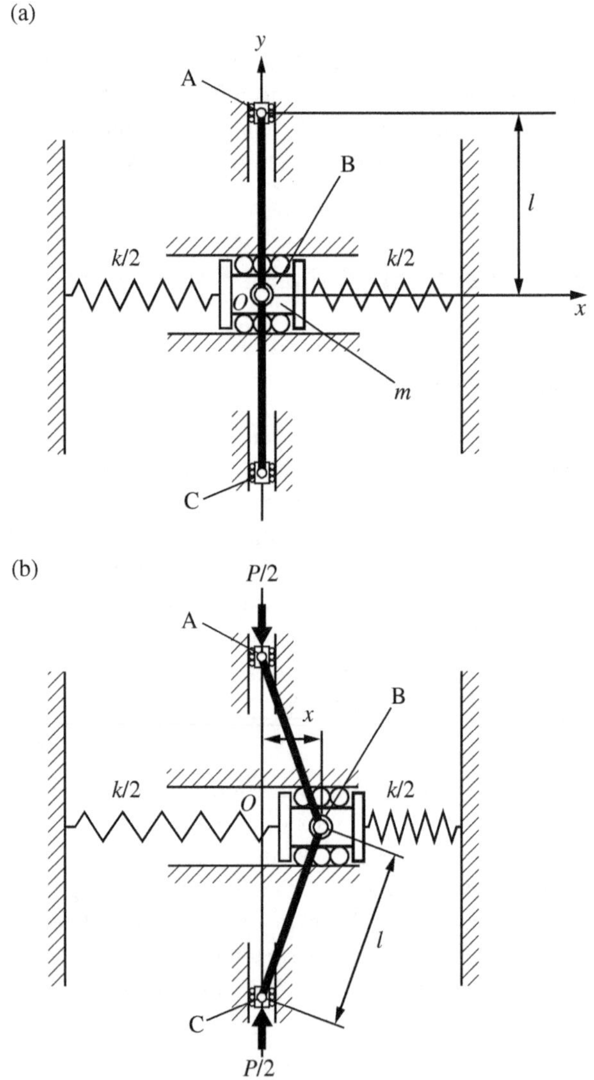

Figure 5.2 Two-link system subject to compressive forces. (a) Straight equilibrium position. (b) Motion of mass due to the compressive forces.

It should be noted that the effect of the compressive forces appears as a nonlinear force with respect to x. Indeed, the compressive forces affect the motion of the mass in x-direction through the links thus, the resulting forces causing the motion of the mass is not equal to the given compressive forces.

Problem 5.1.1 Derive Eq. (5.2).

Ans: First we consider the potential energy U in the case when point A moves from $\boldsymbol{r}_1 = l\boldsymbol{e}_y$ to $\boldsymbol{r}_2 = \sqrt{l^2 - x^2}\boldsymbol{e}_y$ under the conservative force $\boldsymbol{F} = -\frac{P}{2}\boldsymbol{e}_y$. From Eq. (1.33), we obtain U as

$$U = -\int_{\boldsymbol{r}_1}^{\boldsymbol{r}_2} -\frac{P}{2}\boldsymbol{e}_y \cdot dy\boldsymbol{e}_y = \frac{P}{2}(\sqrt{l^2 - x^2} - l). \tag{5.3}$$

By using Eq. (1.25) and considering the other compressive force at point C, we obtain the force in x-direction acting on the mass due to the total compressive forces as

$$F_p = -2\frac{\partial U}{\partial x} = P\frac{x}{\sqrt{l^2 - x^2}}. \tag{5.4}$$

Thus, the equation of motion of the mass is expressed as follows:

$$m\frac{d^2x}{dt^2} = -c\frac{dx}{dt} + F_s + F_p = -c\frac{dx}{dt} - kx + P\frac{x}{\sqrt{l^2 - x^2}}, \tag{5.5}$$

where the first term on the right-hand side is an equivalent viscous damping effect acting on the mass ($c > 0$). In the range of $|x/l| \ll 1$, the effect of the compressive forces is expanded as

$$P\frac{x}{\sqrt{l^2 - x^2}} = P\left(\frac{x}{l}\right)\left\{1 - \left(\frac{x}{l}\right)^2\right\}^{-\frac{1}{2}} = P\left(\frac{x}{l}\right) + \frac{P}{2}\left(\frac{x}{l}\right)^3 + O\left(\left(\frac{x}{l}\right)^5\right). \tag{5.6}$$

Considering only the linear term with respect to $\frac{x}{l}$ leads to the linearized equation of motion:

$$m\frac{d^2x}{dt^2} = -c\frac{dx}{dt} - \left(k - \frac{P}{l}\right)x. \tag{5.7}$$

The characteristic equation is

$$m\lambda^2 + c\lambda + \left(k - \frac{P}{l}\right) = 0. \tag{5.8}$$

By increasing the compressive force P, the eigenvalues λ are changed. Then, the trajectory on the phase space is changed as Figure 5.3, where the coordinates of the phase plane are transformed as mentioned in Section 2.4. When the equivalent spring constant $k_{equiv} = k - \frac{P}{l}$ is changed from positive to negative, the point expressed by the combination of the parameters $k' = \frac{k_{equiv}}{m}$ and $c' = \frac{c}{m}$ is passed through c'-axis ($k' = 0$) along arrow (1) in Figure 2.7, hence the static destabilization called *buckling* occurs.

Let us observe the buckling phenomenon produced in a clamped-clamped beam in which an end is movable in the axial direction and subject to a compressive force by a linear motor (voice coil motor) in video (more in detail, see Section 16.2). Increasing the compressive

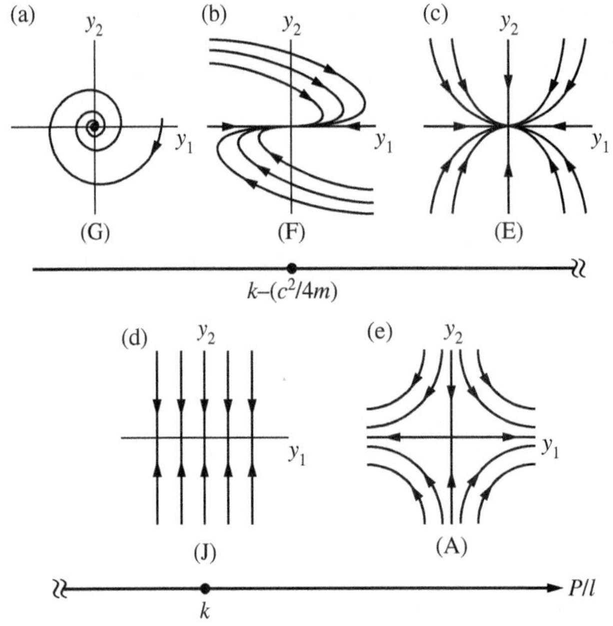

Figure 5.3 Variations of the eigenvalues and the trajectory due to the increase of the compressive force. Capital letters (G), (F), (E), (J), and (A) correspond to those in Figure 2.7.

force, we observe the motion under an initial deflection. In the range of low compressive force (Videoclip 16.2.1(1)), the deflection oscillatory decays, thus the system is described by the phase portrait of Figure 5.3a. If the compressive force is kept increasing (Videoclip 16.2.1(2)), the deflection monotonically decays therefore, the system is described by the phase portrait of Figure 5.3c that is an over damping state. When the compressive force exceeds the critical load (Videoclip 16.2.1(3)), the deflection is not returned to the straight position since the system is statically destabilized as described for the two-link system at $\frac{P}{l} = k$. However, the behavior shown in the video, in the state above the critical load, is different from the theoretical result. The theoretical result predicts above the critical load that the displacement of the mass grows infinitely but, in practice, the displacement is finite. This practical phenomenon is due to the nonlinearity in the system that is neglected in this section. Such a nonlinear phenomenon will be analyzed in Section 7.1.

5.1.2 Effect of a Spring Attached in the Longitudinal Direction

We consider another system described in Figure 5.4 in which a linear spring is attached to the direction parallel to the links in the static equilibrium state ($x = 0$). The relationship between the applied force and the elongation of the spring is linear as in the case of Figure 1.2b. The spring with the initial compression $\Delta l_s = l_{s0} - l_c$ is attached as in Figure 5.4a, where l_{s0} and l_c are the natural length and the length of the spring attached to

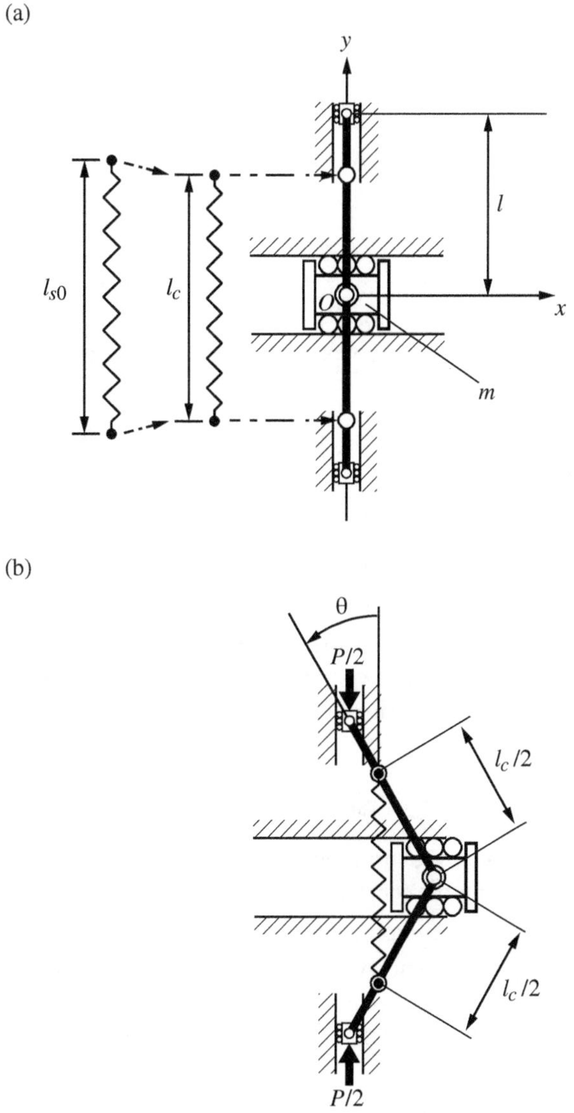

Figure 5.4 Two-link systems with a spring attached in the longitudinal direction. The spring with natural length l_{s0} is initially compressed to length l_c and is attached to the two links.

the links in the static equilibrium state, respectively. The spring force acting on the mass in x-direction is expressed as follows (see Problem 5.1.2):

$$F_{s\text{-}axial} = -\frac{kl_c}{l}\frac{x}{\sqrt{l^2 - x^2}}\left\{\Delta l_s + l_c\left(1 - \frac{\sqrt{l^2 - x^2}}{l}\right)\right\}. \tag{5.9}$$

Unlike the system of Figure 5.2 (see Eq. (5.1)), characterized by springs parallel to the x-direction, the effect of the linear spring perpendicular to such direction appears as a nonlinear force with respect to x. Indeed, the spring force does not act directly on the mass in x-direction but produces a force in x-direction through the links. Such a nonlinearity due to the geometric configuration is generally called *geometric nonlinearity*. Then, the equation of motion of the mass is expressed as follows:

$$m\frac{d^2x}{dt^2} = -c\frac{dx}{dt} + F_{s\text{-}axial} + F_p$$
$$= -c\frac{dx}{dt} - \frac{kl_c}{l}\frac{x}{\sqrt{l^2-x^2}}\left\{\Delta l_s + l_c\left(1 - \frac{\sqrt{l^2-x^2}}{l}\right)\right\} + P\frac{x}{\sqrt{l^2-x^2}}, \tag{5.10}$$

where the first term of the right-hand side is an equivalent viscous damping effect acting on the mass ($c > 0$).

Equation (5.9) is expanded with Taylor series with respect to $\left(\frac{x}{l}\right)$ as

$$F_{s\text{-}axial} = -\frac{kl_c}{l}\left(\frac{x}{l}\right)\left\{1-\left(\frac{x}{l}\right)^2\right\}^{-\frac{1}{2}}\left\{\Delta l_s + l_c\left[1-\left\{1-\left(\frac{x}{l}\right)^2\right\}^{\frac{1}{2}}\right]\right\}$$
$$= -\frac{kl_c}{l}\left(\frac{x}{l}\right)\left\{1+\frac{1}{2}\left(\frac{x}{l}\right)^2+\cdots\right\}\left\{\Delta l_s + \frac{l_c}{2}\left(\frac{x}{l}\right)^2+\cdots\right\}$$
$$= -\frac{kl_c}{l}\left(\frac{x}{l}\right)\left\{\Delta l_s + \frac{1}{2}(\Delta l_s + l_c)\left(\frac{x}{l}\right)^2 + O\left(\left(\frac{x}{l}\right)^4\right)\right\}. \tag{5.11}$$

It is noticed that the linear stiffness depends on the initial compression Δl_s and, in the case without the initial compression, the spring attached in the longitudinal direction does not produce the linear restoring force in x-direction, but only the nonlinear one with respect to x.

Problem 5.1.2 Derive Eq. (5.9).

Ans: Since the spring acts linearly, the restoring force is expressed as $F_s = -k\Delta y$, where Δy represents the compression of the spring. By using the initial compression of the spring $y_1 = \Delta l_s$ and the compression at deflection x, i.e. $y_2 = \Delta l_s + l_c(1-\cos\theta) = \Delta l_s + l_c\left(1 - \frac{\sqrt{l^2-x^2}}{l}\right)$, the potential energy U at deflection x is expressed

$$U = \frac{1}{2}kl_c\left(1 - \frac{\sqrt{l^2-x^2}}{l}\right)\left\{2\Delta l_s + l_c\left(1 - \frac{\sqrt{l^2-x^2}}{l}\right)\right\}. \tag{5.12}$$

From $F_{s\text{-}axial} = -\frac{dU}{dx}$, we obtain Eq. (5.9).

When the third and fourth terms on the right-hand side in Eq. (5.10) are linearized under the assumption of $|x/l| \ll 1$, the equation of motion Eq. (5.10) is linearized as

$$m\frac{d^2x}{dt^2} = -c\frac{dx}{dt} - \left(\frac{kl_c\Delta l_s}{l^2} - \frac{P}{l}\right)x. \tag{5.13}$$

The system features can be represented in the $k' - c'$ plane of Figure 2.7 as a point of coordinates $k' = \frac{k_{equiv}}{m}$ and $c' = \frac{c}{m}$. When the positive equivalent spring constant $k_{equiv} = \frac{kl_c\Delta l_s}{l^2} - \frac{P}{l}$ becomes negative, the (k', c') point associated to the system passes through c'-axis ($k' = 0$) along arrow (1) and *buckling* occurs at the critical buckling load $P = \frac{kl_c\Delta l_s}{l} \stackrel{\text{def}}{=} P_{cr}$. Such critical buckling load increases together with the spring compression

Δl_s at the initial equilibrium state. On the other hand, when $\Delta l_s = 0$, the buckling occurs at $P = 0$ because there is no linear stiffness.

5.2 Spring-Mass-Damper Models in MEMS

5.2.1 Comb-Type MEMS Actuator Devices

Comb-type actuator devices as shown in Figure 5.5 are widely applied as electrostatic actuators in MEMS (Legtenberg et al. 1996; Younis 2011). The mover is supported from the stator through the tethers which make the mover (i) compliant in the desired displacement direction (y-direction) of actuation and (ii) stiff in the respective orthogonal direction (x-direction). A closer look to the working mechanism of the device is provided in Figure 5.6 (zoom on the dotted circle area of Figure 5.5). The input voltage V applied to the fixed

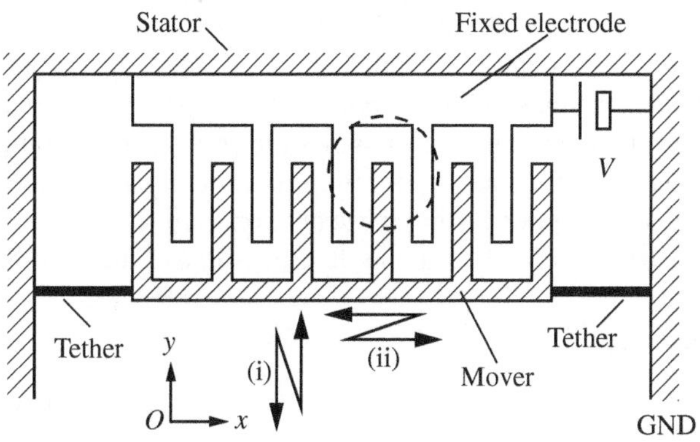

Figure 5.5 Schematic diagram of comb-type actuator. The tethers are compliant in the desired y-direction of actuation as well as stiff in the orthogonal x-direction.

Figure 5.6 Closer look to the part within the dotted circle in Figure 5.5. The y-component of the electrostatic attractive forces provides the actuation while the x-component is balanced to the stiffness of the tethers in x-direction.

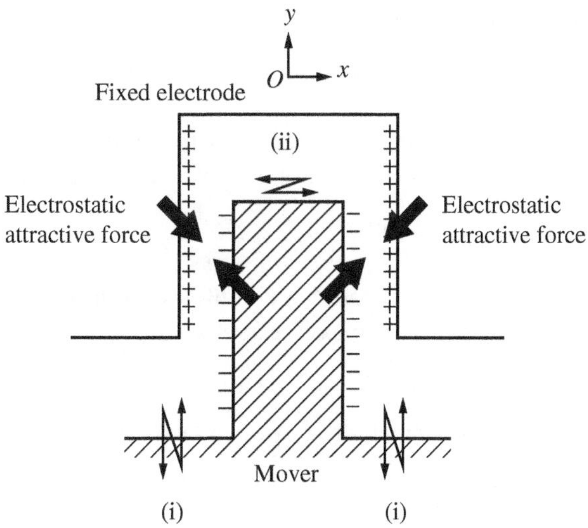

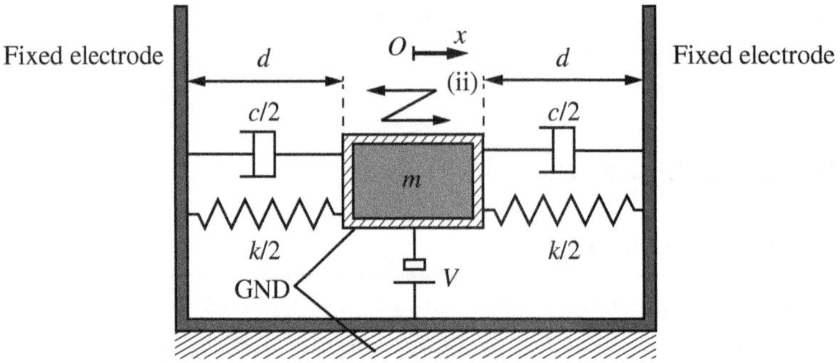

Figure 5.7 Simplified model of comb-type actuator: spring-mass-damper system subject to electrostatic forces.

electrode produces the electrostatic force between the stator and the mover, which is decomposed into its components along x- and y-directions.

Here, we focus on the effect of the attractive electrostatic force on the dynamics of the mover in the relatively stiff direction (x-direction). In Figure 5.7, we introduce a very simplified but essential model, where m is the mass of the mover, the linear spring represents the stiffness of the tethers in x-direction, and the damper shows the equivalent damping effect acting on the mover in x-direction. In the static equilibrium state, in the case without input voltage, the mover is located at the center, having the stators on both left and right sides as fixed electrodes, and the origin O of x-axis is set on the center line of the mass in the static equilibrium state.

The electrostatic force F_e can be assumed as

$$F_e = \frac{\varepsilon A V^2}{x_{gap}^2}, \tag{5.14}$$

where ε is the dielectric constant of gap, A and $x_{gap}(0 \leq x_{gap} \leq d)$ are the overlap area of the fixed electrode and the electrode on the mass, and the distance between two facing electrodes, respectively. The electrostatic force between the right electrode on the mass and the right fixed electrode is expressed as

$$F_{e-r}(x) = \frac{\varepsilon A V^2}{(d-x)^2}, \tag{5.15}$$

where $x(\leq d)$ is the displacement of the mass in Figure 5.7. Similarly, the electrostatic force between the left electrode on the mass and the left fixed electrode is expressed as

$$F_{e-l}(x) = -F_{e-r}(-x) = -\frac{\varepsilon A V^2}{(d+x)^2}. \tag{5.16}$$

These electrostatic forces are nonlinear with respect to x. The sum of these electrostatic forces acting on the mass is expressed as

$$F_e = F_{e-l} + F_{e-r}, \tag{5.17}$$

which does not include any even power terms.

Then, the equation of motion of the mass is expressed as follows:

$$m\frac{d^2x}{dt^2} = -c\frac{dx}{dt} - kx + \varepsilon AV^2\left\{\frac{1}{(d-x)^2} - \frac{1}{(d+x)^2}\right\}. \tag{5.18}$$

The system has the trivial equilibrium state $x = 0$ independent of the input voltage V. Neglecting $O(|x/d|^3)$, Eq. (5.18) is linearized as follows:

$$m\frac{d^2x}{dt^2} = -c\frac{dx}{dt} - \left(k - \frac{4\varepsilon AV^2}{d^3}\right)x, \tag{5.19}$$

which expresses the dynamics near the trivial equilibrium state. Similar to the system of Eqs. (5.7) and (5.13), when the equivalent spring constant $k_{equiv} = k - \frac{4\varepsilon AV^2}{d^3}$ is changed from positive to negative with increasing voltage V, the combination of the parameters $k' = \frac{k_{equiv}}{m}$ and $c' = \frac{c}{m}$ passes through the c'-axis along arrow (1) in Figure 2.7 and the static destabilization occurs. This phenomenon is called electrostatic buckling.

5.2.2 Cantilever-Type MEMS Switch

Figure 5.8 shows a cantilever-type MEMS switch. When the voltage is applied to the electrode, the cantilever is bent and bridges the gap between the two conductors. In particular, Figure 5.9 represents the cross section of the cantilever-type MEMS switch; (a) and (b) show the states of turn-off and on, respectively. We consider the dynamics of the cantilever depending on the input voltage to the electrode. The simplified analytical model for the motion in x-direction is introduced as Figure 5.10; m, k, and c are the equivalent mass, stiffness, and damping of the cantilever, respectively. In the state without input voltage, the gap between the mass and the electrode is denoted as d. The electrostatic force is given by Eq. (5.14). Thus, the equation of motion of the mass m is expressed as

$$m\frac{d^2x}{dt^2} = -c\frac{dx}{dt} - kx + \frac{\varepsilon AV^2}{(d-x)^2}. \tag{5.20}$$

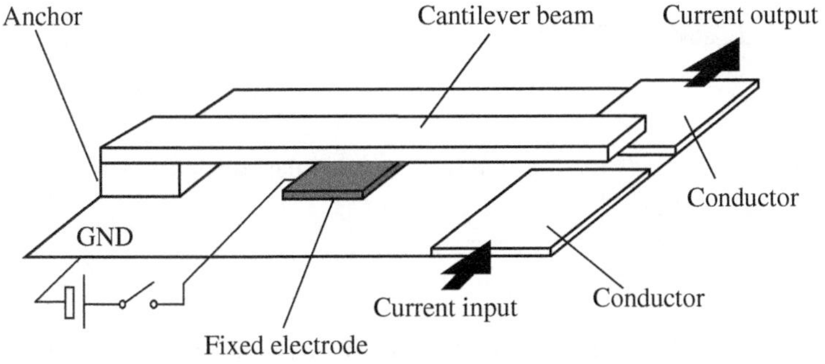

Figure 5.8 Schematic diagram of MEMS switch. By applying sufficient voltage V between the cantilever and the fixed electrode, the cantilever beam is absorptive to the fixed electrode and they are contacted. Then, the switch becomes conductive.

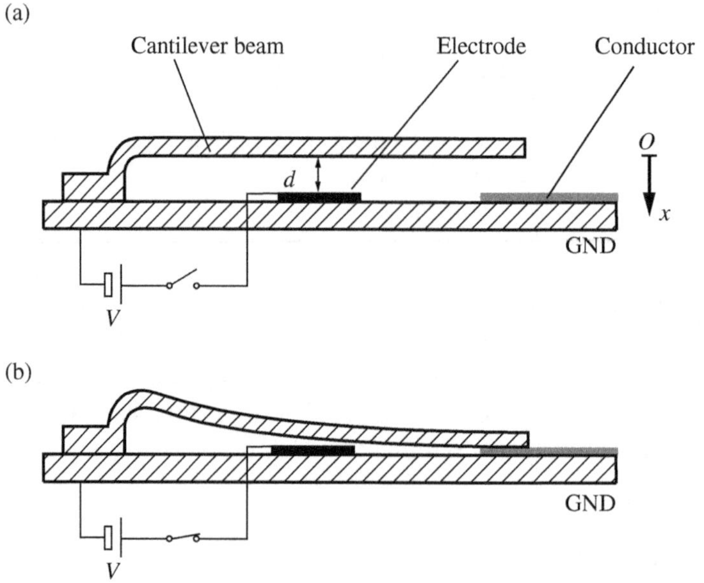

Figure 5.9 The cross section of MEMS Switch. (a) Turn-off state without input voltage. (b) Turn-on with input voltage.

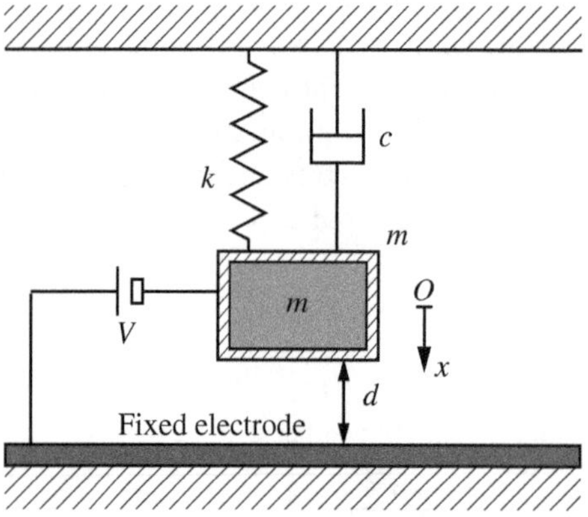

Figure 5.10 Simplified model of MEMS switch.

Introducing the dimensionless time $t^* = \sqrt{\frac{k}{m}}t$ and the dimensionless displacement $x^* = \frac{x}{d}$, we have the dimensionless equation of motion as

$$\frac{d^2x^*}{dt^{*2}} + 2\gamma\frac{dx^*}{dt^*} + x^* = \frac{V^*}{(1-x^*)^2}, \tag{5.21}$$

where $\gamma = \frac{c}{2\sqrt{mk}}$, $V^* = \frac{\varepsilon AV^2}{kd^3}$, and $x^* \leq 1$ because of $x \leq d$. We consider the change in the dynamics of the switch under the increase of dimensionless input voltage V^*. In this section, we perform the linear analysis (see Section 7.7.2 for nonlinear analysis). Neglecting $O(x^{*2})$,

Eq. (5.21) is linearized as

$$\frac{d^2x^*}{dt^{*2}} + 2\gamma\frac{dx^*}{dt^*} + (1 - 2V^*)x^* = V^*. \tag{5.22}$$

It is noted under the application of voltage that the electrostatic force includes the component independent of x^*, which does not exist in the comb-drive equilibrium state. The nontrivial equilibrium state corresponding to the particular solution x_{st} exists as

$$x_{st} = \frac{V^*}{1 - 2V^*}. \tag{5.23}$$

The homogeneous solution $\tilde{x}$ satisfies

$$\frac{d^2\tilde{x}}{dt^{*2}} + 2\gamma\frac{d\tilde{x}}{dt^*} + (1 - 2V^*)\tilde{x} = 0. \tag{5.24}$$

Since the complete solution of Eq. (5.22) is

$$x^* = x_{st} + \tilde{x}, \tag{5.25}$$

the stability of the equilibrium state is obtained from the dynamics of $\tilde{x}$. Regarding $2\gamma > 0$ and $1 - 2V^*$ as c' and k', respectively, in Figure 2.7, we can determine the stability. The relationship between the dimensionless input voltage and the dimensionless displacement of the switch is described as in Figure 5.11.

As increasing V, the equivalent stiffness $k' = 1 - 2V^*$ decreases and it becomes zero for $V^* = \frac{1}{2}$. When $V^* > \frac{1}{2}$, the stiffness is negative. Therefore, the static equilibrium states are described as shown in Figure 5.11, where the solid and dashed lines denote the stable and unstable equilibrium states, respectively, The phase portraits are given in Figure 5.12a,b for $V^* < \frac{1}{2}$ and $V^* > \frac{1}{2}$, respectively. The stable equilibrium state of the switch approaches to the electrode with increasing V^*. When we apply V^* so that the stable equilibrium state is greater than 1, the switch is turned on. By the way, as seen from Figures 5.11 and 5.12b, the unstable equilibrium state existing in $V^* > \frac{1}{2}$ indicates that the switch can go away from the electrode depending on the disturbance. From a physical point of view such phenomenon

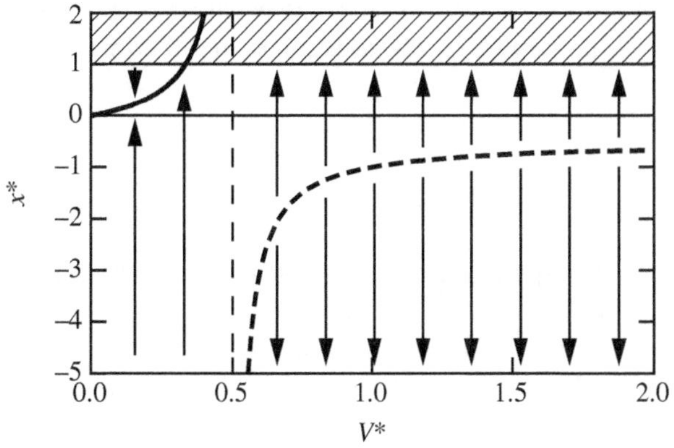

Figure 5.11 Static equilibrium states of MEMS switch.

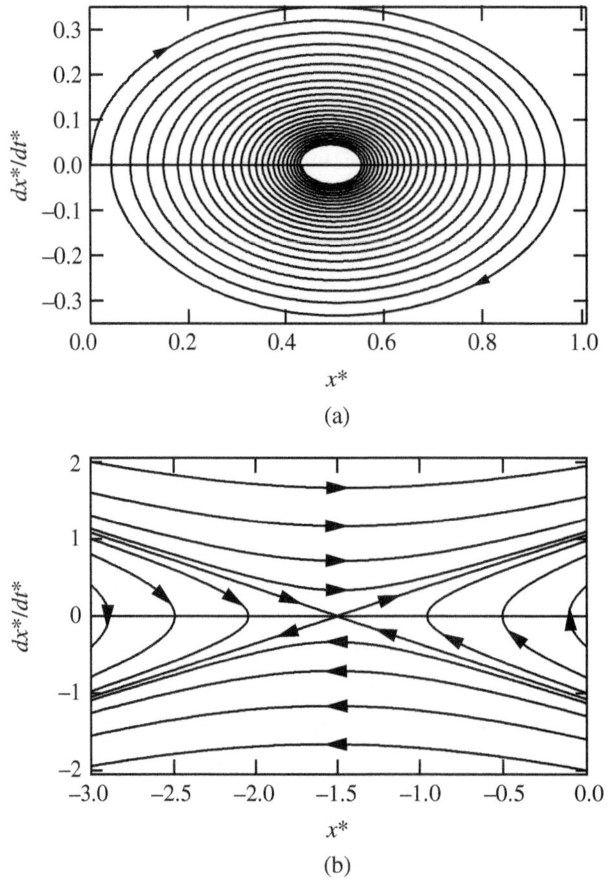

Figure 5.12 Phase portraits of MEMS switch. (a) $V^* < \frac{1}{2}$. (b) $V^* > \frac{1}{2}$.

is unacceptable and may be caused from the linearization of the equation of motion. In Section 7.7.2, the nonlinear analysis is preformed for the equation of motion thus considering the nonlinear effects.

References

Legtenberg, R., A. Groeneveld, and M. Elwenspoek (1996). Comb-drive actuators for large displacements. *Journal of Micromechanics and microengineering* 6(3), 320.

Stoker, J. J. (1992). *Nonlinear Vibrations in Mechanical and Electrical Systems*. Wiley.

Younis, M. I. (2011). *MEMS linear and nonlinear statics and dynamics*, vol. 20. Springer Science & Business Media.

6

Dynamic Instability and Practical Examples

In this chapter, a selection of explanatory and practical examples is discussed in detail to highlight the effects of destabilizing forces, such as nonconservative force fields, on the dynamic response of single- and multi-degree-of-freedom systems. The loss of stability, and the subsequent arising of self-excited oscillations, is investigated for five paradigmatic case studies.

6.1 Self-Excited Oscillation of Belt-Driven Mass-Spring-Damper System

We consider the belt-driven mass-spring-damper system with belt speed of V as shown in Figure 6.1. Let the absolute displacement and velocity of the mass be x and $v = \frac{dx}{dt}$, respectively. The friction force between the mass and belt which contact each other can be schematically assumed as Figure 6.2 (see also Section E.2), where $F_c^{\#}$ and v_r are the friction and the relative velocity of the mass with respect to the belt, respectively. Furthermore, v_r is expressed by the absolute speeds of the belt and the mass, V and $\frac{dx}{dt}$, as

$$v_r = \frac{dx}{dt} - V. \tag{6.1}$$

Since the friction force acting on the mass is $-F_c^{\#}$, the equation of motion of the mass is

$$m\frac{d^2x}{dt^2} = -F_c^{\#}(v_r) - c\frac{dx}{dt} - kx. \tag{6.2}$$

By letting $\frac{d^2x}{dt^2} = \frac{dx}{dt} = 0$, we obtain the equilibrium state. Then, the relative velocity v_{r-st} in the equilibrium state is expressed from Eq. (6.1) as

$$v_{r-st} = 0 - V = -V \tag{6.3}$$

and Eq. (6.1) is rewritten as

$$v_r - v_{r-st} = \frac{dx}{dt}. \tag{6.4}$$

Also, from Eq. (6.2), we have the following equilibrium equation:

$$0 = -F_c^{\#}(v_{r-st}) - kx_{st}. \tag{6.5}$$

Linear and Nonlinear Instabilities in Mechanical Systems: Analysis, Control and Application, First Edition. Hiroshi Yabuno.

Companion website: www.wiley.com/go/yabuno/instabilitiesinmechanicalsystems

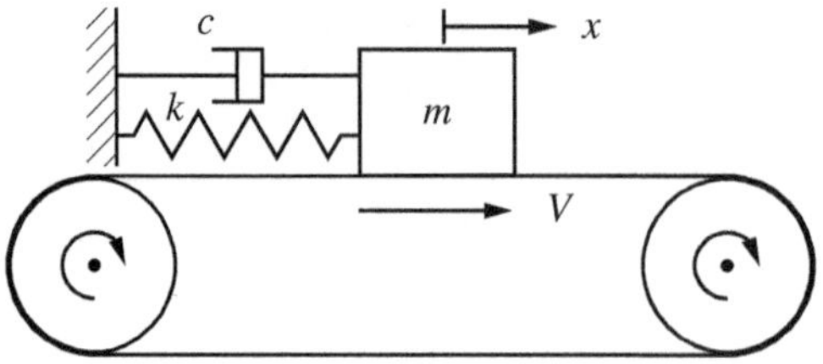

Figure 6.1 Belt-driven mass-spring-damper system, where x is the absolute displacement of mass from the static equilibrium state and V is the absolute velocity of the belt.

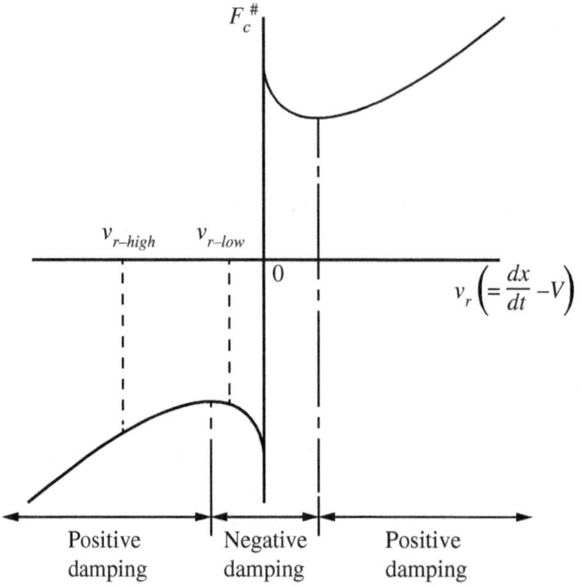

Figure 6.2 Schematic diagram of friction caused by the relative sliding velocity between two contact bodies: v_r is the relative sliding velocity between the contacting surfaces of solid bodies and $F_c^\#$ is the friction force; see also Section E.2.

It is noticed that the equilibrium state x_{st} depends on the belt speed $V(= -v_{r-st})$.

Assuming that the function $F_c^\#$ of v_r is smooth and analytic in the neighborhood of the equilibrium state $v_r = v_{r-st}(= -V)$ or $\frac{dx}{dt} = 0$, we can expand $F_c^\#(v_r)$ with Taylor series as

$$F_c^\#(v_r) = a_0 + a_1(v_r - v_{r-st}) + a_2(v_r - v_{r-st})^2 + \cdots , \tag{6.6}$$

or equivalently

$$F_c^\#(v_r) = a_0 + a_1\frac{dx}{dt} + a_2\left(\frac{dx}{dt}\right)^2 + \cdots , \tag{6.7}$$

where $a_n = \frac{1}{n!}\frac{d^nF_c^\#}{dv_r^n}\bigg|_{v_r=v_{r-st}}$. If Eq. (6.7) has a nonzero radius of convergence, we can find $v_0(v_0 > 0)$ such that (Yoshida (2010))

$$\sup\ |a_n v_0^n| < 1. \tag{6.8}$$

Letting

$$\Delta v_r = \frac{v_r - v_{r-st}}{v_0} = \frac{\frac{dx}{dt}}{v_0}, \tag{6.9}$$

we can rewrite Eqs. (6.6) and (6.7) as

$$F_c^{\#}(v_r) = c_0 + c_1 \Delta v_r + c_2 \Delta v_r^2 + \cdots, \tag{6.10}$$

where $c_n = a_n v_0^n$ (sup $|c_n| < 1$).

Let us investigate the dynamics in the case of $|\Delta v_r| \ll 1$, i.e. when the relative velocity v_r is in the neighborhood of the relative velocity $v_{r-st}(= -V)$ in the equilibrium state of the mass. Then, the friction force $F_c^{\#}(v_r)$ is linearized as

$$F_c^{\#}(v_r) \approx c_0 + c_1 \Delta v_r = F_c^{\#}(v_{r-st}) + \left.\frac{dF_c^{\#}}{dv_r}\right|_{v_r = v_{r-st}} \frac{dx}{dt}. \tag{6.11}$$

Expressing x as

$$x = x_{st} + \Delta x \tag{6.12}$$

and considering the equilibrium equation (6.5), we have the following linearized equation of motion:

$$m\frac{d^2\Delta x}{dt^2} + \left(\left.\frac{dF_c^{\#}}{dv_r}\right|_{v_r = v_{r-st}} + c \right) \frac{d\Delta x}{dt} + k\Delta x = 0. \tag{6.13}$$

The system is equivalent to Eq. (2.114). Because the damping coefficient c is positive, in the case when the friction $F_c^{\#}$ has a positive slope $\left.\frac{dF_c^{\#}}{dv_r}\right|_{v_r = v_{r-st}} > 0$ at $v_r = v_{r-st}(= -V)$ (for example, $v_r = v_{r-high}$ in Figure 6.2), i.e. a positive damping characteristic, the eigenvalues of Eq. (6.13) are located in the first quadrant in Figure 2.7 and the trivial equilibrium state is stable. Decreasing the belt velocity of V, the slope $\left.\frac{dF_c^{\#}}{dv_r}\right|_{v_r = v_{r-st}}$ is decreased and changed from positive to negative (for example, $v_r = v_{r-low}$ in Figure 6.2). Further decreasing makes the equivalent damping $\left(\left.\frac{dF_c^{\#}}{dv_r}\right|_{v_r = v_{r-st}} + c \right)$ negative. Hence, such system parameter exceeds the stability boundary $c' = 0$ along arrow (2) in Figure 2.7 and the system is dynamically destabilized producing a self-excited oscillation. In this critical point, Hopf bifurcation occurs (see Section 7.4.3).

In the wide belt speed range from very high speed to very low one, the root locus is described as Figure 6.3; decreasing the speed, the eigenvalues move according to the arrows. The trajectory on the phase space is changed as Figure 6.4, where the coordinates of the phase plane are transformed as mentioned in Section 2.6. At points from (1) to (7) of Figure 6.3, the phase portraits are changed from (E) to (B) in the same order shown in Figure 6.4, respectively. It is noticed that at a very low speed the dynamically unstable state (self-excited oscillation: phase portrait (D)) is changed to the statically unstable state (phase portrait (B)). The points (F), (H) and (C) on the root locus corresponds to the stable inflected node, center, and unstable inflected node in Figure 6.4, respectively. Before point (F), the phase portrait of the system is the stable node as (E) in Figure 6.4. In the intervals between (F) and (H) and between (H) and (C), it is stable focus as (G) and unstable focus

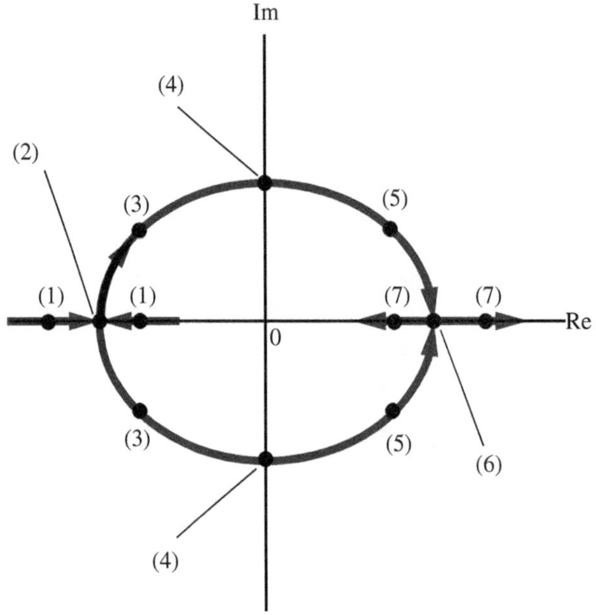

Figure 6.3 Root locus: by decreasing the belt speed, the two roots move in the order of numbers. At points (4) Hopf bifurcation is produced.

as (D), respectively. After point (C), since it is the unstable node as (B), the system becomes statically unstable.

6.2 Flutter of Wing

We consider a simple two-degree-of-freedom model of the cross section of an aircraft wing as in Figure 6.5, where the origin of the static coordinate system is at the mass center G in the static equilibrium state. The heave and pitch motions are denoted with y and θ. The lift force acts on the aerodynamics center C^A and is assumed as $L = C_0V^2\theta$, where C_0 and V are a constant determined by the wing span, the air density, and so on, and a uniform air stream of velocity, respectively. The equivalent restoring force and torsional moment are expressed by the spring and the torsional spring at the elastic center C^E, respectively; the spring force and torsional spring moment are, $F_s = -k(y + a\theta)$ and $M_s = -\delta\theta$, respectively. In general, C^A is in front of G and b is positive. On the other hand, C^E can be located in front or back of G; hence, depending on the sign of a, the dynamics is qualitatively different as will be shown below.

The equations of motion of the heaving and pitch motions with respect to G are expressed as

$$\begin{cases} m\dfrac{d^2y}{dt^2} + ky + (ka - C_0V^2)\theta = 0, & (6.14) \\ I\dfrac{d^2\theta}{dt^2} + kay + (ka^2 + \delta - bC_0V^2)\theta = 0. & (6.15) \end{cases}$$

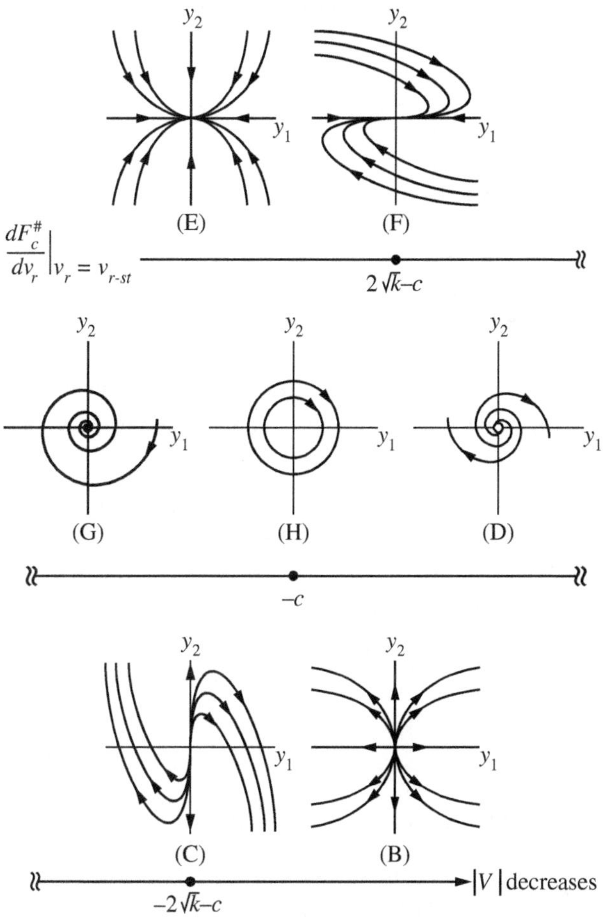

Figure 6.4 Variations of the phase portrait due to the decrease of belt speed. When decreasing the belt speed V, the relative velocity $v_{r-st}(< 0)$ in the equilibrium state is increased. Then, at this state, the slope $\frac{dF_c^{\#}}{dv_r}\Big|_{v_r=v_{r-st}}$ in Figure 6.2 is changed from positive to negative and the absolute value increases. According to this change of friction, the phase portrait is changed as the arrows in this figure. Phase portraits of (E), (F), (G), (H), (D), (C), and (B) correspond to those in Figure 2.7 and are here associated with points from (1) to (7), respectively, on the root locus of Figure 6.3. In cases of (F), (H), and (C), the slope $\frac{dF_c^{\#}}{dv_r}\Big|_{v_r=v_{r-st}}$ is $2\sqrt{mk}-c$, $-c$, and $-2\sqrt{mk}-c$, respectively.

Using b as the representative length and $\sqrt{\frac{I}{\delta+ka^2}}$ as the representative time, we rewrite the equations of motion in nondimensional form as

$$\begin{cases} m^*\dfrac{d^2y^*}{dt^{*2}} + k^*y^* + (k^*a^* - V^*)\theta = 0, & (6.16) \\ \dfrac{d^2\theta}{dt^{*2}} + k^*a^*y^* + (1 - V^*)\theta = 0, & (6.17) \end{cases}$$

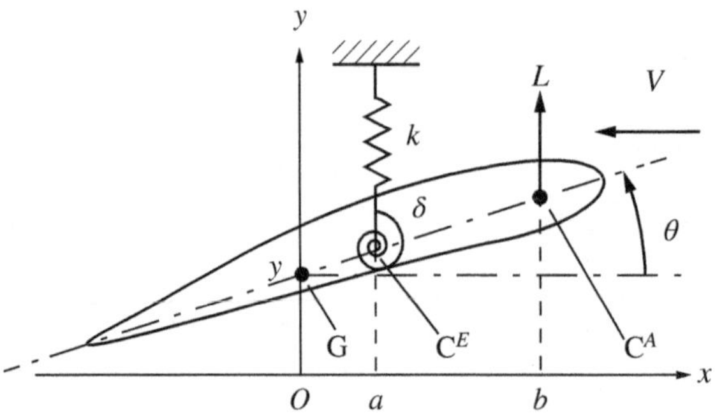

Figure 6.5 Two-dimensional airfoil model.

where $y^* \left(=\frac{y}{b}\right)$ represents the dimensionless lateral motion, and V^* represents the dimensionless velocity of air stream, i.e. the dimensionless wing speed. The nondimensional coefficients are

$$m^* = \frac{mb^2}{I} > 0, \quad k^* = \frac{kb^2}{\delta + ka^2} > 0, \quad a^* = \frac{a}{b}, \quad V^* = \frac{C_0 b V^2}{\delta + ka^2} > 0. \tag{6.18}$$

Equations (6.16) and (6.17) are rewritten in matrix form as

$$\begin{bmatrix} m^* & 0 \\ 0 & 1 \end{bmatrix} \frac{d^2}{dt^{*2}} \begin{bmatrix} y^* \\ \theta \end{bmatrix} + \begin{bmatrix} k^* & k^*a^* - V^* \\ k^*a^* & 1 - V^* \end{bmatrix} \begin{bmatrix} y^* \\ \theta \end{bmatrix} = \begin{bmatrix} 0 \\ 0 \end{bmatrix}. \tag{6.19}$$

The coefficients matrix K' in Eq. (3.1) is

$$K' = \begin{bmatrix} k^* & k^*a^* - V^* \\ k^*a^* & 1 - V^* \end{bmatrix}, \tag{6.20}$$

which is expressed by the sum of the stiffness and circulatory matrices as

$$K = \begin{bmatrix} k^* & k^*a^* - \dfrac{V^*}{2} \\ k^*a^* - \dfrac{V^*}{2} & 1 - V^* \end{bmatrix} \quad \text{and} \quad N = \begin{bmatrix} 0 & -\dfrac{V^*}{2} \\ \dfrac{V^*}{2} & 0 \end{bmatrix}. \tag{6.21}$$

Therefore, if the lift force does not act, i.e. $V^* = 0$, there is no circulatory force $\boldsymbol{F}_N$ and the system is conservative. In other words, the lift force makes the system nonconservative and can produce self-excited oscillation.

Substituting

$$y = A_y e^{\lambda t}, \quad \theta = A_\theta e^{\lambda t} \tag{6.22}$$

into Eq. (6.19) yields

$$L(V^*)\boldsymbol{\Phi} = \boldsymbol{0}, \tag{6.23}$$

where $\boldsymbol{\Phi}$ is the kernel of $L(V^*)$ expressed by

$$\boldsymbol{\Phi} = \begin{bmatrix} A_y \\ A_\theta \end{bmatrix}, \tag{6.24}$$

and the linear operator is expressed by

$$L(V^*) = \begin{bmatrix} m^*\lambda^2 + k^* & k^*a^* - V^* \\ k^*a^* & \lambda^2 + 1 - V^* \end{bmatrix}. \tag{6.25}$$

The condition that there exists nontrivial $\mathbf{\Phi}$ is $|L(V^*)| = 0$, i.e.

$$c_4\lambda^4 + c_2\lambda^2 + c_0 = 0, \tag{6.26}$$

where

$$c_0 = k^*(a^* - 1)V^* + k^*(1 - k^*a^{*2}), \tag{6.27}$$

$$c_2 = -m^*V^* + m^* + k^*, \tag{6.28}$$

$$c_4 = m^*. \tag{6.29}$$

Equation (6.26) is the characteristic equation equivalent to the following biquadratic equation with respect to $\Lambda = \lambda^2$ (see Eq. (4.16)):

$$c_4\Lambda^2 + c_2\Lambda + c_0 = 0. \tag{6.30}$$

The discriminant is

$$D = \{m^*(V^* - 1) + k^*\}^2 + 4m^*k^*a^*(k^*a^* - V^*), \tag{6.31}$$

and two solutions Λ_1 and Λ_2 of Eq. (6.30) have the following relationship:

$$\Lambda_1 + \Lambda_2 = -\frac{c_2}{c_4}, \tag{6.32}$$

$$\Lambda_1\Lambda_2 = \frac{c_0}{c_4}. \tag{6.33}$$

Using the above equations, in Sections 6.2.1 and 6.2.2, we examine the change of stability when increasing flight speed V^*.

6.2.1 Static Destabilization in Case When the Mass Center is Located in Front of the Elastic Center

First, we consider the case when the mass center G is located in front of the elastic center C^E, i.e. $a^* < 0$. Because the discriminant D is positive independent of V^*, the values $\Lambda_i (i = 1, 2)$ are always real. In particular, the solutions of Eq. (6.26) (i) for $\Lambda_i < 0$, are a pair of complex conjugate purely imaginary numbers, $\pm\sqrt{-\Lambda_i}i$, (ii) for $\Lambda_i > 0$, are positive and negative real numbers, $\pm\sqrt{\Lambda_i}$, (iii) for $\Lambda_i = 0$, are double zero.

At $V^* = 0$, c_2 is positive and c_0 is also positive because of $1 - k^*a^{*2} = \frac{\delta}{\delta + ka^2}$. At $V^* = \frac{1-k^*a^{*2}}{1-a^*} \overset{\text{def}}{=} V_1^*$ corresponding to $c_0 = 0$, we obtain

$$c_2 = \frac{k^*(1 + m^*a^{*2}) - a^*(m^* + k^*)}{1 - a^*} > 0. \tag{6.34}$$

At $V^* = 1 + \frac{k^*}{m^*} \overset{\text{def}}{=} V_2^*$ corresponding to $c_2 = 0$, we get

$$c_0 = -\left(\frac{k^{*2}}{m^*} + k^{*2}a^{*2}\right) + a^*\left(k^* + \frac{k^{*2}}{m^*}\right) < 0. \tag{6.35}$$

The schematic diagrams of c_0 and c_2 with respect to V^* are shown in Figure 6.6a.

When $0 \leq V^* < V_1^*$, $\Lambda_i (i = 1, 2)$ are negative; V_1^* is generally different from that in Fig. 6.6(a). The four roots of Eq. (6.26) are two pairs of conjugate purely imaginary numbers and the system is stable in the sense of Liapunov.

When $V^* = V_1^*$, $\Lambda_i (i = 1, 2)$ are zero and negative real numbers. Two roots of Eq. (6.26) are a pair of conjugate purely imaginary numbers and the other two roots are double zero. The system is unstable.

When $V_1^* < V^*$, $\Lambda_i (i = 1, 2)$ are positive and negative real numbers. Two roots of Eq. (6.26) are a pair of conjugate purely imaginary numbers and the other two roots are positive and negative real numbers. The system becomes statically unstable and the buckling occurs.

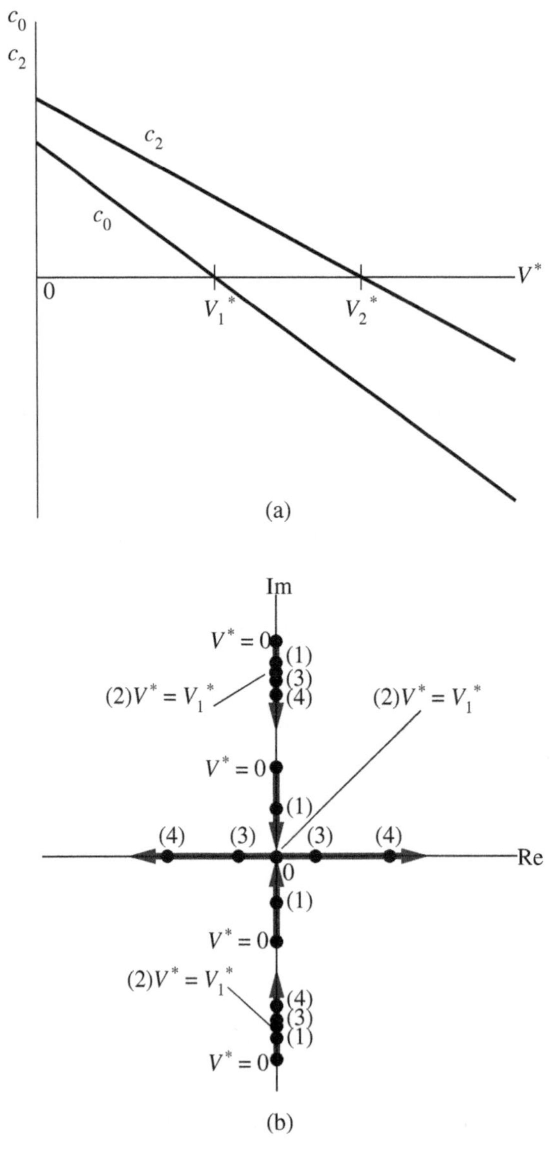

Figure 6.6 Static destabilization of wing. (a) Schematic diagrams of c_0 and c_2. (b) Root locus with respect to V^* where $m^* = 1, k^* = 1/2, a^* = -1$. With increasing V^*, the four eigenvalues move in numerical order: (1) $V^* = 0.18$; (2) $V^* = V_1^* = 0.25$; (3) $V^* = 0.28$; (4) $V^* = 0.50$.

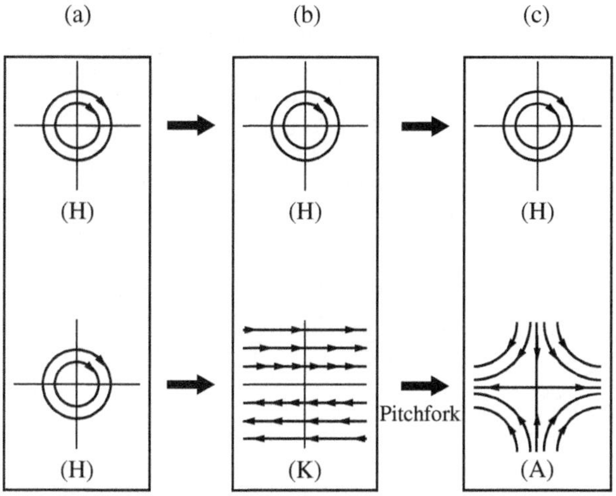

Figure 6.7 Variation of phase portraits due to wing speed ($a^* < 0$): (a) $V^* < V_1^*$; (b) $V^* = V_1^*$; and (c) $V^* > V_1^*$. Capital letters (H), (K), and (A) correspond to those in Figure 2.7.

The root locus with respect to V^* is shown in Figure 6.6b. The variation of the phase portraits by increasing V^* is also schematically shown in Figure 6.7.

6.2.2 Static and Dynamic Destabilization in Case When the Mass Center is Located Behind the Elastic Center

Next, we consider the case when the mass center G is located behind the elastic center C^E, i.e. $a^* > 0$. It is noticed from Eq. (6.31) that the sign of the discriminant can be negative depending on V^*. When it is negative, the values $\Lambda_i (i = 1, 2)$ are a pair of complex conjugate numbers and the four solutions of Eq. (6.26) are a pair of complex conjugate numbers with a positive real part and a pair of complex conjugate numbers with a negative real part (see Section 4.4).

Hereafter, for simplicity, we consider the case that the elastic center coincides with the aerodynamic center, i.e. $a^* = 1$ (Ziegler 2013). Then, Eqs. (6.27), (6.28), and (6.31) are written respectively as

$$c_0 = k^*(1 - k^*), \tag{6.36}$$

$$c_2 = -m^* V^* + m^* + k^*, \tag{6.37}$$

$$D = \{m^*(V^* - 1) + k^*\}^2 + 4m^* k^*(k^* - V^*). \tag{6.38}$$

The discriminant D is expressed by the quadratic function of V^* and it is positive at $V^* = 0$. The coefficient c_2 is zero at $V_2^* = \frac{m^*+k^*}{m^*}$ for the minimum D. Schematic diagrams of c_0, c_2, and D with respect to V^* are described in Figure 6.8a.

Therefore, when $0 \le V^* < V_1^*$, $\Lambda_i (i = 1, 2)$ are negative. The four roots of Eq. (6.26) are two pairs of conjugate purely imaginary numbers and the system is stable in the sense of Liapunov.

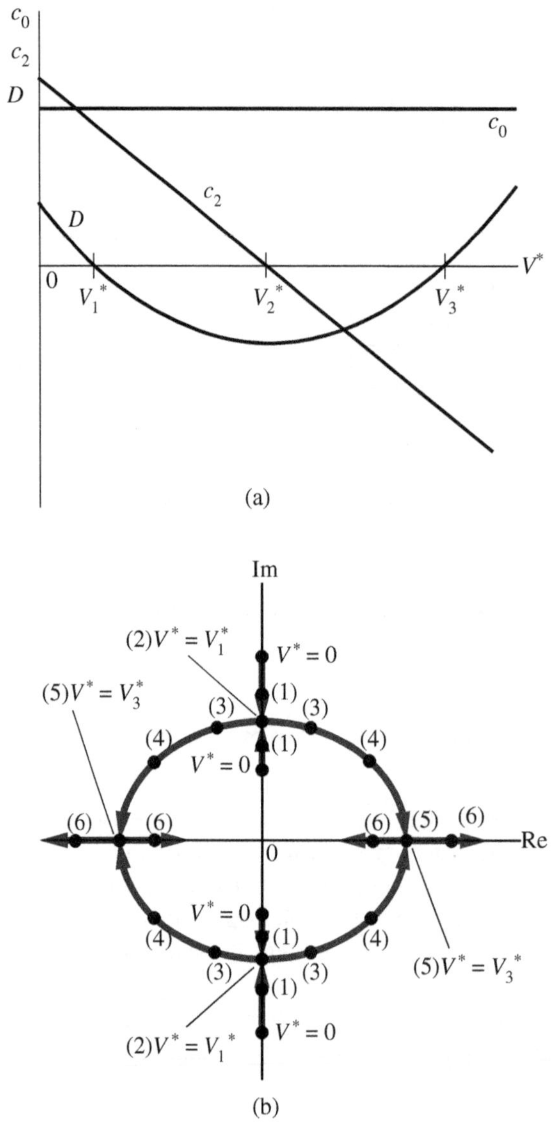

Figure 6.8 Dynamic and static destabilization of wing. (a) Schematic diagrams of c_0, c_2, and D. (b) Root locus with respect to V^* where $m^* = 1, k^* = 1/2, a^* = 1$. With increasing V^*, the four eigenvalues move in numerical order: (1) $V^* = 0.41$; (2) $V^* = V_1^* = 0.50$; (3) $V^* = 0.72$; (4) $V^* = 1.62$; (5) $V^* = V_3^* = 2.50$; (6) $V^* = 2.65$. At $V^* = V_1$ and $V^* = V_3$, Hamiltonian Hopf bifurcation and pitchfork bifurcation are produced, respectively.

When $V^* = V_1^*$, $\Lambda_i (i = 1, 2)$ are repeated negative numbers. The duplicated two roots of Eq. (6.26) are a pair of conjugate purely imaginary numbers. Also in this case, the system is stable in the sense of Liapunov.

When $V_1^* < V^* < V_3^*$, $\Lambda_i (i = 1, 2)$ are a pair of conjugate complex numbers with negative real part. Two roots of Eq. (6.26) are a pair of conjugate complex numbers with negative real part and the other two roots are a pair of conjugate complex numbers with positive real part.

As mentioned above, in $0 < V^* < V_3^*$, a dynamic destabilization is produced by the special change of the eigenvalues. Hence, two pure imaginary eigenvalues merge into a double one at point (2) in Figure 6.8b and after that, the double eigenvalues are separated into

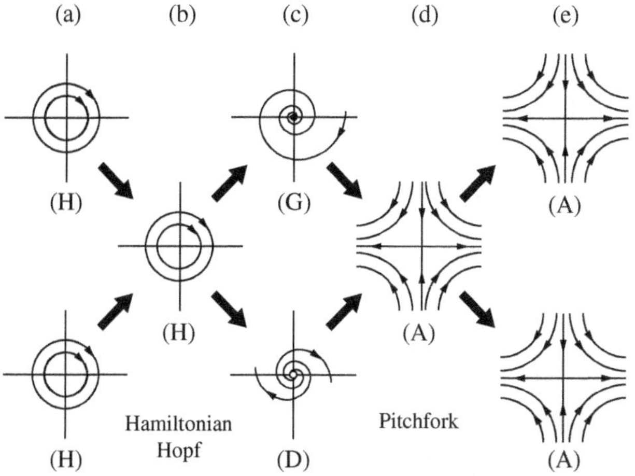

Figure 6.9 Variation of phase portraits due to wing speed ($a^* > 0$): (a) $V^* < V_1^*$; (b) $V^* = V_1^*$; (c) $V_1^* < V^* < V_3^*$; (d) $V^* = V_3^*$; and (e) $V_3^* < V^*$. Capital letters (A), (D), (G), and (H) correspond to those in Figure 2.7. In the ranges of (c) and (e), self-excited oscillation and buckling occur, respectively.

two complex eigenvalues whose real parts are positive and negative. This destabilization is called *Hamiltonian Hopf bifurcation* (Kirillov 2013; Marsden and Ratiu 1994). Self-excited oscillation is then produced.

When $V^* = V_3^*$, Λ_i $(i = 1, 2)$ are multiple positive numbers. Two roots of Eq. (6.26) are a duplicated repeated real positive numbers and the other two roots are repeated duplicated negative real numbers.

When $V^* > V_3^*$, Λ_i $(i = 1, 2)$ are two positive numbers. Two roots of Eq. (6.26) are positive real numbers and the other two roots are negative real numbers.

Increasing the velocity V^*, the system becomes dynamically unstable at $V^* = V_1^*$ and statically unstable at $V^* = V_3^*$. A root locus with respect to V^* is shown in Figure 6.8b. The variation of the phase portraits with increasing V^* is also schematically shown in Figure 6.9.

6.3 Hunting Motion in a Railway Vehicle

A railway vehicle wheelset experiences the problem of self-excited oscillation, which is called *hunting motion* (Wickens 1965) due to a nonconservative contact force (Kalker 1979) between the wheels and rails when the running speed is above a critical speed. In this section, we examine the resonance mechanism, in which the railway vehicle runs on straight rails, by using a simplified two-degree-of-freedom model accounting on the lateral and yaw motions, y and ψ, respectively, as Figure 6.10, where r_0, k_x, k_y, γ_e, and d_0 are the centered wheel rolling radius, longitudinal stiffness, lateral stiffness, wheel tread angle, and half-track gauge, respectively.

The tread angle makes the running performance on curved rails smoother (Figure 6.11a shows the configuration of a wheel set in the static equilibrium state.). For example, let us consider the situation that the wheelset runs on the curved rails as in Figure 6.11b. Then, because the wheelset is automatically shifted outside, the peripheral speed of the contact

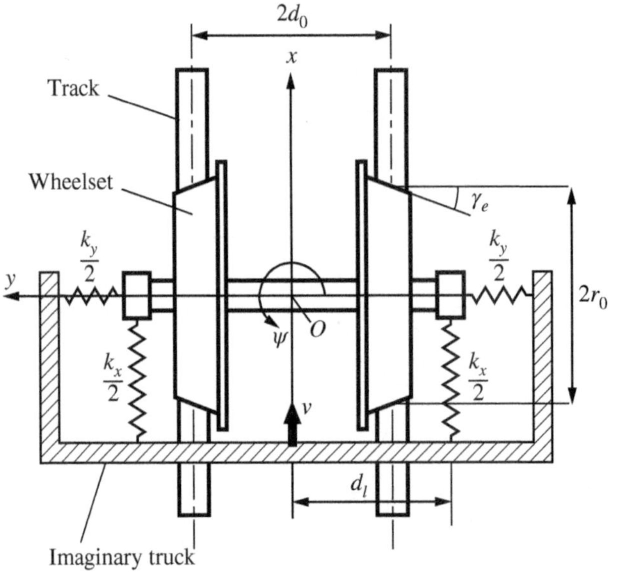

Figure 6.10 Two-degree-of-freedom model of a railway vehicle wheelset with running speed v: x and ψ are the lateral and yaw motions, respectively. The tread angle γ_e on the wheels makes the running performance on curved rails smoother. The axle is supported from the truck in x- and y-directions with the springs.

part of the left wheel becomes greater than that of the right wheel due to the tread angle γ_e, and d_l turns out to be greater than d_r. As a result, the slips between the rails and wheels are minimized, and the running performance on the curved rails rises. On the other hand, this tread angle makes the running performance on the straight rails worse. Considering the situation as Figure 6.11c that the wheelset is moved in y-direction under a certain disturbance, we notice that the peripheral speed of the contact part of the left wheel is higher than that of the right wheel because the diameter of the left wheel at the contact part d_l is larger than that of the right wheel at the contact part d_r. Since the left and right wheels are connected and have the same angular velocity, the slips between the wheels and the rails occur and then, a special contact force called *creep force* (Wickens 2005) acts between the wheels and rails.

Let us consider the behavior of the simple model of Figure 6.10 at a constant running speed v on the straight rails. The origin O of the moving coordinate system $x-y-z$ with a constant speed v in x-direction is set at the center of the axle of the wheelset in the equilibrium state. In this case, the x- and y-axes are parallel to the running direction and the axle of the wheelset, respectively. Letting the mass and the moment of inertia of the wheelset around z-axis be m and I, respectively, we can express the equation of motion for the lateral and yaw motions in the form corresponding to Eq. (3.1) as

$$M\frac{d^2\boldsymbol{x}}{dt^2} + D\frac{d\boldsymbol{x}}{dt} + (K+N)\boldsymbol{x} = \boldsymbol{0}, \tag{6.39}$$

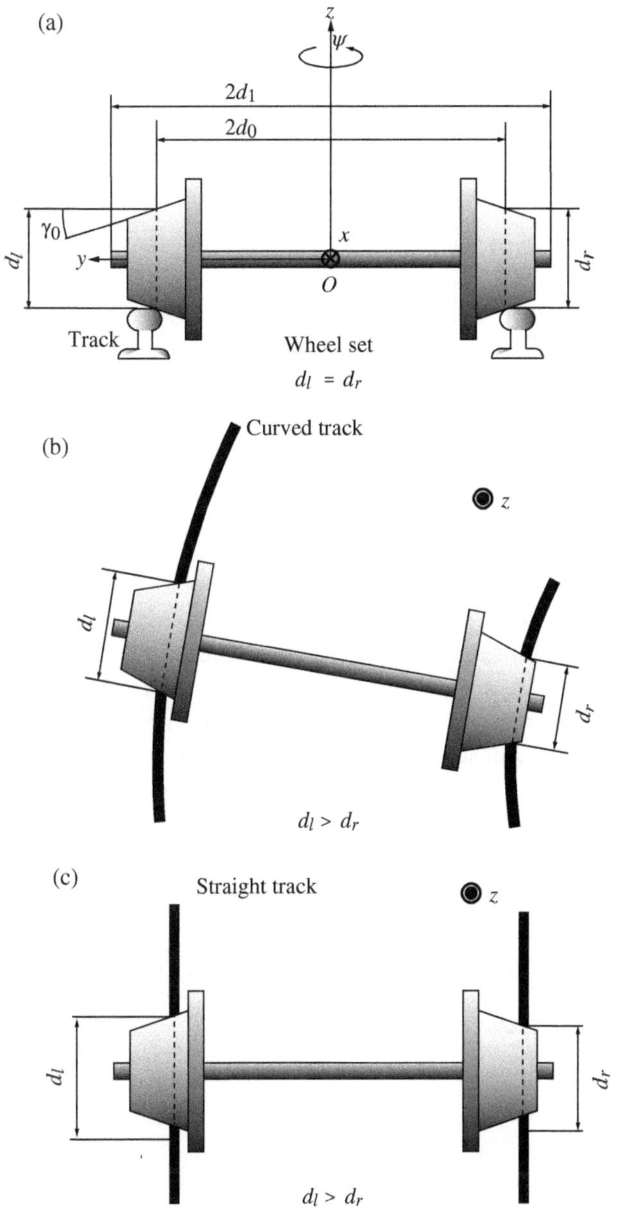

Figure 6.11 Configurations of wheelset. (a) When the wheelset runs on straight rails while keeping the coincidence of the mass center of the wheelset and the center of the rails, d_l and d_r are the same. We here denote d_l as the diameter of the circle which the contact point on the left wheel makes to the rail by the rotation of the axle, and d_r as the diameter of the circle which the contact point on the right wheel makes to the rail by the rotation of the axle. (b) When the wheelset runs on right-hand curved rails, the mass center of the wheelset automatically moves into the outside of the curved rail and d_l turns out to be larger than d_r. Then, since the peripheral speed of the left wheel is higher than that of the right wheel, the slip between the wheels and rails is decreased due to the existence of tread angle γ_e. (c) When the wheelset running on straight rails is disturbed in y direction due to fluctuation noise, d_l becomes larger than d_r. Then, since the peripheral speed of the left wheel is higher than that of the right wheel, the slip between the wheels and rails is caused by the existence of tread angle γ_e.

where

$$\boldsymbol{x} = \begin{bmatrix} y \\ \psi \end{bmatrix}, \quad M = \begin{bmatrix} m & 0 \\ 0 & I \end{bmatrix}, \quad D = \begin{bmatrix} \dfrac{2\kappa_{yy}}{v} & 0 \\ 0 & \dfrac{2\kappa_{xx} d_0^2}{v} \end{bmatrix},$$

$$K = \begin{bmatrix} k_y & -\kappa_{yy} + \dfrac{\kappa_{xx}\gamma_e d_0}{r_0} \\ -\kappa_{yy} + \dfrac{\kappa_{xx}\gamma_e d_0}{r_0} & k_x d_1^2 \end{bmatrix},$$

$$N = \begin{bmatrix} 0 & -\kappa_{yy} - \dfrac{\kappa_{xx}\gamma_e d_0}{r_0} \\ \kappa_{yy} + \dfrac{\kappa_{xx}\gamma_e d_0}{r_0} & 0 \end{bmatrix}.$$

The longitudinal and lateral creep coefficients to characterize the contact force between the rails and wheels are κ_{xx} and κ_{yy}. While a gyroscopic force does not exit, a circulatory force is included in Eq. (6.39). Therefore, the system is nonconservative and the self-excited oscillation due to the mode coupling can be produced. If there is no contact force, i.e. $\kappa_{xx} = \kappa_{yy} = 0$, the circulatory matrix N becomes a zero matrix and the self-excited oscillation is not produced. However, if there is no contact force, the railway vehicle cannot run. The circulatory matrix N is deduced under some physical assumptions and derivation of Eqs. (6.39) and (6.40) is shown in Appendix H.

The natural frequencies of the lateral and yaw motions are $\omega_y = \sqrt{\frac{k_y}{m}}$ and $\omega_\psi = \sqrt{\frac{k_x d_1^2}{I}}$, respectively. Using the half-track gauge d_0 as the representative length and the inverse value of the natural frequency in the yaw motion $\frac{1}{\omega_\psi}$ as the representative time, we rewrite the equations of motion in the following dimensionless form (Iwnicki 2006; Wickens 1965, 2005):

$$\begin{cases} \ddot{y}^* + \dfrac{d_{11}}{v^*}\dot{y}^* + k_{11}y^* + k_{12}\psi = 0, & (6.40) \\ \ddot{\psi} + \dfrac{d_{22}}{v^*}\dot{\psi} + k_{21}y^* + \psi = 0, & (6.41) \end{cases}$$

where the dot denotes the derivative with respect to the dimensionless time $t^*(= \omega_\psi t)$, $y^* \left(= \frac{y}{d_0}\right)$ is the dimensionless lateral motion, and $v^* \left(= \frac{v}{d_0 \omega_\psi}\right)$ is the dimensionless running speed. The dimensionless coefficients are

$$d_{11} = \frac{2\kappa_{yy}}{md_0\omega_\psi^2}\ (2.40), \quad k_{11} = \frac{\omega_y^2}{\omega_\psi^2}\ (0.075), \quad k_{12} = -\frac{2\kappa_{yy}}{md_0\omega_\psi^2}\ (-2.40),$$

$$d_{22} = \frac{2\kappa_{xx}d_0}{I\omega_\psi^2}\ (4.15), \quad k_{21} = \frac{2d_0^2\kappa_{xx}\gamma_e}{Ir_0\omega_\psi^2}\ (0.141), \qquad (6.42)$$

where the values in the parenthesis are those of the experiments shown in Section 16.3. Let the solutions of Eqs. (6.40) and (6.41) be

$$y^* = A_y e^{\lambda t^*}, \quad \psi = A_\psi e^{\lambda t^*}. \tag{6.43}$$

Substituting these into Eqs. (6.40) and (6.41) yields

$$L(v^*)\mathbf{\Phi} = \mathbf{0}, \tag{6.44}$$

where $L(v^*)$ is the linear operator as follows:

$$L(v^*) = \begin{bmatrix} \lambda^2 + \dfrac{d_{11}}{v^*}\lambda + k_{11} & k_{12} \\ k_{21} & \lambda^2 + \dfrac{d_{22}}{v^*}\lambda + 1 \end{bmatrix}. \tag{6.45}$$

The vector $\mathbf{\Phi}$ is the kernel of L:

$$\mathbf{\Phi} = \begin{bmatrix} A_y \\ A_\psi \end{bmatrix}. \tag{6.46}$$

The condition that there exists nontrivial $\boldsymbol{x}$, i.e. $|L| = 0$, is

$$\lambda^4 + \frac{d_{11} + d_{22}}{v^*}\lambda^3 + \left(\frac{d_{11}d_{22}}{v^{*2}} + k_{11} + 1\right)\lambda^2 + \frac{d_{11} + d_{22}k_{11}}{v^*}\lambda + k_{11} - k_{12}k_{21} = 0. \tag{6.47}$$

Increasing the dimensionless running speed v^*, the four characteristic roots of λ vary along the arrows in Figure 6.12. The variation of the subspaces according to the root locus is described in Figure 6.13. The phase space consists of two kinds of two-dimensional subspaces. One keeps the stable subspace independent of the running speed. The other changes from stable subspace to the unstable one and the self-excited oscillation is produced through Hopf bifurcation (see Section 7.4.3).

At the critical speed $v^* = v^*_{cr}$, whose phase portrait is Figure 6.13d, λ includes a pair of purely imaginary values, $\pm i\omega$, and Eq. (6.45) is expressed as

$$L(v^*_{cr}) = \begin{bmatrix} k_{11} - \omega^2 + i\dfrac{d_{11}\omega}{v^*_{cr}} & k_{12} \\ k_{21} & 1 - \omega^2 + i\dfrac{d_{22}\omega}{v^*_{cr}} \end{bmatrix}. \tag{6.48}$$

The associated eigenmode is

$$\mathbf{\Phi}_0 = \begin{bmatrix} k_{12} \\ -k_{11} + \omega^2 - i\dfrac{d_{11}\omega}{v^*_{cr}} \end{bmatrix}, \tag{6.49}$$

which satisfies

$$L(v^*_{cr})\mathbf{\Phi}_0 = \mathbf{0}. \tag{6.50}$$

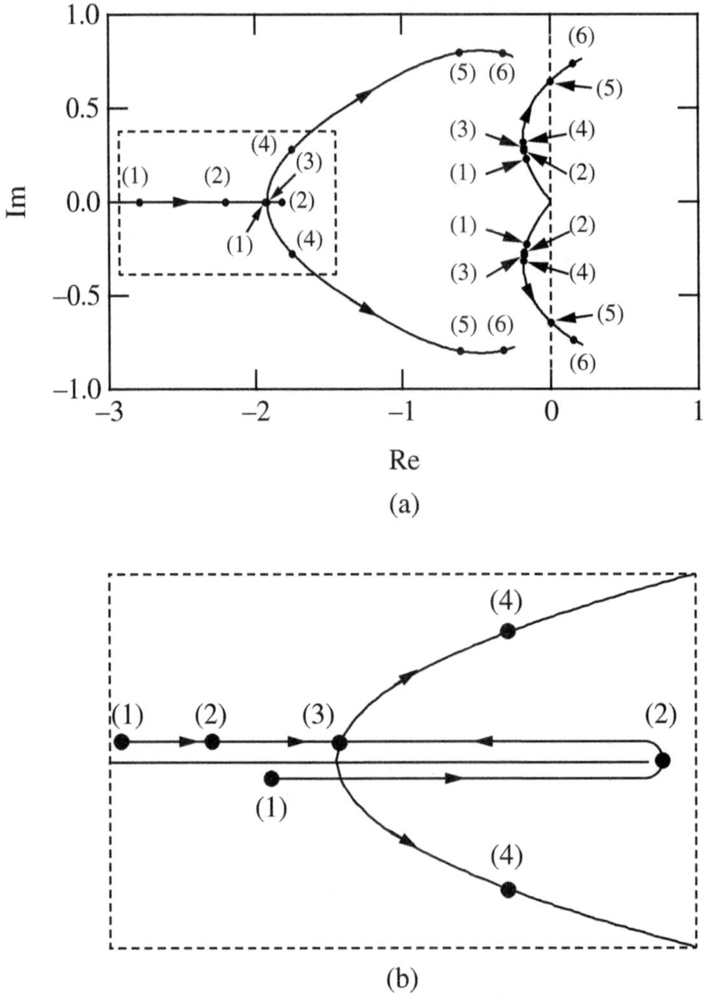

Figure 6.12 Root locus. (a) Root locus under the increase of running speed v^*; (b) Zoom on the part within the dotted square in (a). With increasing the running speed, the four roots move in the numerical order; the self-excited oscillation occurs at (5) through Hopf bifurcation.

Also, from Eq. (6.47), ω satisfies

$$\omega^4 - i\frac{d_{11}+d_{22}}{v_{cr}^*}\omega^3 - \left(\frac{d_{11}d_{22}}{v_{cr}^{*2}} + k_{11} + 1\right)\omega^2 + i\frac{d_{11}+d_{22}k_{11}}{v_{cr}^*}\omega + k_{11} - k_{12}k_{21} = 0. \tag{6.51}$$

Separating the real and imaginary parts of Eq. (6.51) leads to

$$\begin{cases} \omega^4 - \left(\dfrac{d_{11}d_{22}}{v_{cr}^{*2}} + k_{11} + 1\right)\omega^2 + k_{11} - k_{12}k_{21} = 0, & (6.52) \\ (d_{11}+d_{22})\omega^2 - (d_{11}+d_{22}k_{11}) = 0. & (6.53) \end{cases}$$

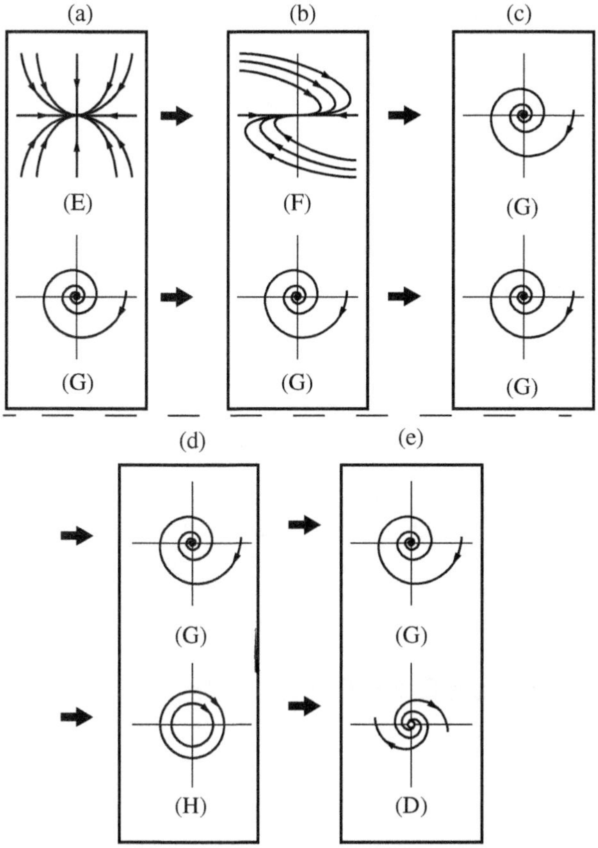

Figure 6.13 Variation of two two-dimensional subspaces. The dynamics consist of two two-dimensional subspaces for any running velocity. The trivial steady state in the upper two-dimensional subspace is stable for any speeds. The trivial steady state in the lower two-dimensional subspace is changed from stable to unstable at (d), i.e. $v^* = v^*_{cr}$.

From Eq. (6.53), we obtain the response frequency ω at $v^* = v^*_{cr}$ as

$$\omega = \sqrt{\frac{d_{11} + d_{22}k_{11}}{d_{11} + d_{22}}}. \tag{6.54}$$

Using this, we obtain v^*_{cr} from Eq. (6.52) as

$$v^*_{cr} = \sqrt{\frac{d_{11}d_{22}\omega^2}{\omega^4 - (k_{11} + 1)\omega^2 + k_{11} - k_{12}k_{21}}}. \tag{6.55}$$

Figure 6.14 shows the dependence of the critical speed on k_{21} related to the tread angle γ_e. Increasing the tread angle, i.e. k_{21}, which makes the running performance on curved rails smoother, decreases the critical running speed v^*_{cr}, and degrades the running performance on the straight rails. As a result, the running performances of this railway vehicle on curved

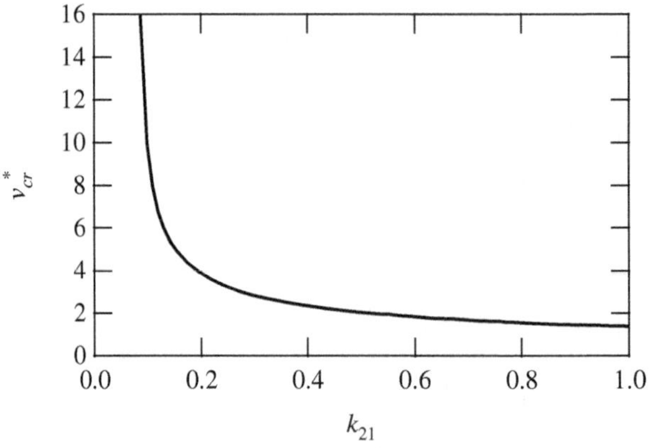

Figure 6.14 Decrease of the critical running speed on the straight rails by the increase of k_{21}, i.e. the tread angle γ_e.

and straight rails are incompatible. As a method to break through the problem, an application of gyroscopic damper is proposed (Yabuno et al. 2008; Yoshino et al. 2015) and the experimental investigation is shown in the video in Section 16.4.

6.4 Dynamic Instability in Jeffcott Rotor Due to Internal Damping

6.4.1 Fundamental Rotor Dynamics

We consider rotor dynamics about the simplified model as shown in Figure 6.15. The disk is mounted at the center of a rotating massless shaft with an angular velocity (or rotational speed) Ω. The shaft has a circular cross section and there is no direction difference in the elasticity. The shaft intercepts the disk at the elastic center M. In general, the mass center of the disk C deviates slightly from M of a quantity e as shown in Figure 6.15b; this deviation is called *eccentricity*. The described rotor system, the Jeffcott rotor, is a extremely simplified model, but retains the essential characteristics of rotor dynamics. Since the lateral deflection and the inclination of the rotor are decoupled in this system (Ishida and Yamamoto (2013)), we can deal with the system as a two-degree-of-freedom system on $x-y$ plane in Figure 6.15, where z-axis coincides with the bearing center line.

The equation of motion of the mass center of the disk is

$$m\frac{d^2\boldsymbol{r}_c}{dt^2} = \boldsymbol{F}_{elastic}, \tag{6.56}$$

where $\boldsymbol{r}_c$ is the position vector of the mass center. $\boldsymbol{F}_{elastic}$ is the restoring force of the elastic shaft expressed as

$$\boldsymbol{F}_{elastic} = -k\boldsymbol{r}, \tag{6.57}$$

where k is the linear stiffness of the shaft. The vector $\boldsymbol{r}$ is the position vector of M as

$$\boldsymbol{r} = x\boldsymbol{e}_x + y\boldsymbol{e}_y. \tag{6.58}$$

Figure 6.15 Jeffcott rotor. M is the elastic center of the disk, C the mass center of the disk, and e the eccentricity. $x - y$ is a static coordinate and $\xi - \eta$ is a rotational coordinate whose angular velocity is equal to the rotational speed of the shaft Ω.

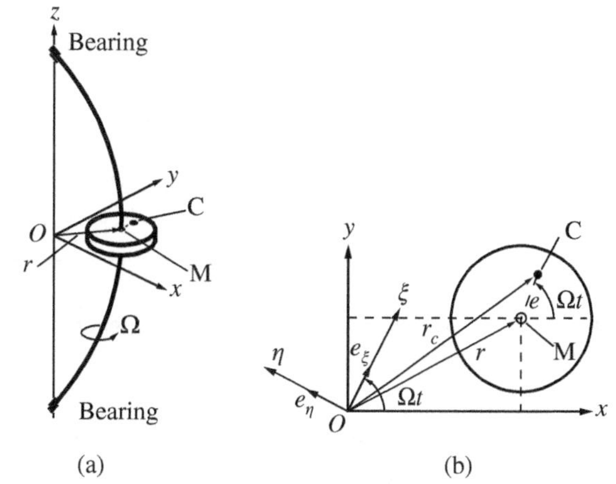

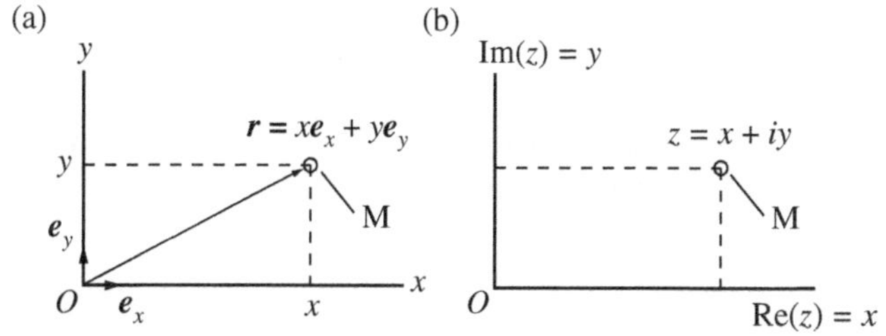

Figure 6.16 Position of elastic center M expressed by (a) vector $\boldsymbol{r}$ and (b) complex value z.

The position vector $\boldsymbol{r}_c$ is expressed using $\boldsymbol{r}$ as

$$\boldsymbol{r}_c = \boldsymbol{r} + e(\cos \Omega t \boldsymbol{e}_x + \sin \Omega t \boldsymbol{e}_y). \tag{6.59}$$

Substituting Eqs. (6.57)–(6.59) into Eq. (6.56) yields

$$M\frac{d^2\boldsymbol{r}}{dt^2} + K\boldsymbol{r} = \boldsymbol{F}_{ex}(t), \tag{6.60}$$

where

$$\boldsymbol{r} = \begin{bmatrix} x \\ y \end{bmatrix}, \quad M = \begin{bmatrix} m & 0 \\ 0 & m \end{bmatrix}, \quad K = \begin{bmatrix} k & 0 \\ 0 & k \end{bmatrix},$$

$$\boldsymbol{F}_{ex}(t) = me\Omega^2 \begin{bmatrix} \cos \Omega t \\ \sin \Omega t \end{bmatrix}. \tag{6.61}$$

This is regarded as a system with conservative restoring forces, expressed by the symmetric stiffness matrix K, and subject to the external harmonic excitations expressed by the term on the right-hand side. This equation is equivalent to

$$\begin{cases} \dfrac{d^2x}{dt^2} + \omega^2 x = e\Omega^2 \cos \Omega t, & (6.62) \\ \dfrac{d^2y}{dt^2} + \omega^2 y = e\Omega^2 \sin \Omega t, & (6.63) \end{cases}$$

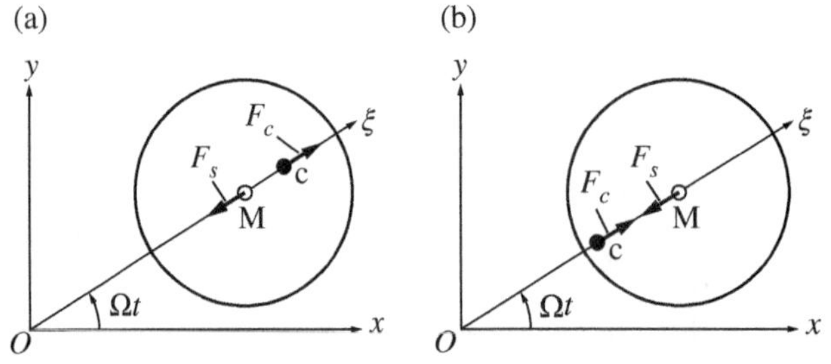

Figure 6.17 Relative position of the mass center and the elastic center depending on rotational speed: (a) $\Omega < \omega$; (b) $\Omega > \omega$.

where $\omega = \sqrt{\frac{k}{m}}$. Different from a pair of decoupled oscillators, in rotor systems, the phase difference of the external harmonic excitations is necessarily $\pi/2$. Therefore, the system can be written also in a complex form as

$$\frac{d^2z}{dt^2} + \omega^2 z = e\Omega^2 e^{i\Omega t}, \tag{6.64}$$

where $z = x + iy$ (see Figure 6.16). This expression will be used in Section 6.4.3. In this section, we consider Eqs. (6.62) and (6.63). A particular solution is

$$\begin{cases} X_p = \dfrac{e\Omega^2}{-\Omega^2 + \omega^2} \cos \Omega t, \\ Y_p = \dfrac{e\Omega^2}{-\Omega^2 + \omega^2} \sin \Omega t. \end{cases} \tag{6.65}$$

These correspond to the steady-state response after the transient state when the homogeneous solution is decayed due to the external damping generally present in a practical system, but here not considered in Eqs. (6.62) and (6.63) (for the consideration of the external damping term, see Section 6.4.3). Then, the position vector of the elastic center is

$$\boldsymbol{r} = X_p \boldsymbol{e}_x + Y_p \boldsymbol{e}_y = \frac{e\Omega^2}{\omega^2 - \Omega^2}(\cos \Omega t \boldsymbol{e}_x + \sin \Omega t \boldsymbol{e}_y). \tag{6.66}$$

At the critical rotational speed $\Omega = \omega$, in the analysis ignoring damping effect, it is concluded that the amplitude becomes infinity, but, in the practical case, the external damping effect keeps the resonance amplitude finite (see Section 6.4.3). From Eq. (6.59), the position vector of the mass center is

$$\begin{aligned} \boldsymbol{r}_c &= \boldsymbol{r} + e(\cos \Omega t \boldsymbol{e}_x + \sin \Omega t \boldsymbol{e}_y) \\ &= \frac{e\omega^2}{\omega^2 - \Omega^2}(\cos \Omega t \boldsymbol{e}_x + \sin \Omega t \boldsymbol{e}_y). \end{aligned} \tag{6.67}$$

At very high rotational speed $\frac{\omega}{\Omega} \ll 1$, $\boldsymbol{r}_c$ becomes approximately a zero vector, and the mass center is almost located on the bearing center line. This phenomenon is known as the *self-centering* (e.g. Ishida and Yamamoto 2013).

By the way, the relative positional relation between points C and M is

$$|\boldsymbol{r}_c| - |\boldsymbol{r}| = e\frac{\omega^2 - \Omega^2}{|\omega^2 - \Omega^2|}, \tag{6.68}$$

and the configurations between the mass center C and the elastic center M are shown for $\Omega < \omega$ and $\Omega > \omega$ in Figure 6.17a,b, respectively. Different from the low speed range of $\Omega < \omega$, in the high speed range $\Omega > \omega$, the centrifugal force F_c acting on the mass center C and the restoring force F_s acting on the elastic center M are facing each other as shown in Figure 6.17b, and these forces can be regarded as compressive forces in a sense. Due to such a relative position, it was concluded that in the high speed range, above the critical speed $\Omega > \omega$, the operation was impossible as described in the pioneering research on rotor dynamics by Rankin (for more details, see (Ishida and Yamamoto 2013)). In Section 6.4.2, even in such a situation, the operation is possible in accordance with the result of Eq. (6.65).

6.4.2 Effects of the Centrifugal Force and the Coriolis Force on Static Stability

We investigate the dynamics using the polar coordinate system $\xi - \eta$ rotating with angular velocity Ω as shown in Figure 6.15b: the base vectors are

$$\begin{cases} \boldsymbol{e}_\xi = \cos\Omega t \boldsymbol{e}_x + \sin\Omega t \boldsymbol{e}_y, & (6.69) \\ \boldsymbol{e}_\eta = -\sin\Omega t \boldsymbol{e}_x + \cos\Omega t \boldsymbol{e}_y, & (6.70) \end{cases}$$

and their derivatives with respect to time are

$$\begin{cases} \dfrac{d\boldsymbol{e}_\xi}{dt} = \Omega \boldsymbol{e}_\eta, & (6.71) \\ \dfrac{d\boldsymbol{e}_\eta}{dt} = -\Omega \boldsymbol{e}_\xi. & (6.72) \end{cases}$$

Using these equations, we can express the position, velocity, and acceleration vectors of the elastic center M, $\boldsymbol{r}$, $\frac{d\boldsymbol{r}}{dt}$, and $\frac{d^2\boldsymbol{r}}{dt^2}$, respectively, as follows:

$$\boldsymbol{r} = x\boldsymbol{e}_x + y\boldsymbol{e}_y = \xi\boldsymbol{e}_\xi + \eta\boldsymbol{e}_\eta, \tag{6.73}$$

$$\frac{d\boldsymbol{r}}{dt} = \left(\frac{d\xi}{dt} - \Omega\eta\right)\boldsymbol{e}_\xi + \left(\frac{d\eta}{dt} + \Omega\xi\right)\boldsymbol{e}_\eta, \tag{6.74}$$

$$\frac{d^2\boldsymbol{r}}{dt^2} = \left(\frac{d^2\xi}{dt^2} - 2\Omega\frac{d\eta}{dt} - \Omega^2\xi\right)\boldsymbol{e}_\xi + \left(\frac{d^2\eta}{dt^2} + 2\Omega\frac{d\xi}{dt} - \Omega^2\eta\right)\boldsymbol{e}_\eta. \tag{6.75}$$

The acceleration vector of the mass center C is

$$\begin{aligned} \frac{d^2\boldsymbol{r}_c}{dt^2} &= \frac{d^2\boldsymbol{r}}{dt^2} + e\frac{d^2\boldsymbol{e}_\xi}{dt^2} \\ &= \left(\frac{d^2\xi}{dt^2} - 2\Omega\frac{d\eta}{dt} - \Omega^2\xi - e\Omega^2\right)\boldsymbol{e}_\xi + \left(\frac{d^2\eta}{dt^2} + 2\Omega\frac{d\xi}{dt} - \Omega^2\eta\right)\boldsymbol{e}_\eta. \end{aligned} \tag{6.76}$$

The elastic force of the shaft $\boldsymbol{F}_{elastic}$ is expressed as

$$\boldsymbol{F}_{elastic} = -k\xi\boldsymbol{e}_\xi - k\eta\boldsymbol{e}_\eta. \tag{6.77}$$

Substituting Eqs. (6.76) and (6.77) into Eq. (6.56) yields the equation of motion in the polar coordinate as

$$\begin{cases} m\dfrac{d^2\xi}{dt^2} - 2m\Omega\dfrac{d\eta}{dt} + (k - m\Omega^2)\xi = me\Omega^2, & (6.78) \\ m\dfrac{d^2\eta}{dt^2} + 2m\Omega\dfrac{d\xi}{dt} + (k - m\Omega^2)\eta = 0. & (6.79) \end{cases}$$

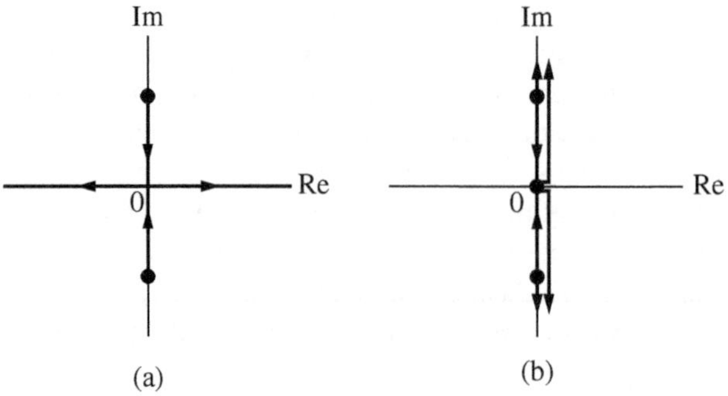

Figure 6.18 Root loci by increasing rotational speed. (a) Without considering Coriolis force. (b) With considering Coriolis force.

These particular solution is

$$\xi_p = \frac{e\Omega^2}{\omega^2 - \Omega^2}, \quad \eta_p = 0, \tag{6.80}$$

which corresponds to the steady-state response of Eq. (6.65). In addition, the particular solutions ξ_p and η_p are, respectively, equal to the solutions ξ_{st} and η_{st} of the following equilibrium equations:

$$\begin{cases} (k - m\Omega^2)\xi_{st} = me\Omega^2, & (6.81) \\ (k - m\Omega^2)\eta_{st} = 0. & (6.82) \end{cases}$$

In order to determine the stability of the equilibrium state, substituting $\xi = \xi_{st} + \Delta\xi$ and $\eta = \eta_{st} + \Delta\eta$ into Eqs. (6.78) and (6.79) and considering Eqs. (6.81) and (6.82) yields

$$M\frac{d^2\Delta\boldsymbol{r}}{dt^2} + G\frac{d\Delta\boldsymbol{r}}{dt} + K\Delta\boldsymbol{r} = 0, \tag{6.83}$$

where

$$\Delta\boldsymbol{r} = \begin{bmatrix} \Delta\xi \\ \Delta\eta \end{bmatrix}, \quad M = \begin{bmatrix} m & 0 \\ 0 & m \end{bmatrix}, \quad G = \begin{bmatrix} 0 & -2m\Omega \\ 2m\Omega & 0 \end{bmatrix},$$
$$K = \begin{bmatrix} k - m\Omega^2 & 0 \\ 0 & k - m\Omega^2 \end{bmatrix}. \tag{6.84}$$

Equation (6.83) is equivalent to the homogeneous equations for Eqs. (6.78) and (6.79). As seen from the component in stiffness matrix K, if $\Omega > \omega \left(= \sqrt{\frac{k}{m}}\right)$, the centrifugal force produces a positional force with a negative stiffness. Without considering the gyroscopic effect in Eq. (6.83), the root locus is described in Figure 6.18a. When the rotational speed is above the critical speed, i.e. $\Omega > \omega$, the equivalent stiffness $k - m\Omega^2$ becomes negative. Therefore, the phase portrait is changed with increasing the rotational speed as shown in Figure 6.19, and it seems that the rotor system is buckled and cannot be operated with a rotational speed above the critical speed as the before-mentioned pioneering research. However, the operation is possible even above the critical speed since the gyroscopic effect $G\frac{d\Delta\boldsymbol{r}}{dt}$ caused by Coriolis force avoids the buckling. This fact is proved by using the characteristic

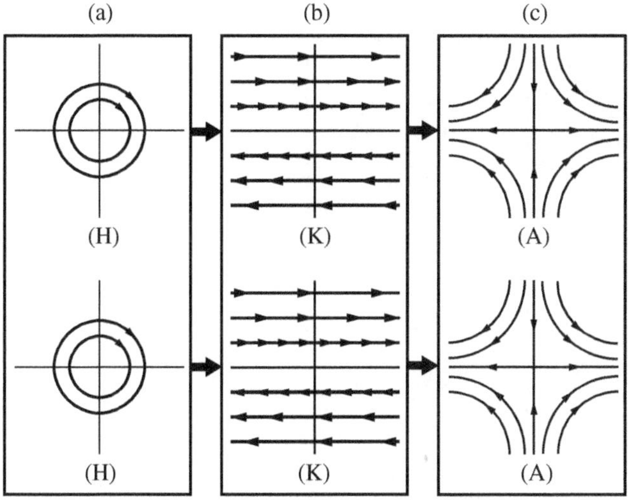

Figure 6.19 Phase portraits without considering Coriolis force.

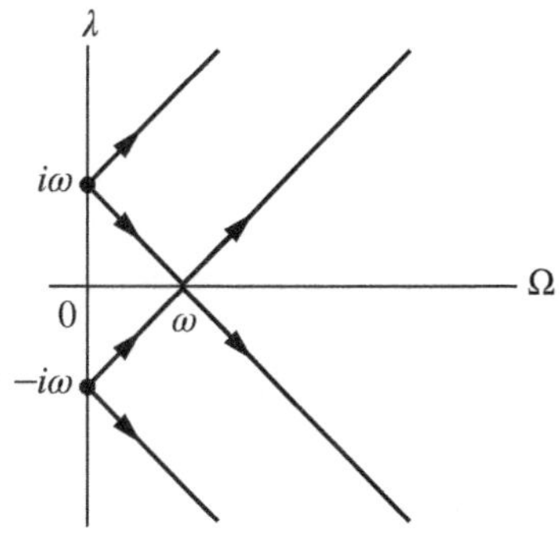

Figure 6.20 Campbell diagram. Due to the Coriolis force, all eigenvalues are purely imaginary for any rotational speed except for $\Omega = \omega$.

equation depending on the rotational speed:

$$\lambda^4 + 2(\omega^2 + \Omega^2)\lambda^2 + (\omega^2 - \Omega^2)^2 = 0. \tag{6.85}$$

The solutions are

$$\lambda = \pm(\omega + \Omega)i, \ \pm|\omega - \Omega|i. \tag{6.86}$$

Therefore, the root locus is described as in Figure 6.18b. All eigenvalues are purely imaginary independent of rotational speed except for $\Omega = \omega$. With increasing the rotational speed, the eigenvalues are changed along the arrow in Figure 6.20, and the phase portraits are changed into those of Figure 6.21. Therefore, even above the critical speed, the rotor systems can be operated. Hence, the gyroscopic force avoids the static instability above the critical rotational speed due to the centrifugal force. Such a stabilization could be seen in a monorail (Cousins 1913).

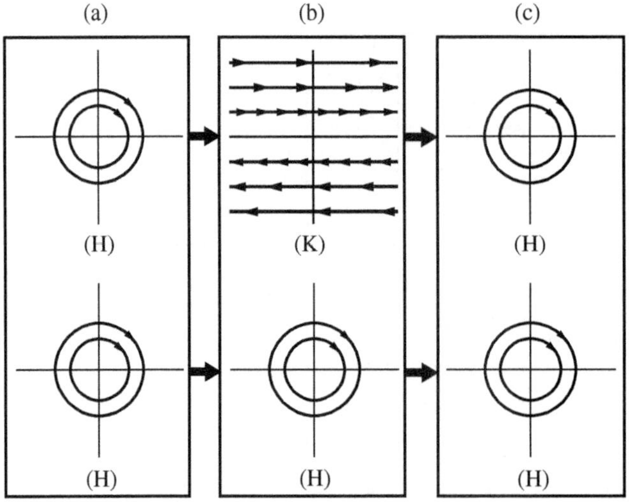

Figure 6.21 Phase portraits with considering Coriolis force.

6.4.3 Effect of External Damping

In general, two kinds of damping forces can act on rotor systems. One is the external damping force, which is, for example, friction with surrounding static fluid proportional to the absolute velocity on $x-y$ plane. Assuming that it is isotropic, we can express the damping forces in x- and y-directions as

$$D_x = -c_e\frac{dx}{dt}, \quad D_y = -c_e\frac{dy}{dt}, \tag{6.87}$$

where the external damping coefficient c_e is a positive constant. The external damping is described by the dampers set in the transversal direction to the shaft express as shown in Figure 6.24a. Then, these dissipative forces corresponding to F_D in Eq. (3.7) are added to Eq. (6.60) as

$$M\frac{d^2\boldsymbol{r}}{dt^2} + D\frac{d\boldsymbol{r}}{dt} + K\boldsymbol{r} = \boldsymbol{F}_{ex}(t), \tag{6.88}$$

where D is a positive definite symmetric matrix as

$$M = \begin{bmatrix} m & 0 \\ 0 & m \end{bmatrix}, \quad D = \begin{bmatrix} c_e & 0 \\ 0 & c_e \end{bmatrix}, \quad K = \begin{bmatrix} k & 0 \\ 0 & k \end{bmatrix}. \tag{6.89}$$

Equation (6.88) is equivalent to

$$\begin{cases} \dfrac{d^2x}{dt^2} + 2\gamma_e\omega\dfrac{dx}{dt} + \omega^2 x = e\Omega^2\cos\Omega t, & (6.90) \\ \dfrac{d^2y}{dt^2} + 2\gamma_e\omega\dfrac{dy}{dt} + \omega^2 y = e\Omega^2\sin\Omega t, & (6.91) \end{cases}$$

where $\gamma_e = \frac{c_e}{2\sqrt{mk}}$. Referring to the solution of Eq. (C.1) equivalent to these equations, we obtain the steady-state response that is the particular solution as

$$\begin{cases} x_p = a_{pc-x}\cos\Omega t + a_{ps-x}\sin\Omega t, & (6.92) \\ y_p = a_{pc-y}\cos\Omega t + a_{ps-y}\sin\Omega t, & (6.93) \end{cases}$$

where a_{pc-x}, a_{ps-x}, a_{pc-y}, and a_{ps-y} are

$$a_{pc-x} = \frac{e\Omega^2(\omega^2 - \Omega^2)}{(\omega^2 - \Omega^2)^2 + 4\gamma_e^2\omega^2\Omega^2}, \tag{6.94}$$

$$a_{ps-x} = \frac{2e\gamma_e\omega\Omega^3}{(\omega^2 - \Omega^2)^2 + 4\gamma_e^2\omega^2\Omega^2}, \tag{6.95}$$

$$a_{pc-y} = \frac{-2e\gamma_e\omega\Omega^3}{(\omega^2 - \Omega^2)^2 + 4\gamma_e^2\omega^2\Omega^2}, \tag{6.96}$$

$$a_{ps-y} = \frac{e\Omega^2(\omega^2 - \Omega^2)}{(\omega^2 - \Omega^2)^2 + 4\gamma_e^2\omega^2\Omega^2}. \tag{6.97}$$

As a result, the external damping avoids the infinite response amplitude at the resonance point, i.e. the critical speed $\Omega = \omega$. In case without eccentricity, i.e. $e = 0$, the dynamics is described by the homogeneous equations (6.90) and (6.91) with zero right-hand side, i.e.

$$\begin{cases} \frac{d^2x}{dt^2} + 2\gamma_e\omega\frac{dx}{dt} + \omega^2 x = 0, & (6.98) \\ \frac{d^2y}{dt^2} + 2\gamma_e\omega\frac{dy}{dt} + \omega^2 y = 0, & (6.99) \end{cases}$$

and the whirling motion is decayed with time. In Section 6.4.4, we investigate the effect of internal damping on such a rotor system without eccentricity.

Before going to the next section, we mention the analysis based on a complex form, which can directly characterize whirling motions in rotor dynamics. Let us consider the rotor dynamics governed by Eqs. (6.90) and (6.91). We introduce the complex value as $z = x + iy$ as shown in Figure 6.16. Then, since z is equivalent to $\boldsymbol{r} = x\boldsymbol{e}_x + y\boldsymbol{e}_y$, adding Eq. (6.90) to Eq. (6.91) multiplied by the imaginary unit i, yields the equation of motion in complex form as

$$\frac{d^2z}{dt^2} + 2\gamma_e\omega\frac{dz}{dt} + \omega^2 z = e\Omega^2 e^{i\Omega t}. \tag{6.100}$$

First, we consider the homogeneous equation, that is Eq. (6.100) with zero term on the right-hand side. Substituting $z = \mathrm{e}^{\lambda t}$ into Eq. (6.100) yields

$$\lambda^2 + 2\gamma_e\omega\lambda + \omega^2 = 0. \tag{6.101}$$

In the case of $0 < \gamma_e < 1$, the eigenvalues are $\lambda = -\gamma_e\omega \pm \sqrt{1-\gamma_e^2}\omega i$. The homogeneous solution is

$$\tilde{z} = A_f \mathrm{e}^{-\gamma_e\omega t}\mathrm{e}^{i\sqrt{1-\gamma_e^2}\omega t} + A_b \mathrm{e}^{-\gamma_e\omega t}\mathrm{e}^{-i\sqrt{1-\gamma_e^2}\omega t}, \tag{6.102}$$

where the constant complex A_f and A_b are expressed as $A_f = a_f\mathrm{e}^{i\phi_f}$ and $A_b = a_b\mathrm{e}^{-i\phi_b}$ using the real constant values, a_f, ϕ_f, a_b, and ϕ_b, determined by the initial condition. Then,

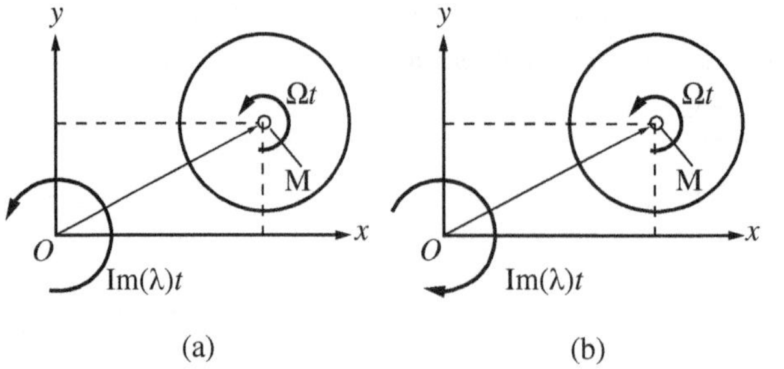

Figure 6.22 Forward and backward whirling motions. (a) Forward whirling motion ($\mathrm{Im}(\lambda) > 0$). (b) Backward whirling motion ($\mathrm{Im}(\lambda) < 0$).

Eq. (6.102) is expressed as

$$\tilde{z} = a_f e^{-\gamma_e \omega t} e^{i(\sqrt{1-\gamma_e^2}\omega t + \phi_f)} + a_b e^{-\gamma_e \omega t} e^{-i(\sqrt{1-\gamma_e^2}\omega t + \phi_b)}. \tag{6.103}$$

The first term related to the eigenvalue with a positive imaginary part expresses the *forward whirling motion* , and the second term related to the eigenvalue with a negative imaginary part expresses the backward whirling motion, as shown in Figure 6.22a,b, respectively.

6.4.4 Dynamics Instability Due to Internal Damping

The damping other than the external damping is the internal damping, which is related to the strain of the shaft fiber. According to Gasch et al. (2006), we investigate the characteristics of the internal damping paying attention to the difference with respect to the external damping. Figure 6.23 schematically describes the configurations of the rotor: part (a) shows the case without damping and the rotor describes a whirling motion with natural frequency ω under the rotational speed Ω; part (b) represents the case of the rotor system subject to external and internal damping. In this section, in order to focus on the essential characteristics of external and internal damping, we consider the case without eccentricity *e*.

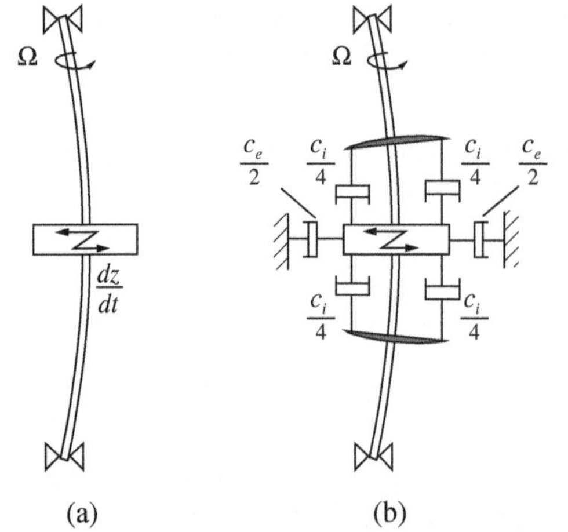

Figure 6.23 Jeffcott rotor subject to external and internal damping. (a) Case without damping, where $\frac{dz}{dt}$ and Ω express the velocity of the whirling motion and the angular velocity of the shaft, respectively. (b) Case with external and internal damping, where c_e and c_i are external and internal damping coefficients, respectively.

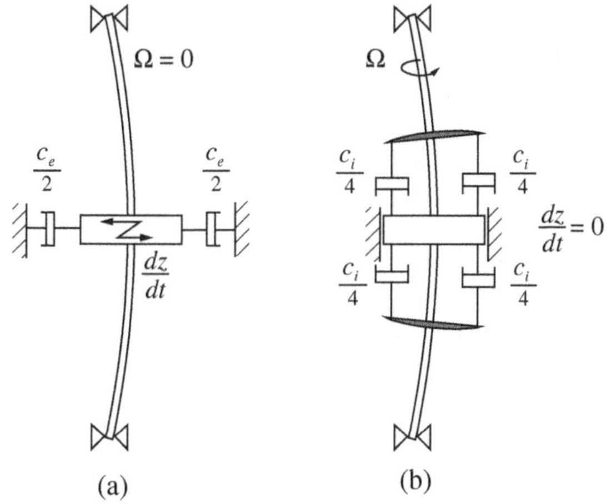

Figure 6.24 External and internal damping effects. (a) External damping: this damping is caused from the whirling motion of the disk, but not related to the rotation of the shaft; the external damping is schematically described by the dampers set in the transversal direction to the shaft. (b) Internal damping: this damping is caused from the rotation of the shaft, but not related to the whirling motion of the disk; the internal damping is schematically described by the dampers set in the parallel direction to the shaft.

First, as mentioned in Section 6.4.3, the external damping is schematically shown with dampers as in Figure 6.24a. Independent of the rotation of the shaft, the viscous damping force is caused by the velocity of M on $x - y$ plane and is proportional to the velocity $\frac{dz}{dt}$ as defined by Eq. (6.87). Hence, the external damping can be produced even in the whirling motion without rotational speed ($\Omega = 0$). On the other hand, the internal damping is caused by the rotation of the shaft even in the case without its whirling motion $\frac{dz}{dt} = 0$.

Let us consider Figure 6.24b. The shaft is fixed with a curved configuration. Then, the rotation of the shaft is possible, but the whirling motion is impossible. Even in such a situation, the strain of the shaft fiber is alternatively changed with the rotation and causes the so called internal damping, which is represented by the dampers in Figure 6.24b. The damping effect is related to the difference between the whirling angular velocity and the rotational speed Ω; in Figure 6.24b, whirling angular velocity is zero, but the rotational speed is not zero. Therefore, the internal damping effect is expressed using the rotational frame $\xi - \eta$ as $-c_i(\dot{\xi}\boldsymbol{e}_\xi + \dot{\eta}\boldsymbol{e}_\eta)$. As a result, in the case with internal damping, the equations of motion (6.78) and (6.79) in the rotational frame are changed into

$$\begin{cases} m\ddot{\xi} - 2m\Omega\dot{\eta} + (k - m\Omega^2)\xi + c_i\dot{\xi} = 0, & (6.104) \\ m\ddot{\eta} + 2m\Omega\dot{\xi} + (k - m\Omega^2)\eta + c_i\dot{\eta} = 0. & (6.105) \end{cases}$$

Furthermore, because of $-c_i(\dot{\xi}\boldsymbol{e}_\xi + \dot{\eta}\boldsymbol{e}_\eta) = -c_i\{(\dot{x} + \Omega y)\boldsymbol{e}_x + (\dot{y} - \Omega x)\boldsymbol{e}_y\}$, these equations of motion are expressed in the static frame as

$$\begin{cases} m\ddot{x} + c_i\dot{x} + kx + c_i\Omega y = 0, & (6.106) \\ m\ddot{y} + c_i\dot{y} - c_i\Omega x + ky = 0, & (6.107) \end{cases}$$

or equivalently

$$M\frac{d^2\boldsymbol{r}}{dt^2} + D\frac{d\boldsymbol{r}}{dt} + (K + N)\boldsymbol{r} = \boldsymbol{0}, \tag{6.108}$$

where the coefficient matrices are

$$M = \begin{bmatrix} m & 0 \\ 0 & m \end{bmatrix}, \quad D = \begin{bmatrix} c_i & 0 \\ 0 & c_i \end{bmatrix}, \quad K = \begin{bmatrix} k & 0 \\ 0 & k \end{bmatrix}, \quad N = \begin{bmatrix} 0 & c_i\Omega \\ -c_i\Omega & 0 \end{bmatrix}. \tag{6.109}$$

The matrices, D and K, are positive definite symmetric. The internal damping effect appears as a velocity-dependent damping effect in D and produces the circulatory force expressed by N, which makes the system nonconservative.

Next, considering also the external damping, we investigate the effect of the internal damping on the stability of rotor system. The external damping effect appears in matrix D in Eq. (6.109) as

$$D = \begin{bmatrix} c_i + c_e & 0 \\ 0 & c_i + c_e \end{bmatrix}. \tag{6.110}$$

Introducing $z = x + iy$, the equation of motion is expressed in a complex form as

$$\ddot{z} + 2(\gamma_e + \gamma_i)\omega\dot{z} + \omega^2 z - 2i\gamma_i\omega\Omega z = 0, \tag{6.111}$$

where $\gamma_e = \frac{c_e}{2\sqrt{mk}}$, $\gamma_i = \frac{c_i}{2\sqrt{mk}}$, $\omega = \sqrt{\frac{k}{m}}$. Substituting

$$z = Ae^{\lambda t} \tag{6.112}$$

into Eq. (6.111) yields

$$\lambda^2 + 2(\gamma_e + \gamma_i)\omega\lambda + \omega^2 - 2i\gamma_i\omega\Omega = 0. \tag{6.113}$$

Under the assumption of $\gamma_e \ll 1$, $\gamma_i \ll 1$, by neglecting $O(\gamma_e^2)$ and $O(\gamma_i^2)$, the solutions can be approximately expressed as

$$\begin{cases} \lambda_f = \gamma_i\Omega - (\gamma_e + \gamma_i)\omega + \omega i, & (6.114) \\ \lambda_b = -\{\gamma_i\Omega + (\gamma_e + \gamma_i)\omega\} - \omega i, & (6.115) \end{cases}$$

which are related to the forward and backward whirling motions, respectively. As a result, we obtain the whirling motion as

$$z = A_f e^{\lambda_f t} + A_b e^{\lambda_b t} = a_f e^{\{\gamma_i\Omega - (\gamma_e+\gamma_i)\omega\}t} e^{i(\omega t + \phi_f)} + a_b e^{-\{\gamma_i\Omega + (\gamma_e+\gamma_i)\omega\}t} e^{-i(\omega t + \phi_b)}, \tag{6.116}$$

where $A_f = a_f e^{i\phi_f}$ and $A_b = a_b e^{i\phi_b}$; a_f, ϕ_f, a_b, and ϕ_b are constants obtained from the initial condition. The first term proportional to $e^{i\omega t}$, related to λ_f with positive imaginary part, and the second term proportional to $e^{-i\omega t}$, related to λ_b with negative imaginary part, are the forward and backward whirling motions, respectively. The radii of the forward and backward whirling motions are respectively

$$a_f e^{\{\gamma_i\Omega - (\gamma_e+\gamma_i)\omega\}t} \quad \text{and} \quad a_b e^{-\{\gamma_i\Omega + (\gamma_e+\gamma_i)\omega\}t}. \tag{6.117}$$

The backward whirling motion decays with time for any rotational speed since the radius decays with time due to $-\{\gamma_i\Omega + (\gamma_e + \gamma_i)\omega\} < 0$. The forward whirling motion is dynamically stable when $\Omega < \frac{\gamma_e+\gamma_i}{\gamma_i}\omega$ since the radius decays with time, but dynamically unstable when $\Omega > \frac{\gamma_e+\gamma_i}{\gamma_i}\omega$ since the radius grows with time. As a result, $\Omega = \frac{\gamma_e+\gamma_i}{\gamma_i}\omega$ is the critical rotational speed, and when $\Omega > \frac{\gamma_e+\gamma_i}{\gamma_i}\omega$, the internal damping causes a self-excited forward whirling motion through Hopf bifurcation.

6.5 Dynamic Instability in Fluid-Conveying Pipe Due to Follower Force

In this section, we briefly introduce dynamic instability due to follower force. Different from the compressive force to produce the buckling, the follower force is defined as the force whose direction is kept tangent to the axial direction of the structure (Herrmann 1971). The follower force acts on the structure as a circulatory force.

We investigate the self-excited oscillation in a fluid conveying pipe. The system is well known as an essential model producing self-excited oscillation due to circulatory force and has attracted many researchers for a long time (Paidoussis 1998). We consider the simplified two-degree-of-freedom model as shown in Figure 6.25, which is first introduced by Benjamin (1962) and consists of two rigid pipes with length of $\frac{l}{2}$, where V is a fluid velocity which is assumed constant, m and m_f are the mass per unit length of the pipe and fluid, respectively, l is the length of half the pipe, and k is the stiffness coefficient of the torsional springs at the hinges. The fluid is discharged at the end of the pipe with velocity V.

The dynamics is described by φ_1 and φ_2. According to the literature (Seyranian et al. 1994; Sugiyama and Noda 1981), the nondimensional equations of motion are

$$\begin{bmatrix} \frac{4}{3} & \frac{1}{2} \\ \frac{1}{2} & \frac{1}{3} \end{bmatrix} \begin{bmatrix} \ddot{\varphi}_1 \\ \ddot{\varphi}_2 \end{bmatrix} + \begin{bmatrix} \eta V^* & 2\eta V^* \\ 0 & \eta V^* \end{bmatrix} \begin{bmatrix} \dot{\varphi}_1 \\ \dot{\varphi}_2 \end{bmatrix} + \begin{bmatrix} 2 - V^{*2} & V^{*2} - 1 \\ -1 & 1 \end{bmatrix} \begin{bmatrix} \varphi_1 \\ \varphi_2 \end{bmatrix} = \begin{bmatrix} 0 \\ 0 \end{bmatrix}, \tag{6.118}$$

where the nondimensional fluid speed, the nondimensional time, and the mass ratio are expressed by the dimensional parameters respectively as

$$V^* = Vl\sqrt{\frac{m_f}{kl}}, \quad t^* = \sqrt{\frac{k}{(m+m_f)l^3}}t, \quad \eta = \sqrt{\frac{m_f}{m+m_f}}. \tag{6.119}$$

The dot denotes the derivative with respect to t^*. The system has the gyroscopic and circulatory terms due to the fluid speed V^*.

First, we consider the case when the fluid mass is very small, i.e. η is assumed to be zero. The gyroscopic terms can be neglected. In case with stationary fluid ($V^* = 0$), since

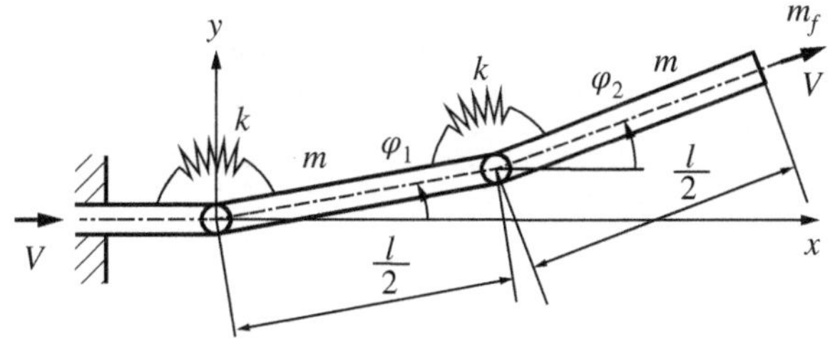

Figure 6.25 Two-degree-of-freedom model of fluid conveying pipe consisting of two rigid pipes with length of $\frac{l}{2}$. The restoring moments with stiffness coefficient k act through the torsional springs at the hinges. The flow is discharged at velocity V.

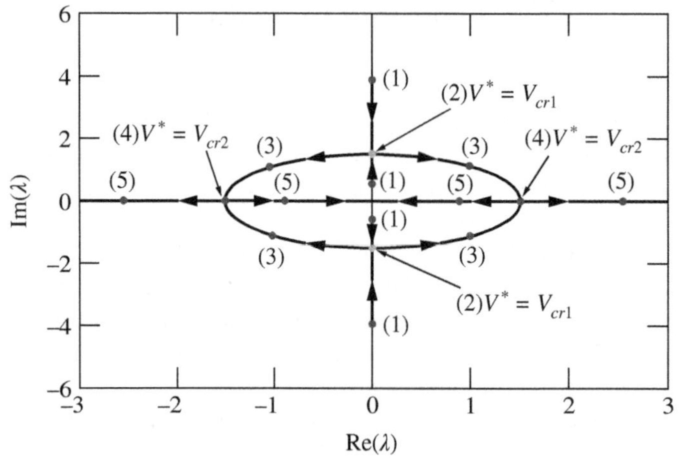

Figure 6.26 Root locus in the case when fluid mass can be neglected. With increasing V^*, the four eigenvalues move in numerical order: (1) $V^* = 0$; (2) $V^* = V^*_{cr1} = 1.59$; (3) $V^* = 1.90$; (4) $V^* = V^*_{cr2} = 2.15$; (5) $V^* = 2.30$. At V^*_{cr1}, the system produces self-excited oscillation through Hamiltonian Hopf bifurcation. At V^*_{cr2}, the system is buckled.

the matrix K' related to the positional force is symmetric, i.e. $N = 0$ (see Section 3.1), the system is conservative and the behavior corresponds to that in Section 4.3. In the case with flow speed, the system becomes nonconservative due to circulatory terms. With increasing the flow speed, the eigenvalues move along the arrow in Figure 6.26. The root locus is qualitatively the same as that of flutter shown in Figure 6.8b. At $V^* = V^*_{cr1}$, the self-excited oscillation starts through Hamiltonian Hopf bifurcation. When increasing the fluid speed to higher levels, instead of the self-excited oscillation, the system is buckled at $V^* = V^*_{cr2}$.

Next, we consider the case when the fluid mass cannot be neglected. In case with stationary fluid ($V^* = 0$), the system is conservative again. With increasing the flow speed, the eigenvalues move along the arrow in Figure 6.27. At $V^* = V^*_{cr1}$, the eigenvalues related to one of two modes are changed from the complex conjugate pair with negative real part to two negative real numbers. At $V^* = V^*_{cr2}$, the complex conjugate eigenvalues with negative real part in the other mode is changed to the complex conjugate eigenvalues with positive real part. Therefore, at this point, the self-excited oscillation is produced through Hopf bifurcation. With increasing the flow speed again, this mode producing the self-excited oscillation exhibits buckling at $V^* = V^*_{cr3}$. The corresponding experiments are mentioned in Sections 4.5 and 16.5.

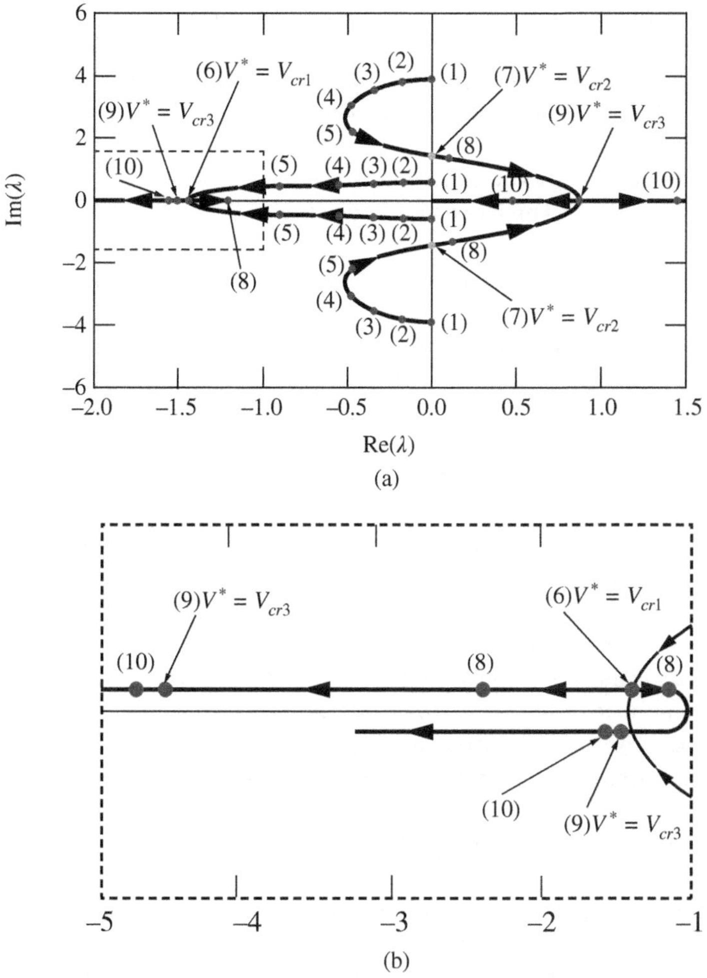

Figure 6.27 Root locus in case when fluid mass cannot be neglected. (a) Root locus under the increase of fluid speed V^*. (b) Zoom on the part with in the dashed square in (a). With increasing V^*, the four eigenvalues move in numerical order: (1) $V^* = 0$; (2) $V^* = 0.4$; (3) $V^* = 0.8$; (4) $V^* = 1.2$; (5) $V^* = 1.6$; (6) $V^* = V^*_{cr1} = 1.84$; (7) $V^* = V^*_{cr2} = 1.90$; (8) $V^* = 1.96$; (9) $V^* = V^*_{cr3} = 2.47$; (10) $V^* = 2.55$. At V^*_{cr1}, the system has complex conjugate pair with negative real part and repeated negative real eigenvalues. At V^*_{cr2}, the system has are pure imaginary conjugate pair and Hopf bifurcation occurs. At V^*_{cr3}, the system has repeated positive and two negative real eigenvalues.

References

Benjamin, T. B. (1962). Dynamics of a system of articulated pipes conveying fluid-I. Theory. Proceedings of the Royal Society of London. Series A *Mathematical and Physical Sciences* 261(1307), 457–486.

Cousins, H. (1913). The stability of gyroscopic single track vehicles. *Engineering* 2, 678–681.

Gasch, R., Nordmann, R., and Pfützner, H. (2006). *Rotordynamik*. Springer-Verlag.

Herrmann, G. (1971). *Dynamics and Stability of Mechanical Systems with Follower Forces.*

Ishida, Y. and Yamamoto, T. (2013). *Linear and Nonlinear Rotordynamics: A Modern Treatment with Applications.* John Wiley & Sons.

Iwnicki, S. (2006). *Handbook of Railway Vehicle Dynamics.* CRC Press.

Kalker, J. (1979). Survey of wheel—rail rolling contact theory. *Vehicle System Dynamics* 8(4), 317–358.

Kirillov, O. N. (2013). Nonconservative Stability Problems of Modern Physics, vol. 14. Walter de Gruyter.

Marsden, J. E. and Ratiu, T.S. (1994). Introduction to mechanics and symmetry: a basic exposition of classical mechanical systems. In: vol. 17. Springer.

Paidoussis, M. P. (1998). Fluid-Structure Interactions: Slender Structures and Axial Flow, vol. 1. Academic Press.

Seyranian, A. P., Lund, E., and Olhoff, N. (1994). Multiple eigenvalues in structural optimization problems. *Structural and Multidisciplinary Optimization* 8(4), 207–227.

Sugiyama, Y. and Noda, T. (1981). Studies on stability of two-degree-of-freedom articulated pipes conveying fluid: effect of an attached mass and damping. *Bulletin of JSME* 24(194), 1354–1362.

Wickens, A. (1965). The dynamic stability of a simplified four-wheeled railway vehicle having profiled wheels. *International Journal of Solids and Structures* 1(4): 385–406.

Wickens, A. (2005). *Fundamentals of Rail Vehicle Dynamics.* CRC Press.

Yabuno, H., Takano, H., and Okamoto, H. (2008). Stabilization control of hunting motion of railway vehicle wheelset using gyroscopic damper. *Journal of Vibration and Control* 14(12), 209–230.

Yoshida, Z. (2010). *Nonlinear Science: the Challenge of Complex Systems.* Springer Science & Business Media.

Yoshino, H., Hosoya, T., Yabuno, H., Lin, S., and Suda, Y. (2015). Theoretical and experimental analyses on stabilization of hunting motion by utilizing the traction motor as a passive gyroscopic damper. *Proceedings of the Institution of Mechanical Engineers, Part F: Journal of Rail and Rapid Transit* 229(4), 395–401.

Ziegler, H. (2013). Principles of Structural Stability, vol. 35. Birkhäuser.

7

Local Bifurcations

As the parameters in a dynamical system are varied, the stability of equilibrium states may be changed or the equilibrium states may be created or destroyed. Such changes depending on the parameters are called *bifurcations*. The parameter values at which the bifurcations occur are called *bifurcation points* (Strogatz 1994; Thomsen 2003). In this chapter, we focus on local bifurcations produced in the neighborhood of a trivial equilibrium state by the variation of one parameter. Such bifurcations are called *codimension one bifurcation* (for higher codimension bifurcations, see Guckenheimer and Holmes (1983)). Some of the destabilization phenomena due to change of one parameter were already encountered in Section 2.6. These phenomena occur when the eigenvalues of Jacobian matrix matrix include zero or a conjugate pair of pure imaginary eigenvalues while the single parameter changes. The stability of the trivial steady state is theoretically clarified by invoking the so-called *small motions assumption*, i.e. ignoring the nonlinear terms of degree two and higher in dependent variables to describe the motions. When the displacement becomes large, the assumption is violated. The nonlinear effects are taken into account in the equations of motion and it is necessary to carry out nonlinear analysis to solve them.

The dynamics of the two-link model presented in Section 5.1 is investigated again under the consideration of the nonlinearity of the springs. This is a simple but essential model to exhibit the so-called pitchfork bifurcation phenomenon. The nonlinear characteristics are analytically investigated in detail and compared with experimental results. Also, some other codimension one bifurcations are introduced and characterized by nonlinear analysis.

7.1 Nonlinear Analysis of a Two-Link Model Subject to Compressive Forces

We analyze the nonlinear dynamics of the system in Figure 5.2. In Section 5.1, the nonlinear characteristics of the spring force and the compression are neglected, but we here deal with the equation of motion without linearization.

Linear and Nonlinear Instabilities in Mechanical Systems: Analysis, Control and Application, First Edition. Hiroshi Yabuno.

Companion website: www.wiley.com/go/yabuno/instabilitiesinmechanicalsystems

7.1.1 Nonlinearity of Equivalent Spring Stiffness

Taking into account the nonlinear component of the spring stiffness (see Appendix A), we express the left and right spring forces as

$$F_{sl}^{\#}(x) = \frac{1}{2}(k_1 x + k_2 x^2 + k_3 x^3 + \cdots), \tag{7.1}$$

and

$$F_{sr}^{\#} = -F_{sl}^{\#}(-x) = \frac{1}{2}(k_1 x - k_2 x^2 + k_3 x^3 + \cdots), \tag{7.2}$$

respectively, where x corresponds to Δx in Appendix A. Equation (5.1) is replaced under the consideration of cubic nonlinearity as

$$F_s^{\#} = F_{sl}^{\#} + F_{sr}^{\#} = k_1 x + k_3 x^3 \overset{\text{def}}{=} -F_s. \tag{7.3}$$

From Eq. (5.2), the compressive force P acts on the mass as the force in x-direction

$$F_p = P\frac{x}{\sqrt{l^2 - x^2}}. \tag{7.4}$$

The stiffness characteristic is

$$F_p^{\#} = -F_p = -P\frac{x}{\sqrt{l^2 - x^2}}. \tag{7.5}$$

Introducing the representative length and time as $L = l$ and $T = \sqrt{\frac{m}{k_1}}$, respectively, $F_s^{\#}$ and $F_p^{\#}$ are nondimensionalized as

$$F_s^{\#*} = \frac{T^2}{mL}F_s^{\#} = \frac{T^2}{mL}(k_1 x + k_3 x^3 + \cdots) = x^* + k_3^* x^{*3} + O(x^{*5}) \tag{7.6}$$

and

$$F_p^{\#*} = \frac{T^2}{mL}F_p^{\#} = -\frac{T^2}{mL}P\frac{x}{\sqrt{l^2 - x^2}} = -px^* - \frac{p}{2}x^{*3} + O(x^{*5}), \tag{7.7}$$

respectively. Here the dimensionless parameters are $k_3^* = \frac{k_3 l^2}{k_1}$ and $p = \frac{P}{k_1 l}$, while the dimensionless displacement $x^* = \frac{x}{l}$ is assumed to be $|x^*| \ll 1$. Therefore, the spring has positive linear stiffness since the coefficient of x^* is 1 and its nonlinear characteristic is hardening when $k_3^* > 0$ $(k_3 > 0)$ and softening when $k_3^* < 0$ $(k_3 < 0)$. On the other hand, the effect of the compressive force $p > 0$ $(P > 0)$ has a spring characteristic with negative linear stiffness and softening nonlinearity because of negative coefficients of x^* and x^{*3}. The equivalent nonlinear spring characteristic due to the spring and the compression is

$$F_{equiv}^{\#*} = F_s^{\#*} + F_p^{\#*} = (1 - p)x^* + \left(k_3^* - \frac{p}{2}\right)x^{*3} + O(x^{*5}). \tag{7.8}$$

The linear stiffness is positive when $p < 1$ and negative when $p > 1$. The nonlinearity is hardening for when $k_3^* - \frac{p}{2} > 0$ and softening for $k_3^* - \frac{p}{2} < 0$, respectively. As a result, the dimensionless equivalent nonlinear spring force acting on the mass is

$$F_{equiv}^{*} = -F_{equiv}^{\#*} = -(1 - p)x^* - \left(k_3^* - \frac{p}{2}\right)x^{*3} + O(x^{*5}). \tag{7.9}$$

7.1.2 Equilibrium States and Their Stability

As clarified by the linear analysis in Section 5.1, increasing the compressive force p (P in the dimensional form), the trivial equilibrium state is destabilized at $p = 1$ ($P = lk_1$), where the linear stiffness is zero. In order to focus on the neighborhood of the critical load $p = 1$, we set the dimensionless compressive load as

$$p = 1 + \epsilon \quad (|\epsilon| \ll 1). \tag{7.10}$$

Then, neglecting ϵx^{*3} due to its higher order smallness, we can rewrite Eqs. (7.8) and (7.9) as

$$F^{\#*}_{equiv} = -\epsilon x^* + \alpha_3 x^{*3} + O(x^{*5}), \tag{7.11}$$

and

$$F^*_{equiv} = \epsilon x^* - \alpha_3 x^{*3} + O(x^{*5}), \tag{7.12}$$

respectively, where $\alpha_3 = k_3^* - \frac{1}{2}$. From Eq. (7.11), the equivalent nonlinear stiffness may be hardening or softening in accordance with the positive or negative sign of α_3, respectively. Hence, the relatively strong hardening nonlinearity of an attached practical spring, i.e. $k_3^* > \frac{1}{2}$, makes the equivalent nonlinear stiffness hardening ($\alpha_3 > 0$). When $k_3^* < \frac{1}{2}$, i.e. the attached spring is weakly hardening or softening, the equivalent nonlinear characteristic is softening ($\alpha_3 < 0$). When $k_3^* = \frac{1}{2}$, the equivalent nonlinear stiffness disappears.

Finally, from Eq. (5.5), we obtain the nondimensional equation of motion in the third order approximation with respect to x^* as

$$\begin{aligned} \frac{d^2x^*}{dt^{*2}} &= -2\gamma \frac{dx^*}{dt^*} + F^*_{equiv} \\ &= -2\gamma \frac{dx^*}{dt^*} + \epsilon x^* - \alpha_3 x^{*3} + O(x^{*5}). \end{aligned} \tag{7.13}$$

The dimensionless conservative force F^*_{equiv} has the potential energy:

$$U^* = -\int F^*_{equiv} dx^* = -\int (\epsilon x^* - \alpha_3 x^{*3}) dx^* = -\frac{\epsilon}{2} x^{*2} + \frac{\alpha_3}{4} x^{*4}, \tag{7.14}$$

where the integral constant is 0. The equilibrium states x_{st} satisfy the following equilibrium equation:

$$F^*_{equiv}(x_{st}) = -\left.\frac{dU^*}{dx^*}\right|_{x^*=x_{st}} = (\epsilon - \alpha_3 x_{st}^2) x_{st} = 0. \tag{7.15}$$

First, we have the trivial equilibrium state $x_{st} = 0$ for any ϵ as also obtained from the linear analysis in Section 5.1. In addition, there exist the nontrivial equilibrium states which are found to be

$$x_{st} = \pm\sqrt{\frac{\epsilon}{\alpha_3}}. \tag{7.16}$$

In cases of $\alpha_3 > 0$ and $\alpha_3 < 0$, the nontrivial equilibrium states are in the regions of $\epsilon > 0$ and $\epsilon < 0$, respectively. In case of $\alpha_3 = 0$, the equivalent spring is linear, and Eq. (7.15) does not provide any nontrivial steady state.

The stability of such equilibrium states is examined by introducing $x_1 = x^*$ and $x_2 = \frac{dx^*}{dt^*}$ in Eq. (7.13):

$$\frac{d}{dt^*}\begin{bmatrix} x_1 \\ x_2 \end{bmatrix} = \begin{bmatrix} x_2 \\ -2\gamma x_2 - \dfrac{dU^*}{dx_1} \end{bmatrix}. \tag{7.17}$$

The solutions of $\frac{dx_1}{dt^*} = \frac{dx_2}{dt^*} = 0$ are the equilibrium states $\boldsymbol{x}_{st} = [x_{1st}, x_{2st}]^T$ and satisfy

$$\begin{cases} 0 = x_{2st}, & (7.18) \\ 0 = -2\gamma x_{2st} - \left.\dfrac{dU^*}{dx_1}\right|_{x_1 = x_{1st}}, & (7.19) \end{cases}$$

which is equivalent to Eq. (7.15) because of $x_{1st} = x_{st}$.

Setting $x_1 = x_{1st} + \Delta x_1$ and $x_2 = x_{2st} + \Delta x_2$ and neglecting the nonlinear terms with respect to Δx_1 and Δx_2, we have the equations for the time-variation of Δx_1 and Δx_2 as

$$\frac{d}{dt^*}\begin{bmatrix} \Delta x_1 \\ \Delta x_2 \end{bmatrix} = D\boldsymbol{f}(\boldsymbol{x}_{st})\begin{bmatrix} \Delta x_1 \\ \Delta x_2 \end{bmatrix}, \tag{7.20}$$

where the Jacobian matrix $D\boldsymbol{f}(\boldsymbol{x}_{st})$ is expressed as

$$D\boldsymbol{f}(\boldsymbol{x}_{st}) = \left.\begin{bmatrix} 0 & 1 \\ -\dfrac{d^2U^*}{dx_1^2} & -2\gamma \end{bmatrix}\right|_{x_1 = x_{1st}}. \tag{7.21}$$

Equation (7.21) provides the following characteristic equation

$$\lambda^2 + 2\gamma\lambda + \left.\frac{d^2U^*}{dx_1^2}\right|_{x_1 = x_{1st}} = 0, \tag{7.22}$$

so that the eigenvalues are found to be

$$\lambda = -\gamma \pm \sqrt{\gamma^2 - \left.\frac{d^2U^*}{dx_1^2}\right|_{x_1 = x_{1st}}}, \tag{7.23}$$

where $\gamma > 0$. In case of $\left.\frac{d^2U^*}{dx_1^2}\right|_{x_1 = x_{1st}} \geq 0$, the two eigenvalues λ do not include a positive real number or a complex number with positive real part. In case of $\left.\frac{d^2U^*}{dx_1^2}\right|_{x_1 = x_{1st}} < 0$, the two eigenvalues are positive and negative real numbers. Therefore, the equilibrium states are stable and unstable in cases of $\left.\frac{d^2U^*}{dx_1^2}\right|_{x_1 = x_{1st}} \geq 0$ and $\left.\frac{d^2U^*}{dx_1^2}\right|_{x_1 = x_{1st}} < 0$, respectively. In particular, in the case of $\left.\frac{d^2U^*}{dx_1^2}\right|_{x_1 = x_{1st}} > 0$, the equilibrium states are asymptotically stable. Let us determine the stability of the trivial and nontrivial equilibrium states. The second derivative of U^* with respect to x_1 at the equilibrium states is

$$\left.\frac{d^2U^*}{dx_1^2}\right|_{x_1 = x_{1st}} = -\epsilon + 3\alpha_3 x_{1st}^2. \tag{7.24}$$

For the trivial equilibrium state:

$$\left.\frac{d^2U^*}{dx_1^2}\right|_{x_1 = x_{1st}} = -\epsilon. \tag{7.25}$$

The trivial equilibrium state is stable when $\epsilon < 0$ and unstable when $\epsilon > 0$ as also shown for the linear analysis in Section 5.1.

For the nontrivial equilibrium states of Eq. (7.16):

$$\left.\frac{d^2U^*}{dx_1^2}\right|_{x_1=x_{1st}} = 2\epsilon. \tag{7.26}$$

In case $\alpha_3 > 0$, the nontrivial equilibrium states are stable since they are defined in the region of $\epsilon > 0$. In case $\alpha_3 < 0$, the nontrivial equilibrium states are unstable because they exist in the region of $\epsilon < 0$.

While changing the parameter ϵ, an eigenvalue of Jacobian matrix with respect to the trivial equilibrium state becomes zero and the stability is lost. At the critical point, the nontrivial equilibrium states of Eq. (7.16) emerge. Such a bifurcation is called a *pitchfork bifurcation*. In particular, a pitchfork bifurcation in which the unstable trivial equilibrium state and the stable nontrivial equilibrium states coexist at the same value of ϵ is called a *supercritical pitchfork bifurcation*. On the other hand, a pitchfork bifurcation showing that the stable trivial equilibrium state and the unstable nontrivial equilibrium states coexist at the same value of ϵ is called a *subcritical pitchfork bifurcation*.

Let us investigate in detail the dynamics when $\alpha_3 > 0$. The bifurcation diagram expressing the relationship between the parameter ϵ and the equilibrium states $x_{st} = x_{1st}$ is shown in Figure 7.1a, where the solid and the dashed lines denote the stable equilibrium state and the unstable one, respectively. The potential energy curves obtained from Eq. (7.14)

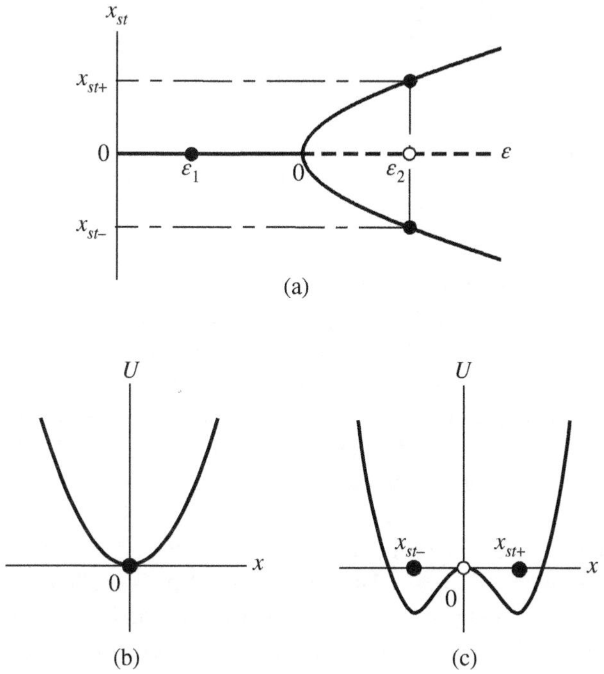

Figure 7.1 Supercritical pitchfork bifurcation ($\alpha_3 > 0$). (a) is bifurcation diagram, where the solid and dashed lines denote the stable and unstable equilibrium states, respectively. In case of $\epsilon < 0$ ($p < 1$), there is only the stable trivial equilibrium state obtained by the linear analysis. In case of $\epsilon > 0$ ($p > 1$), in addition to the unstable trivial equilibrium state obtained by the linear analysis, there are stable nontrivial equilibrium states. (b) is the potential energy curve in $\epsilon = \epsilon_1 < 0$. (c) is the potential energy curve in $\epsilon = \epsilon_2 > 0$.

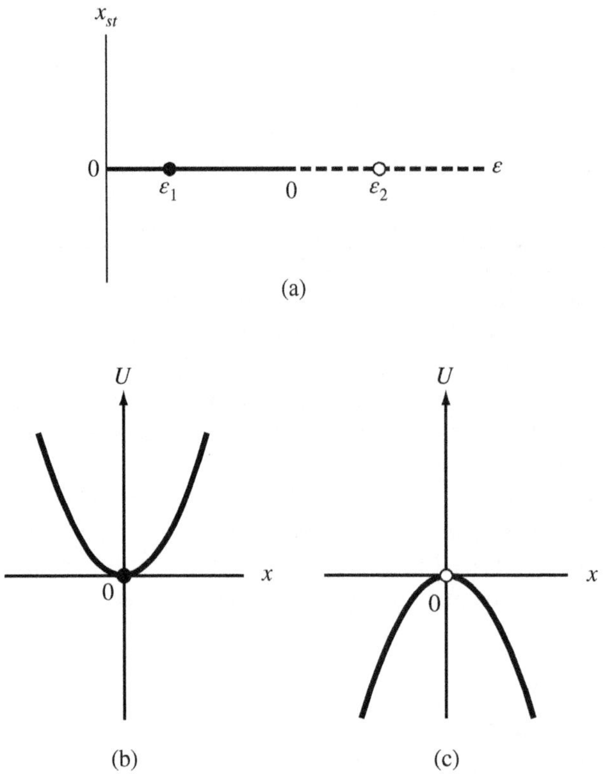

Figure 7.2 Case without nonlinearity ($\alpha_3 = 0$). (a) is bifurcation diagram, where the solid and dashed lines denote the stable equilibrium state and unstable one, respectively. (b) is the potential energy curve in $\epsilon = \epsilon_1 < 0$. (c) is the potential energy curve in $\epsilon = \epsilon_2 > 0$.

are shown as Figure 7.1b,c when the compressive force is below and above the critical load $\epsilon = 0$ ($p = 1$), i.e. $\epsilon < 0$ ($p < 1$) and $\epsilon > 0$ ($p > 1$), respectively. In particular, the stable nontrivial equilibrium states that cannot be predicted by the linear theory in Section 5.1 are derived and their existence is in accordance with the phenomenon experimentally observed in the video showing the bucking of a slender beam (see the videos referred to in Section 5.1.1). The postbuckling behavior in the video can be theoretically explained only by the nonlinear analysis for the equation of motion considering the nonlinearity of the system.

When $\alpha_3 = 0$, the bifurcation diagram is represented in Figure 7.2a, which corresponds to the result in the linear analysis (see Section 5.1). There are no nontrivial equilibrium states. Figure 7.2b,c report the potential energy curves in the ranges below and above the critical load $\epsilon = 0$. In $\epsilon > 0$, there are no minimum points expressing the stable equilibrium states.

Next, we consider the case of $\alpha_3 < 0$. The bifurcation diagram is provided in Figure 7.3a, where the solid and the dashed lines denote the stable equilibrium state and the unstable one, respectively. The potential energy curves obtained are shown as Figure 7.3b,c when the compressive force is below and above the critical load $\epsilon = 0$ ($p = 1$), respectively. In the case

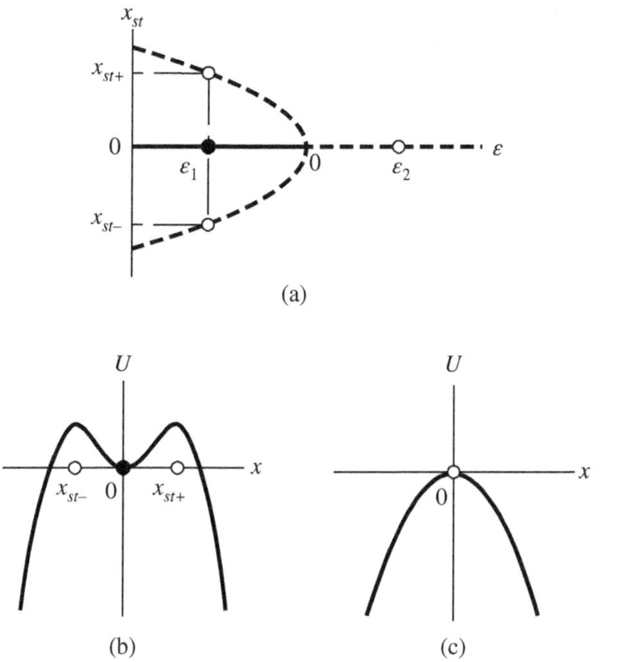

Figure 7.3 Subcritical pitchfork bifurcation ($\alpha_3 < 0$). (a) is bifurcation diagram, where the solid and dashed lines denote the stable equilibrium state and unstable one, respectively. (b) is the potential energy curve in $\epsilon_1 = \epsilon < 0$. (c) is the potential energy curve in $\epsilon_2 = \epsilon > 0$.

$\epsilon < 0$, the unstable nontrivial equilibrium states exist. The unstable nontrivial equilibrium states indicate that even if the compressive force is below the critical load $\epsilon = 0$ by the linear theory, the displacement can grow depending on the initial displacement and velocity. Then, the higher order nonlinear analysis for the equation of motion considering the nonlinear terms of degree of fifth and higher of x^* or x_1 is required to determine whether the displacement grows infinitely or converges to a finite equilibrium state.

7.2 Reduction of Dynamics Near a Critical Point

First, we consider only the linear term in Eq. (7.13) as

$$\frac{d^2x^*}{dt^{*2}} = -2\gamma\frac{dx^*}{dt^*} + \epsilon x^*, \tag{7.27}$$

and investigate the linear dynamics depending on the compressive force $p = 1 + \epsilon$ or ϵ. The matrix A corresponding to Eq. (2.110) is expressed as

$$A = \begin{bmatrix} 0 & 1 \\ \epsilon & -2\gamma \end{bmatrix}. \tag{7.28}$$

As increasing the compressive force p, the eigenvalues and the phase spaces are changed as in Figure 5.3, where the axes of the phase spaces are transformed into $y_1 - y_2$ by the transformation matrix P introduced in Section 2.4. Hereafter, we focus on the phase spaces, (E), (J), and (A), in the neighborhood of the critical load $p_{cr} = 1$. Equation (7.13) is expressed as follows:

$$\frac{d}{dt^*}\begin{bmatrix} x_1 \\ x_2 \end{bmatrix} = A\begin{bmatrix} x_1 \\ x_2 \end{bmatrix} - \alpha_3 x_1^3 \begin{bmatrix} 0 \\ 1 \end{bmatrix}. \tag{7.29}$$

Because of $|\epsilon| \ll 1$, the eigenvalues of A are

$$\begin{cases} \lambda_1 = -\gamma + \sqrt{\gamma^2 + \epsilon} = \dfrac{\epsilon}{2\gamma} + O(\epsilon^2), & (7.30) \\ \lambda_2 = -\gamma - \sqrt{\gamma^2 + \epsilon} = -2\gamma + O(\epsilon). & (7.31) \end{cases}$$

By substituting

$$\begin{bmatrix} x_1 \\ x_2 \end{bmatrix} = P\begin{bmatrix} y_1 \\ y_2 \end{bmatrix}, \quad P = \begin{bmatrix} 1 & 1 \\ \lambda_1 & \lambda_2 \end{bmatrix}, \tag{7.32}$$

into Eq. (7.29) and premultiplying the result by P^{-1} yields

$$\frac{d}{dt^*}\begin{bmatrix} y_1 \\ y_2 \end{bmatrix} = \begin{bmatrix} \lambda_1 & 0 \\ 0 & \lambda_2 \end{bmatrix}\begin{bmatrix} y_1 \\ y_2 \end{bmatrix} + \frac{\alpha_3}{\lambda_2 - \lambda_1}(y_1 + y_2)^3 \begin{bmatrix} 1 \\ -1 \end{bmatrix}, \tag{7.33}$$

whose linear part is diagonalized.

Of course, the linearized system

$$\frac{d}{dt^*}\begin{bmatrix} y_1 \\ y_2 \end{bmatrix} = \begin{bmatrix} \lambda_1 & 0 \\ 0 & \lambda_2 \end{bmatrix}\begin{bmatrix} y_1 \\ y_2 \end{bmatrix} \approx \begin{bmatrix} \dfrac{\epsilon}{2\gamma} & 0 \\ 0 & -2\gamma \end{bmatrix}\begin{bmatrix} y_1 \\ y_2 \end{bmatrix} \tag{7.34}$$

has the phase spaces of Figure 5.3(E), (J) and (A) depending on ϵ $(= p - 1)$, where y_1 and y_2 axes are the invariant subspaces corresponding to the eigenvalues λ_1 and λ_2, respectively. As increasing p, i.e. ϵ, y_1-axis changes from the stable subspace ($p < 1$, $\epsilon < 0$: (E)) to the unstable one ($p > 1$, $\epsilon > 0$: (A)) through the center subspace ($p = 1$, $\epsilon = 0$: (J)) depending on the change of the sign of λ_1, while y_2-axis maintains a stable subspace because the corresponding eigenvalue λ_2 is negative independently of p, i.e. ϵ. Also, y_2-component of the flow is much faster than y_1-component and tends exponentially to zero because of $\lambda_2 = -2\gamma = O(1)$. Therefore, we can focus on the dynamics on y_1-axis for the stability analysis. Substituting $y_2 = 0$, $\lambda_1 = \frac{\epsilon}{2\gamma}$, and $\lambda_2 = -2\gamma$ into Eq. (7.33) yields

$$\frac{dy_1}{dt^*} = \frac{\epsilon}{2\gamma}y_1 - \frac{\alpha_3}{2\gamma}y_1^3, \tag{7.35}$$

where $O(\epsilon y_1^3)$ is neglected due to the higher order smallness. Because of $y_1 = x_1 - y_2 \approx x_1 = x^*$ from Eq. (7.32), this equation leads to

$$\frac{dx^*}{dt^*} = \frac{\epsilon}{2\gamma}x - \frac{\alpha_3}{2\gamma}x^{*3}, \tag{7.36}$$

which is referred to as the normal form of *pitchfork bifurcation* in Section 7.3. It should be noticed that this equation is also derived by neglecting the inertia effect of $\frac{d^2x^*}{dt^{*2}}$ in Eq. (7.13).

7.3 Pitchfork Bifurcation

We consider the equation including cubic nonlinearity

$$\frac{dx}{dt} = \epsilon x - \alpha_3 x^3, \tag{7.37}$$

which is the normal form of pitchfork bifurcation. We examine the qualitative change of the dynamics depending on the variation of parameter ϵ under the constant parameter α_3; the targeted parameter as ϵ is called a *control parameter*. Equation (7.35) has the same form by replacing $y_1^* \to x$, $\frac{\epsilon}{2\gamma} \to \epsilon$ and $\frac{\alpha_3}{2\gamma} \to \alpha_3$. Let us examine the nonlinear characteristics. The equilibrium equation is

$$(\epsilon - \alpha_3 x_{st}^2) x_{st} = 0. \tag{7.38}$$

There are the trivial equilibrium state $x_{st} = 0$ and the nontrivial ones

$$x_{st} = \pm\sqrt{\frac{\epsilon}{\alpha_3}}, \tag{7.39}$$

which exist only in the condition of $\frac{\epsilon}{\alpha_3} > 0$; they exist in the region of $\epsilon > 0$ and of $\epsilon < 0$ when $\alpha_3 > 0$ and $\alpha_3 < 0$, respectively. Let us determine the stability of the equilibrium states. We analyze the time-evolution of the deviation from the equilibrium state Δx under the condition of $|\Delta x| \ll 1$. Substituting $x = x_{st} + \Delta x$ into Eq. (7.37) and taking only the linear term with respect to Δx yields the equation for Δx

$$\frac{d\Delta x}{dt} = (\epsilon - 3\alpha_3 x_{st}^2)\Delta x. \tag{7.40}$$

The solution is

$$\Delta x = c e^{(\epsilon - 3\alpha_3 x_{st}^2)t}, \tag{7.41}$$

where c is the integral constant determined by the initial condition. The stability of the above equilibrium states is determined as follows. With respect to the trivial solution, Eq. (7.41) becomes

$$\Delta x = c e^{\epsilon t}. \tag{7.42}$$

Therefore, in the regions of $\epsilon < 0$ and of $\epsilon > 0$, the trivial equilibrium state $x_{st} = 0$ is stable and unstable, respectively. With respect to the nontrivial equilibrium states of Eq. (7.39), Eq. (7.41) becomes

$$\Delta x = c e^{-2\epsilon t}. \tag{7.43}$$

Thus, the nontrivial equilibrium states in the regions of $\epsilon > 0$ and of $\epsilon < 0$ are stable and unstable, respectively.

The relationship between the parameter ϵ and the equilibrium states x_{st} is described as Figures 7.1 and 7.3 for $\alpha_3 > 0$ and $\alpha_3 < 0$, respectively. These bifurcations are called a *pitchfork bifurcation*. In particular, the former is called a *supercritical pitchfork bifurcation*. Exceeding the bifurcation point $\epsilon = 0$, the stable nontrivial equilibrium states emerges. On the other hand, the latter bifurcation phenomenon is called a *subcritical pitchfork bifurcation*. Exceeding the bifurcation point $\epsilon = 0$, the unstable equilibrium states existing in $\epsilon < 0$ disappear as shown in Figure 7.3a.

7.4 Other Codimension One Bifurcations

7.4.1 Saddle-Node Bifurcation

Next, we consider a bifurcation due to the quadratic nonlinearity, which is governed with

$$\frac{dx}{dt} = \epsilon - \alpha_2 x^2. \tag{7.44}$$

This equation is the normal form of *saddle-node bifurcation*. We examine the bifurcation when the parameter ϵ is varied and α_2 is constant. There are the nontrivial equilibrium states as follows:

$$x_{st} = \pm\sqrt{\frac{\epsilon}{\alpha_2}}. \tag{7.45}$$

The equilibrium states exist in the region of $\epsilon > 0$ and $\epsilon < 0$ when $\alpha_2 > 0$ and $\alpha_2 < 0$, respectively. Setting $x = x_{st} + \Delta x$ yields the linearized equation with respect to Δx as follows:

$$\frac{d\Delta x}{dt} = -2\alpha_2 x_{st}\Delta x. \tag{7.46}$$

The solution is

$$\Delta x = ce^{-2\alpha_2 x_{st} t}, \tag{7.47}$$

where c is the integral constant determined by the initial condition. The bifurcation digram is described as Figure 7.4a,b. A practical example of the saddle-node bifurcation is the dynamics of MEMS-switch which will be mentioned in Section 7.7.2.

7.4.2 Transcritical Bifurcation

Next, we consider another bifurcation due to the quadratic nonlinearity that is governed with

$$\frac{dx}{dt} = \epsilon x - \alpha_2 x^2. \tag{7.48}$$

This equation is the normal form of *transcritical bifurcation*. We examine the bifurcation when the parameter ϵ is varied and α_2 is constant. The equilibrium equation is

$$0 = (\epsilon - \alpha_2 x_{st})x_{st}. \tag{7.49}$$

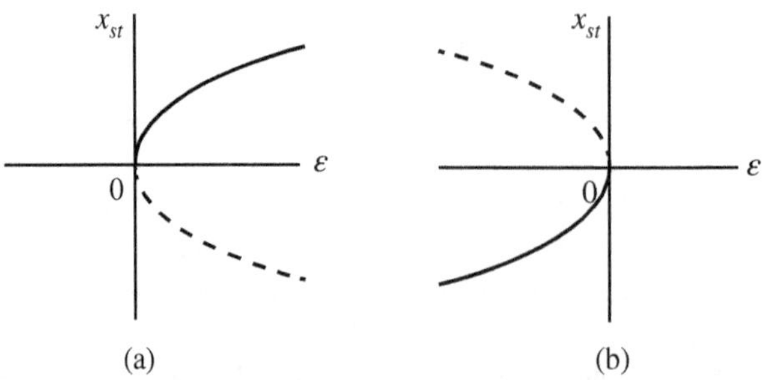

Figure 7.4 Saddle-node bifurcation, where the solid and dashed lines denote stable and unstable equilibrium states, respectively: (a) $\alpha_2 > 0$, (b) $\alpha_2 < 0$.

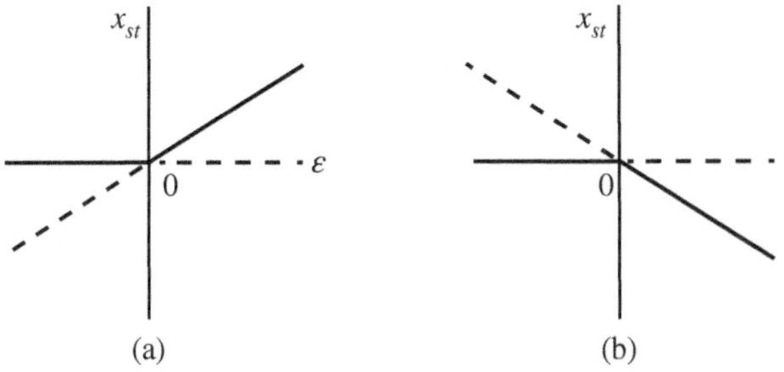

Figure 7.5 Transcritical bifurcation, where the solid and dashed lines denote stable and unstable equilibrium states, respectively: (a) $\alpha_2 > 0$, (b) $\alpha_2 < 0$.

In addition to the trivial equilibrium states $x_{st} = 0$, there are the nontrivial equilibrium states as follows:

$$x_{st} = \frac{\epsilon}{\alpha_2}. \tag{7.50}$$

Setting $x = x_{st} + \Delta x$ yields the linearized equation with respect to Δx as follows:

$$\frac{d\Delta x}{dt} = (\epsilon - 2\alpha_2 x_{st})\Delta x. \tag{7.51}$$

The solution for the trivial equilibrium state is

$$\Delta x = ce^{\epsilon t}. \tag{7.52}$$

The solution for the nontrivial equilibrium states is

$$\Delta x = ce^{-\epsilon t}. \tag{7.53}$$

The bifurcation diagrams for $\alpha_2 > 0$ and $\alpha_2 < 0$ are described as Figure 7.5a,b, respectively. These phenomena are called a *transcritical bifurcation*. In this bifurcation, the trivial and nontrivial equilibrium states coexist for any ϵ, and their stability is always different.

7.4.3 Hopf Bifurcation

In Section 2.6, the static and dynamic destabilizations are investigated for the linear system, depending on the variations of the parameters, k' and c', in Figure 2.7. In Sections 7.3 and 7.4.2, it is clarified that the static destabilization of the trivial equilibrium state according to the arrow (1) in Figure 2.7 is related to the pitchfork and transcritical bifurcations by nonlinear analysis.

In this section, we consider the effect of nonlinearity on the dynamic destabilization of the trivial equilibrium state according to the arrow (2) in Figure 2.7. The eigenvalues of the Jacobian matrix $\boldsymbol{Df}(\boldsymbol{0})$ at the trivial equilibrium state are a pair of complex conjugate

$$\lambda = \epsilon \pm i\omega, \tag{7.54}$$

where ϵ and ω are real, and the real part ϵ is changed from negative to positive. The governing equation for such a system is expressed as follows:

$$\frac{d}{dt}\begin{bmatrix} x_1 \\ x_2 \end{bmatrix} = B \begin{bmatrix} x_1 \\ x_2 \end{bmatrix} + O(2), \tag{7.55}$$

where $O(2)$ denotes higher order terms than the second order of x_1 and x_2, i.e. nonlinear terms; they are proved in cubic nonlinearity in Section 8.2.4. B corresponds to Eq. (2.80)

$$B = \begin{bmatrix} \epsilon & \omega \\ -\omega & \epsilon \end{bmatrix}. \tag{7.56}$$

The system is in the critical state at $\epsilon = 0$ because the eigenvalues are a conjugate pair of pure imaginary eigenvalues. Increasing ϵ, the phase portrait is changed as Figure 2.5. The result has been derived by invoking the so-called *small motions assumption*, i.e. ignoring the higher order terms in Eq. (7.55). In case of $\epsilon > 0$ (Figure 2.5c), when the amplitude of the oscillation becomes large, the assumption is violated. Thus, there are the cases where the linear analysis cannot reveal the dynamics with not small motion. Also for $\epsilon < 0$ (Figure 2.5a) the linear analysis cannot determine, in case of a large disturbance, whether the amplitude of the oscillation decreases. To clarify the dynamics with not small motion, the nonlinear effects that is the higher order terms neglected in the linear analysis must be taken into account in the governing equations and they need to be solved by using nonlinear analysis as in the buckling problem in Section 7.1.

Let us consider the typical polynomial nonlinearity as

$$\frac{d}{dt}\begin{bmatrix} x_1 \\ x_2 \end{bmatrix} = B\begin{bmatrix} x_1 \\ x_2 \end{bmatrix} - (x_1^2 + x_2^2)\begin{bmatrix} \alpha_3 & -\beta_3 \\ \beta_3 & \alpha_3 \end{bmatrix}\begin{bmatrix} x_1 \\ x_2 \end{bmatrix}, \tag{7.57}$$

where the linear part is the matrix B in Eq. (7.56). Then, it is known that the first or lowest order nonlinear terms are cubic and all quadratic terms are removable by normalization (see Section 8.2.4 and Nayfeh (1993) for the detail).

We introduce the polar coordinates as

$$\begin{cases} x_1 = a(t)\cos\Phi(t), & (7.58) \\ x_2 = a(t)\sin\Phi(t), & (7.59) \end{cases}$$

where the radius a is the distance between the origin O and each point on the trajectory and denotes the amplitude of oscillation, while Φ is the argument and denotes the phase of oscillation. Then, Eq. (7.57) is transformed into

$$\begin{cases} \dfrac{da}{dt} = \epsilon a - \alpha_3 a^3, & (7.60) \\ a\dfrac{d\Phi}{dt} = -\omega a - \beta_3 a^3, & (7.61) \end{cases}$$

or in case of $a \neq 0$ into

$$\begin{cases} \dfrac{da}{dt} = \epsilon a - \alpha_3 a^3, & (7.62) \\ \dfrac{d\Phi}{dt} = -\omega - \beta_3 a^2. & (7.63) \end{cases}$$

Similarly to the analysis of equilibrium state in Chapter 2, we first seek for time-invariant amplitudes, which are the so-called steady-state amplitudes. By letting $\frac{da}{dt} = 0$, we obtain the steady-state amplitudes:

$$a_{st} = 0, \tag{7.64}$$

and

$$a_{st} = \sqrt{\frac{\epsilon}{\alpha_3}}. \tag{7.65}$$

In the case without nonlinear terms ($\alpha_3 = 0$), the system exhibits the behavior of the linear system with complex eigenvalues in Section 2.4.4. The nontrivial steady-state amplitude exists in the range of $\epsilon > 0$ and $\epsilon < 0$ when $\alpha_3 > 0$ and $\alpha_3 < 0$, respectively. Substituting $a(t) = a_{st} + \Delta a(t)$ into Eq. (7.60) and taking only the linear term with respect to Δa yields the equation for Δa as

$$\frac{d\Delta a}{dt} = (\epsilon - 3\alpha_3 a_{st}^2)\Delta a. \tag{7.66}$$

The solution is

$$\Delta a = ce^{(\epsilon - 3\alpha_3 a_{st}^2)t}, \tag{7.67}$$

where c is the integral constant determined by the initial condition. Independently of the sign of α_3, the trivial equilibrium state $a_{st} = 0$ is stable or unstable for $\epsilon < 0$ or $\epsilon > 0$, respectively. The nontrivial equilibrium state $a_{st} = \sqrt{\frac{\epsilon}{\alpha_3}}$ is stable when it exists in $\epsilon > 0$, and is unstable when it exists in $\epsilon < 0$. The phase of the nontrivial equilibrium states is obtained from Eq. (7.61):

$$\Phi = -(\omega + \beta_3 a_{st}^2)t + \Phi_0, \tag{7.68}$$

where from Eqs. (7.58) and (7.59), Φ_0 is the initial phase determined by the initial condition of $x_1(0)$ and $x_2(0)$ as

$$\begin{cases} \cos\Phi_0 = \dfrac{x_1(0)}{a_0}, & (7.69) \\ \sin\Phi_0 = \dfrac{x_2(0)}{a_0}, & (7.70) \end{cases}$$

with $a_0 = \sqrt{x_1(0)^2 + x_2(0)^2}$ if the initial condition is on the limit cycle.

First, we investigate the case of $\alpha_3 > 0$. The phase portraits are described as Figure 7.6. Figure 7.6b for $\epsilon > 0$ includes the steady-state oscillation with a stable steady-state amplitude a_{st}, which is unpredictable by the linear analysis in Section 2.4.4 and is called a *stable*

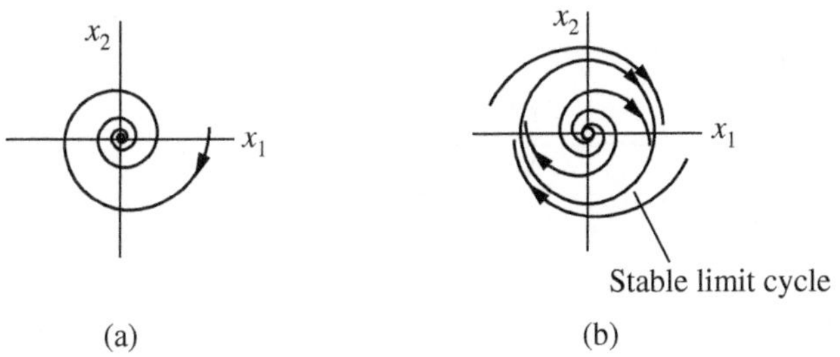

Figure 7.6 Change of phase portrait of supercritical Hopf bifurcation ($\alpha_3 > 0$): (a) $\epsilon < 0$, (b) $\epsilon > 0$.

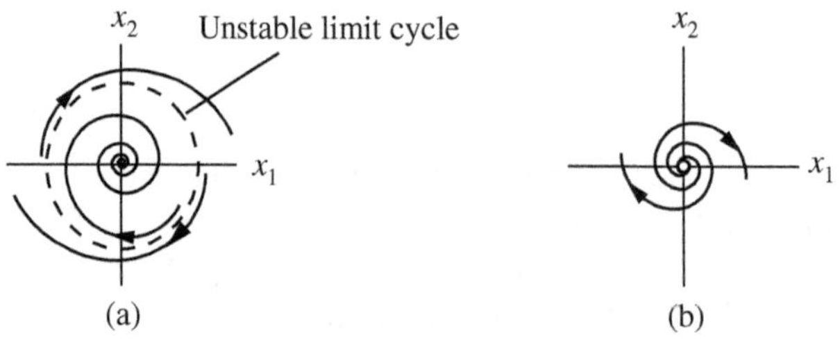

Figure 7.7 Change of phase portrait of subcritical Hopf bifurcation ($\alpha_3 < 0$): (a) $\epsilon < 0$, (b) $\epsilon > 0$.

limit cycle. In both cases, when the initial state is inside and outside of this stable limit cycle, the behavior converges to the oscillation with the nontrivial steady-state amplitude of a_{st}. The bifurcation diagram is described as Figure 7.8a. At $\epsilon = 0$, the stable steady-state oscillation is born. As increasing ϵ, the steady-state amplitude increases.

Next, we investigate the case of $\alpha_3 < 0$. The phase portrait is described as Figure 7.7. Figure 7.7a for $\epsilon < 0$ includes the steady-state oscillation with an unstable steady-state amplitude a_{st}, which is unpredictable by means of the linear theory in Section 2.4.4, and is called an *unstable limit cycle*. In the case when the initial state is inside this unstable limit cycle, the behavior converges to the stable trivial equilibrium state $a_{st} = 0$ as predicted by the linear analysis. However, in the case when the initial state is outside this unstable limit cycle, the amplitude grows with time, in contrast to the result provided by the linear analysis. In order to clarify whether the amplitude grows infinitely or converges to a stable limit cycle, higher order nonlinear analysis is required for the governing equation considering the effects of the higher order nonlinearity. The bifurcation diagram is described as Figure 7.8b. As increasing ϵ, the amplitude of the unstable steady-state oscillation decreases. At $\epsilon = 0$, the unstable steady-state oscillation disappears.

The bifurcations described by the diagrams of Figure 7.8 are called a *Hopf bifurcation*. In particular, the Hopf bifurcation in Figure 7.8a, showing that the unstable trivial equilibrium state and the stable steady-state oscillation coexist at the same value of ϵ, is called a *supercritical Hopf bifurcation*. The Hopf bifurcation in Figure 7.8b, characterized by the coexistence of the stable trivial equilibrium state and the unstable steady-state oscillation at the same value of ϵ, is called a *subcritical Hopf bifurcation*.

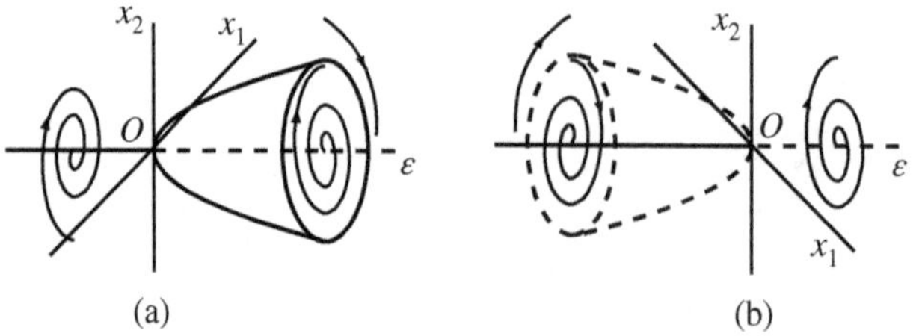

Figure 7.8 Hopf bifurcation: (a) supercritical, (b) subcritical.

7.5 Perturbation of Pitchfork Bifurcation

7.5.1 Bifurcation Diagram

We consider the case where the two link system subject to the compressive force is set with an angle θ from the horizontal direction as shown in Figure 7.9.

The dimensionless equation of motion takes into account the dimensionless gravity effect of $\sigma = -\frac{mg}{k_1 l} \sin\theta$ and Eq. (7.13) turns out to be

$$\frac{d^2x^*}{dt^{*2}} = -2\gamma\frac{dx^*}{dt^*} + \epsilon x^* - \alpha_3 x^{*3} + \sigma. \tag{7.71}$$

This system does not have the trivial equilibrium state. The nontrivial equilibrium states x^*_{st} satisfy

$$\epsilon x_{st} - \alpha_3 x_{st}^3 + \sigma = 0. \tag{7.72}$$

The relationship between the equilibrium state x_{st} and the parameter ϵ is expressed as follows:

$$\epsilon = \alpha_3 x_{st}^2 - \frac{\sigma}{x_{st}}. \tag{7.73}$$

The bifurcation diagrams for $\alpha_3 > 0$ and $\alpha_3 < 0$ are described as Figure 7.10a,b, respectively. The inverse of the slope of the bifurcation diagrams is

$$\frac{d\epsilon}{dx_{st}} = 2\alpha_3 x_{st} + \frac{\sigma}{x_{st}^2}. \tag{7.74}$$

Let us determine the stability of the equilibrium states. Substituting $x = x_{st} + \Delta x$ into Eq. (7.71) and considering Eq. (7.74) yields the linearized equation for Δx as

$$\frac{d^2\Delta x}{dt^{*2}} = -2\gamma\frac{d\Delta x}{dt^*} - \left(\frac{d\epsilon}{dx_{st}}x_{st}\right)\Delta x. \tag{7.75}$$

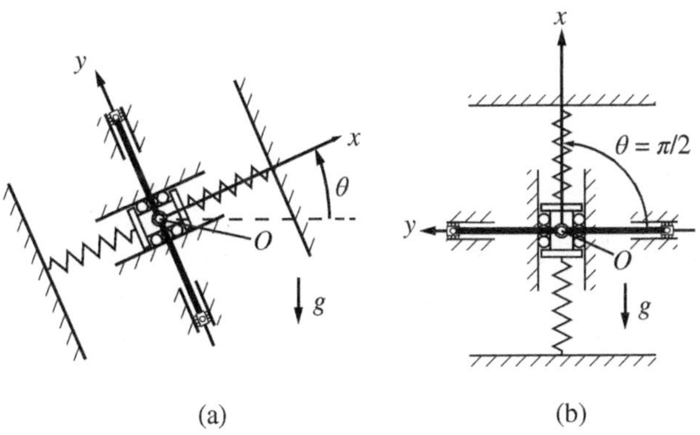

Figure 7.9 Two link system subject to gravity effect. The case of $\theta = 0$ corresponds to the two-link system of Figure 5.2.

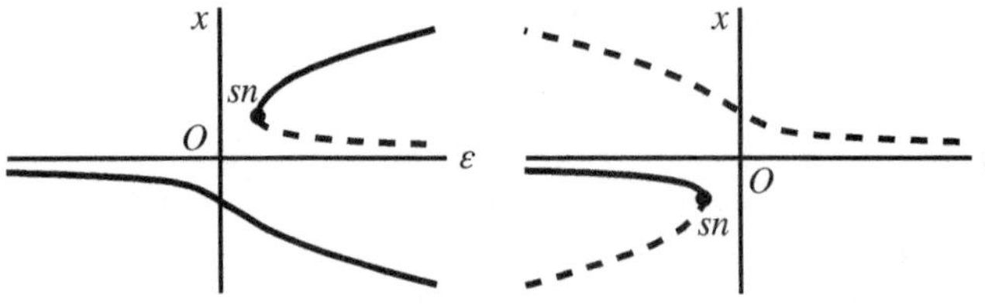

Figure 7.10 Imperfection of pitchfork bifurcation ($\sigma < 0$). Solid and dashed lines denote stable and unstable equilibrium states, respectively: (a) $\alpha_3 > 0$, (b) $\alpha_3 < 0$.

Since $2\gamma > 0$ and $\frac{d\epsilon}{dx_{st}}x_{st}$ are regarded as c' and k' in Figure 2.7, the equilibrium states are stable and unstable, for $\frac{d\epsilon}{dx_{st}}x_{st} > 0$ and $\frac{d\epsilon}{dx_{st}}x_{st} < 0$, respectively. Considering the slope of the branch and the sign of the equilibrium states, we can determine the stability of the equilibrium states as in Figure 7.10. The bifurcation phenomena described by Figure 7.10 is call a *perturbed* pitchfork bifurcation or an *incomplete* pitchfork bifurcation, while the bifurcation phenomena described by Figures 7.1a and 7.3a are called a *complete pitchfork bifurcation*.

7.5.2 Analysis of Bifurcation Point

We characterize the dynamics in the neighborhood of the bifurcation point *sn* in Figure 7.10. At this point (ϵ_{sn}, x_{sn}), because of $\frac{d\epsilon}{dx_{st}} = 0$, we have the following equations from Eqs. (7.73) and (7.74):

$$\begin{cases} \epsilon_{sn} = \alpha_3 x_{sn}^2 - \dfrac{\sigma}{x_{sn}}, & (7.76) \\ 2\alpha_3 x_{sn} + \dfrac{\sigma}{x_{sn}^2} = 0. & (7.77) \end{cases}$$

These equations lead to

$$\epsilon_{sn} - 3\alpha_3 x_{sn}^2 = 0. \tag{7.78}$$

Substituting $\epsilon = \epsilon_{sn} + \epsilon'$ and $x(t) = x_{sn} + x'(t)$ into Eq. (7.71) and considering Eq. (7.78) yields

$$\frac{d^2x'}{dt^2} + 2\gamma\frac{dx'}{dt} - x_{sn}\epsilon' - \epsilon' x' + 3\alpha_3 x_{sn} x'^2 + \alpha_3 x'^3 = 0. \tag{7.79}$$

We consider the case of $\left|\frac{\epsilon'}{\epsilon_{sn}}\right| \ll 1$ and $\left|\frac{x'}{x_{sn}}\right| \ll 1$. Balancing the terms of $x_{sn}\epsilon'$ and $3\alpha_3 x_{sn} x'^2$ leads to the order estimation as $x' = O(\epsilon'^{1/2})$. Then, $x_{sn}\epsilon'$ and $3\alpha_3 x_{sn} x'^2$ are $O(\epsilon')$. Because $\epsilon' x'$ and x'^3 are $O(\epsilon'^{3/2})$, they can be neglected. As a result, Eq. (7.79) is approximated to

$$\frac{d^2x'}{dt^2} + 2\gamma\frac{dx'}{dt} - x_{sn}\epsilon' + 3\alpha_3 x_{sn} x'^2 = 0. \tag{7.80}$$

Furthermore, neglecting the inertia term as Eq. (7.36) leads to

$$\frac{dx'}{dt} = \frac{x_{sn}}{2\gamma}\epsilon' - \frac{3\alpha_3 x_{sn}}{2\gamma}x'^2. \tag{7.81}$$

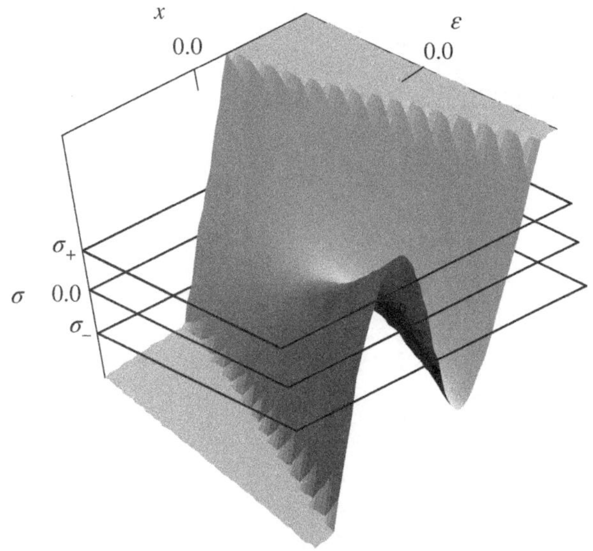

Figure 7.11 Equilibrium surface ($\beta > 0$). All points on the two-dimensional curved surface in $\epsilon - \sigma - x$ space are the equilibrium states satisfying Eq. (7.73).

Because this equation corresponds to the normal form of saddle-node bifurcation of Eq. (7.44), the bifurcation at the point *sn* is a saddle-node.

7.5.3 Equilibrium Surface and Bifurcation Diagrams

Figure 7.11 shows the equilibrium states satisfying Eq. (7.73) in the $\epsilon - \sigma - x$ space, which is called an *equilibrium surface*. The intersection between the equilibrium surface with the plane $\sigma = 0$ is shown in Figure 7.12 and corresponds to the complete pitchfork bifurcation diagram of Figure 7.1a. The intersection between the equilibrium surface with the plane $\sigma = 0.01 > 0$ is shown as Figure 7.13 and corresponds to the bifurcation diagram under positive σ. The intersection between the equilibrium surface with the plane $\sigma = -0.01 < 0$ is shown in Figure 7.14 and corresponds to the perturbed pitchfork bifurcation under negative σ as shown in Figure 7.10a.

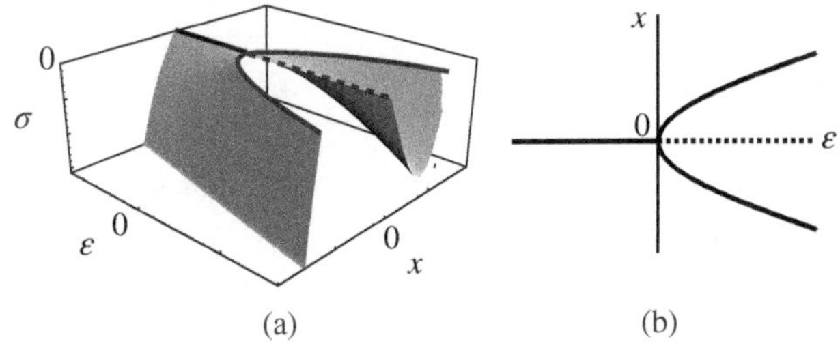

Figure 7.12 Complete pitchfork bifurcation on equilibrium surface ($\sigma = 0$). Source: Reprinted with permission Yabuno (2004).

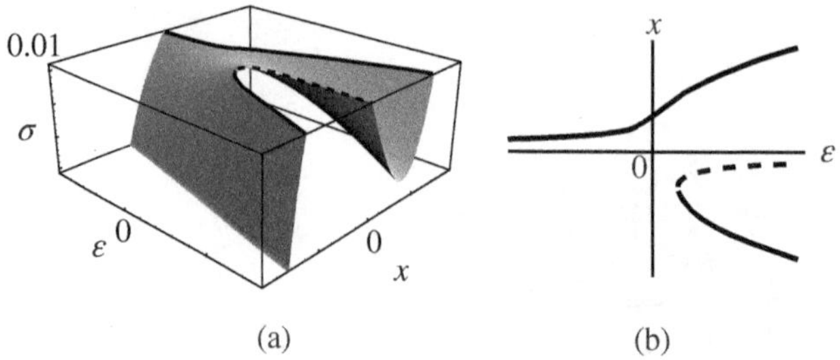

Figure 7.13 Incomplete or perturbed pitchfork bifurcation on equilibrium surface ($\sigma = \sigma_+ > 0$). Source: Reprinted with permission Yabuno (2004).

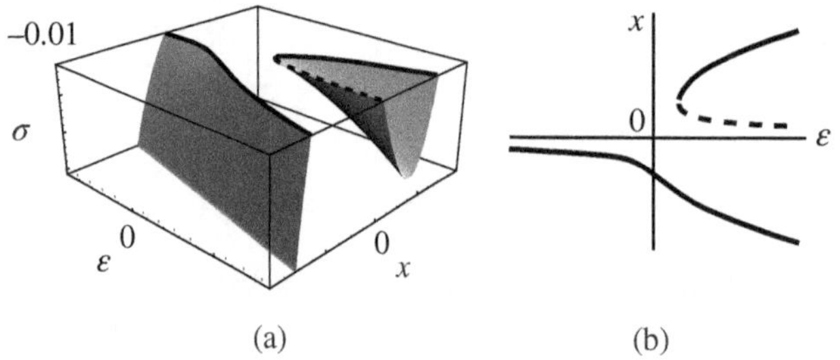

Figure 7.14 Incomplete or perturbed pitchfork bifurcation on equilibrium surface ($\sigma = \sigma_- < 0$). Source: Reprinted with permission Yabuno (2004).

7.6 Effect of Coulomb Friction on Pitchfork Bifurcation

We deal with the two link system in Figure 5.2, where the cubic nonlinearity of the spring is taken into account as in Section 7.1. In this section, we analyze the effect of Coulomb friction (see Appendix E) acting on the nonlinear dynamics. By using Eqs. (5.5) and (7.3), the equation of motion is expressed as

$$m\frac{d^2x}{dt^2} + \left(k_1 - \frac{P}{l}\right)x + \left(k_3 - \frac{P}{2l^3}\right)x^3 = F_c, \tag{7.82}$$

where the viscous damping is neglected but the cubic nonlinearity is taken into account as in Eq. (7.3). F_c denotes the Coulomb friction (see Section E.2). By introducing nondimensional displacement and time $x^* = \frac{x}{l}$ and $t^* = \sqrt{\frac{k_1}{m}}t$, this equation is expressed in the nondimensional form as

$$\frac{d^2x^*}{dt^{*2}} + (1-p)x^* + \left(k_3^* - \frac{p}{2}\right)x^{*3} = F_c^*, \tag{7.83}$$

where $p = \frac{P}{k_1 l}$ and $k_3^* = \frac{k_3 l^2}{k_1}$ are the same nondimensional parameters as in Eq. (7.8). If we focus on the neighborhood of linear buckling point

$$P = P_{cr}(1+\epsilon) \quad (|\epsilon| \ll 1) \tag{7.84}$$

or equivalently

$$p = 1 + \epsilon \quad (|\epsilon| \ll 1), \tag{7.85}$$

according to Eq. (7.13), Eq. (7.83) can rewritten as

$$\frac{d^2x^*}{dt^{*2}} - \epsilon x^* + \alpha_3 x^{*3} + O(x^{*5}) = F_c^*, \tag{7.86}$$

where $\alpha_3 = k_3^* - \frac{1}{2}$ and $F_c^* = \frac{F_c}{k_1 l}$.

7.6.1 Linear Analysis

Before focusing on the nonlinear analysis, let us examine the following equation in which the nonlinear terms in Eq. (7.82) are neglected:

$$m\frac{d^2x}{dt^2} + c\frac{dx}{dt} + \left(k_1 - \frac{P}{l}\right)x = F_c. \tag{7.87}$$

At the critical compressive force, $P = k_1 l \stackrel{\text{def}}{=} P_{cr}$, the coefficient of x expressing the equivalent stiffness, $k_1 - \frac{P}{l}$, becomes 0, and the buckling occurs (see Section 2.6). The behavior in the positive stiffness case $P < P_{cr}$ is mentioned in Appendix E. The equivalent stiffness is equal to k in this appendix. The final rest position after the transient state is changed depending on the initial condition and the set of the final positions makes a region called *dead zone*. In the cases in which the initial position is in the dead zone and the initial velocity is zero, the system remains at rest at the initial position. Such a region exists even in the case $P \geq P_{cr}$ and for any P, the region, which is called *equilibrium region* hereafter, is obtained from

$$\left|\left(k_1 - \frac{P}{l}\right)x_{st}\right| < F_{cmax} \tag{7.88}$$

or equivalently

$$|k_{equiv} x_{st}| < F_{cmax}, \tag{7.89}$$

where F_{cmax} is the maximum static friction force and $k_1 - \frac{P}{l} \stackrel{\text{def}}{=} k_{equiv}$. Therefore, the equilibrium region is expressed as

$$|x_{st}| < \frac{F_{cmax}}{|k_{equiv}|}. \tag{7.90}$$

The equilibrium region with respect to the compressive force P is described as the hatched region in Figure 7.15, where solid and dashed lines are the stable and unstable trivial equilibrium states, respectively, (see also in Figure 7.2). Figure 7.15 can be regarded as the bifurcation diagram in the case with Coulomb friction.

The phase plane in the case of $P < P_{cr}$, i.e. $k_{equiv} > 0$, is qualitatively the same as Figures E.6b and E.10 for the cases far and near from the buckling point $P = P_{cr}$, respectively.

The phase portrait in the case of $m = 1$ and $k_{equiv} = -0.3 < 0$, i.e. $P > P_{cr}$, is shown as Figure 7.16. The thick dashed line is the set of equilibrium states in which the trivial equilibrium state is included. In the case without Coulomb friction, the trivial equilibrium state is unstable but Coulomb friction stabilizes the trivial equilibrium state and also the equilibrium states near the trivial state in the equilibrium region (Yabuno et al. 1999). If the initial conditions are in the hatched region, the behavior approaches to an equilibrium state in the equilibrium region (see the videoclips in Section 16.2).

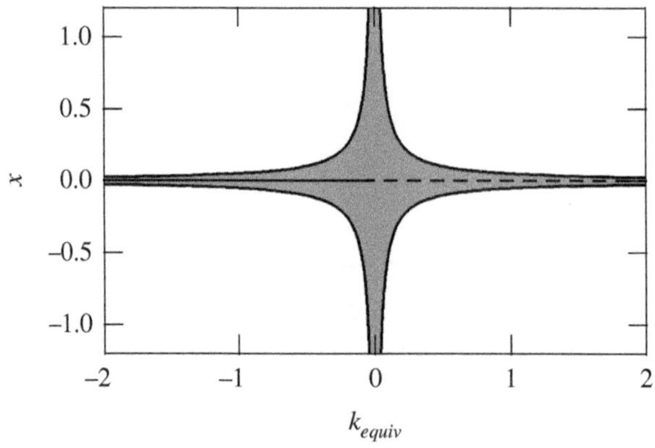

Figure 7.15 Bifurcation diagram with Coulomb friction (linear analysis, $F_{cmax} = 0.05$); solid and dashed lines denote the stable and unstable trivial equilibrium states in the case without Coulomb friction. The hatched region is the equilibrium region whose part for $k_{equiv} > 0$, i.e. $P < P_{cr}$ corresponds to Figure E.9. If the initial position is in this region and the initial velocity is zero, the system keeps this state.

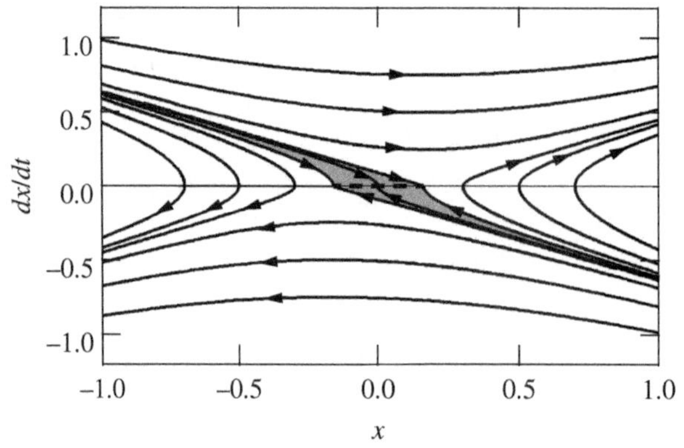

Figure 7.16 Phase portrait in the case of $m = 1$ and $k_{equiv} = -0.3 < 0$, i.e. $P > P_{cr}$ (linear analysis).

7.6.2 Nonlinear Analysis

We analyze the equilibrium states under the Coulomb friction considering the cubic nonlinearity which was neglected in Section 7.6.1. The case without Coulomb friction was already discussed in Section 7.1 leading to the bifurcation diagrams described as Figures 7.1a and 7.3a for hardening and softening cubic nonlinearities, respectively. Here, we focus on the neighborhood of the critical compressive force $P = P_{cr} = k_1 l$ and express p as Eq. (7.85).

According to the method adopted in Section 7.6.1, it is shown from Eq. (7.86) that the equilibrium region satisfies the following equation:

$$| -\epsilon x^*_{st} + \alpha_3 x^{*3}_{st} | < F^*_{cmax}, \tag{7.91}$$

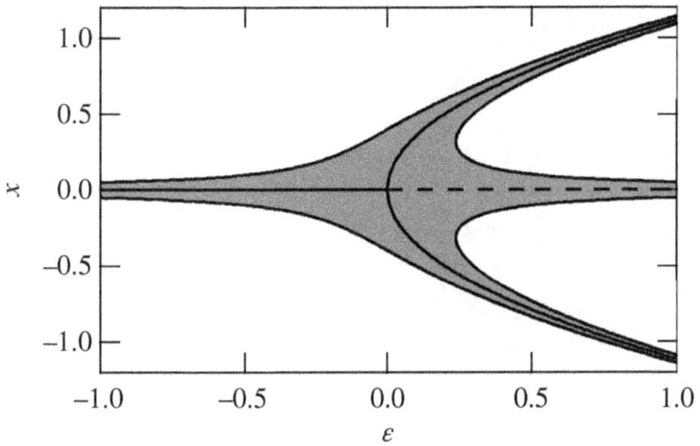

Figure 7.17 Bifurcation diagram with Coulomb friction (in case with hardening cubic nonlinearity); the solid and dashed lines denote the stable and unstable equilibrium states in the case without Coulomb friction, which correspond to those in Figure 7.1a. The parameters are $\alpha_3 = 0.8$ and $F^*_{cmax} = 0.05$.

equivalently

$$-\epsilon x^*_{st} + \alpha_3 x^{*3}_{st} - F^*_{cmax} < 0, \tag{7.92}$$

and

$$-\epsilon x^*_{st} + \alpha_3 x^{*3}_{st} + F^*_{cmax} > 0, \tag{7.93}$$

where $F^*_{cmax} = \frac{F_{cmax}}{k_1 l}$ is the nondimensional maximum static friction force.

First, we show the equilibrium region in the case with hardening cubic nonlinearity, i.e. $\alpha_3 = 0.8 > 0$, in Figure 7.17.

Since F^*_{cmax} is constant, the boundaries in the equilibrium region consist of the combination of the perturbed pitchfork bifurcation diagrams in Figures 7.13b and 7.14b. Similar to the linear analytical result, the equilibrium region is produced as including the stable and unstable equilibrium states in which the Coulomb friction does not exist. In the case when the compressive force is relatively far from the buckling point $\epsilon = 0$ and the equilibrium region is departed into three regions, for example, at $\epsilon = 0.3$ in Figure 7.17, the phase portrait is described as in Figure 7.18. The thick solid and dashed lines denote the equilibrium regions including the stable and unstable equilibrium states, respectively, in which the Coulomb friction does not exist. Even if the equilibrium region includes the unstable equilibrium state as above described when the initial position is in the equilibrium region and the initial velocity is zero, the state keeps at rest at that position. Moreover, when the initial condition is included in the hatched region, the behavior converges to a position in the associated equilibrium region along the arrow in Figure 7.18.

Secondly, we show the equilibrium region in the case with softening cubic nonlinearity, i.e. $\alpha_3 = -0.8 < 0$, in Figure 7.19. Similar to the case with hardening cubic nonlinearity, the equilibrium region is produced including the stable and unstable equilibrium states in which the Coulomb friction does not exist. In the case when the compressive force is relatively far from the buckling point $\epsilon = 0$ and the equilibrium region is departed into three regions, for example, at $\epsilon = -0.3$ in Figure 7.19, the phase portrait is described as in Figure 7.20. As before, the thick solid and dashed lines denote the equilibrium regions

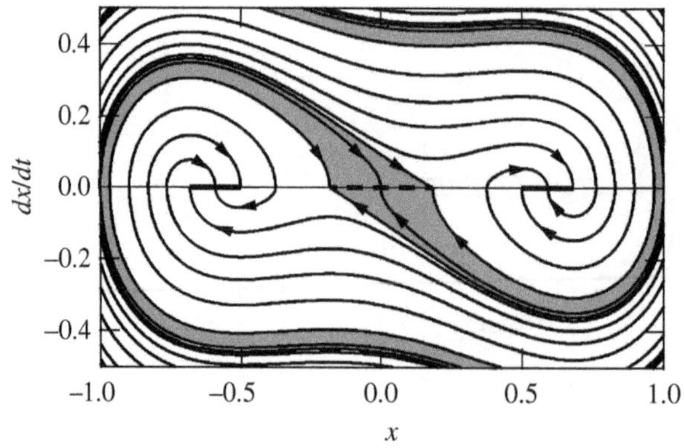

Figure 7.18 Phase portrait with Coulomb friction (in case with hardening cubic nonlinearity). This portrait is for the compressive force in which the equilibrium states are separated into three regions; for example, at $\epsilon = 0.3$ in Figure 7.17.

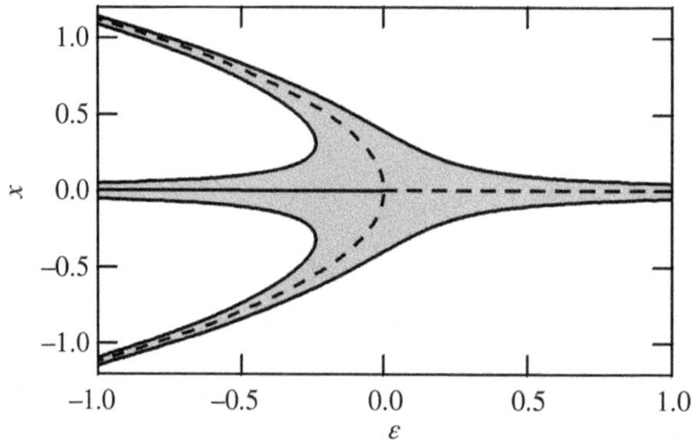

Figure 7.19 Bifurcation diagram with Coulomb friction (in case with softening cubic nonlinearity); the solid and dashed lines denote the stable and unstable equilibrium states in the case without Coulomb friction, which correspond to those in Figure 7.3a. The parameters are $\alpha_3 = -0.8$ and $F^*_{cmax} = 0.05$.

including the stable and unstable equilibrium states, respectively in which the Coulomb friction does not exist. The softening cubic nonlinearity produces subcritical pitchfork bifurcation in the case without Coulomb friction, where the nontrivial equilibrium states are unstable. In the case with Coulomb friction, the equilibrium region is produced including the nontrivial equilibrium states which result to be stabilized in such region. When the initial condition is included in the hatched region, the behavior converges to a position in the associated equilibrium region along the arrow in Figure 7.20. The stability of the equilibrium region can be theoretically investigated by the framework of measure differential inclusions. However, because it is out of level in this manuscript, we just give references (for example, (Glocker 2013; Hetzler 2014; Leine and Van de Wouw 2007)),

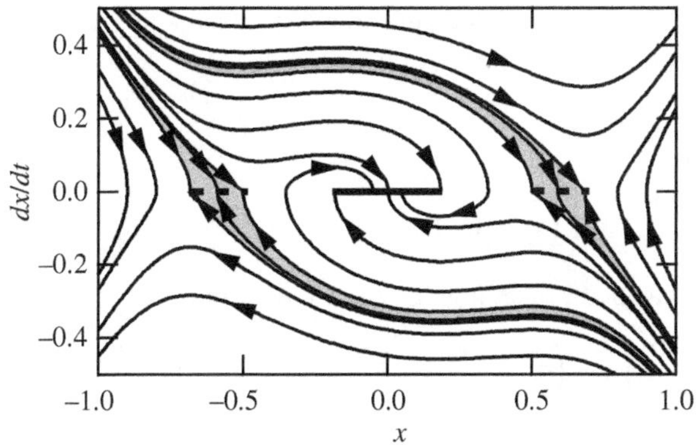

Figure 7.20 Phase portrait with Coulomb friction (in case with softening cubic nonlinearity). This portrait is for the compressive force in which the equilibrium states are separated into three regions; for example, at $\epsilon = -0.3$ in Figure 7.19.

7.7 Nonlinear Characteristics of Static Instability in Spring-Mass-Damper Models of MEMS

We investigated the dynamics of spring-mass-damper models of MEMS devices in Section 5.2 by taking into account their nonlinearity. The comb-type MEMS undergoes electrostatic buckling at a critical input voltage and the trivial steady state is statically destabilized. The linear analysis presented in Section 5.2.1 could determine the critical voltage but did not clarify whether the stable steady states exist instead of destabilized trivial steady states.

More in detail, as previously discussed, the cantilever-type MEMS switch turns on by increasing the input voltage. The linear analysis in Section 5.2.2 provides the lowest voltage needed for the turn-on, but yields the physically unacceptable results above the dimensionless input voltage $V^* = 0.5$. In this section, for these systems, we present the nonlinear equations of motion derived by considering the nonlinear components of the electrostatic and restoring forces acting on the body. Then, we perform nonlinear analyses to clarify the above-mentioned problems which were unsolved through linear analysis.

7.7.1 Pitchfork Bifurcation in Comb-Type MEMS Actuator Device

We investigated the system subject to electrostatic forces in Section 5.2.1, where the electromagnetic force was linearized and the stiffness of the tethers in x-direction in Figure 5.6, i.e. the stiffness in the model shown by Figure 5.7, was assumed to be linear. In this section, we take into account the cubic nonlinearity of the stiffness as in Section 7.1.1, thus obtaining

$$F_s^{\#} = F_{sl}^{\#} + F_{sr}^{\#} = k_1 x + k_3 x^3 \overset{\text{def}}{=} -F_s, \tag{7.94}$$

where k_3 can be positive (hard spring) or negative (soft spring) while k_1 is the positive stiffness. The electrostatic force F_e is expressed from Eq. (5.17) as

$$F_e = \epsilon A V^2 \left\{ \frac{1}{(d-x)^2} - \frac{1}{(d+x)^2} \right\}$$
$$= \frac{4\epsilon A V^2}{d^2} \left\{ \left(\frac{x}{d}\right) + 2\left(\frac{x}{d}\right)^3 \right\} + O\left(\left(\frac{x}{d}\right)^5\right). \tag{7.95}$$

Considering the viscous damping, we obtain the following equation of motion:

$$m\frac{d^2x}{dt^2} = -c\frac{dx}{dt} - \left(k_1 - \frac{4\epsilon A V^2}{d^3}\right)x - \left(k_3 - \frac{8\epsilon A V^2}{d^5}\right)x^3. \tag{7.96}$$

As shown in Section 5.2.1, at the input voltage satisfying $k_1 - \frac{4\epsilon AV^2}{d^3} = 0$, the electrostatic buckling occurs.

Introducing the dimensionless time $t^* = \sqrt{\frac{k_1}{m}}t$ and the dimensionless displacement $x^* = \frac{x}{d}$, we have the dimensionless equation of motion as

$$\frac{d^2x^*}{dt^{*2}} + 2\gamma\frac{dx^*}{dt^*} + (1 - V^*)x^* + (k_3^* - 2V^*)x^{*3} = 0, \tag{7.97}$$

where $\gamma = \frac{c}{2\sqrt{mk}}$, $V^* = \frac{4\epsilon AV^2}{kd^3}$, $k_3^* = \frac{k_3 d^2}{k_1}$ and $x^* \leq 1$ because of $x \leq d$. We analyze the nonlinear dynamics in the neighborhood of the electrostatic buckling point $V^* = 1$ by setting V^* as

$$V^* = 1 + \epsilon \quad (|\epsilon| \ll 1), \tag{7.98}$$

where $\epsilon < 0$ and $\epsilon > 0$ denotes pre- and post-buckling, respectively. Then, neglecting ϵx^{*3} due to its higher order smallness, we can rewrite Eqs. (7.97) as

$$\frac{d^2x^*}{dt^{*2}} + 2\gamma\frac{dx^*}{dt^*} - \epsilon x^* + \alpha_3 x^{*3} = 0, \tag{7.99}$$

where $\alpha_3 = k_3^* - 2$. Since this equation is equivalent to Eq. (7.13), the discussion about nonlinear phenomena is applicable. The equivalent nonlinear stiffness may be hardening or softening in accordance with the positive or negative sign of α_3, respectively. Hence, the relatively strong hardening nonlinearity of an attached practical spring, i.e. $k_3^* > 2$, makes the equivalent nonlinear stiffness hardening ($\alpha_3 > 0$). In this case, the supercritical pitchfork bifurcation is produced as shown in Figure 7.1. In the postbuckling state, the nontrivial stable equilibrium states are produced. When $k_3^* < 2$, i.e. the attached spring is weakly hardening or softening, the equivalent nonlinear characteristic is softening ($\alpha_3 < 0$). In this case, the subcritical pitchfork bifurcation is produced as shown in Figure 7.3. In the prebuckling state, the nontrivial unstable equilibrium states are produced. When $\epsilon > 0$, there is no stable equilibrium state. When $k_3^* = 2$, the equivalent nonlinear stiffness disappears and the bifurcation diagram is described as Figure 7.2.

7.7.2 Saddle-Node Bifurcation in MEMS Switch

In this section, the nonlinear term in Eq. (5.21) which was neglected in Section 5.2.2 is taken into account. Hereafter, the symbol * is omitted for simplicity. The equilibrium equation becomes

$$x_{st} - \frac{V}{(1-x_{st})^2} = 0. \tag{7.100}$$

Then, the equilibrium states satisfies

$$V = x_{st}(1-x_{st})^2 \tag{7.101}$$

and

$$\frac{dV}{dx_{st}} = 3x_{st}^2 - 4x_{st} + 1. \tag{7.102}$$

Because of $0 < x < 1$, the equilibrium states exist only in the range of $V < \frac{4}{27} \stackrel{\text{def}}{=} V_{sn}$ and when $V = V_{sn}, x = \frac{1}{3} \stackrel{\text{def}}{=} x_{sn}$. We thus determine the stability of the equilibrium states. Substituting $x = x_{st} + \Delta x$ $(|\Delta x| \ll 1)$ into Eq. (5.21) and neglecting the nonlinear terms of Δx yields

$$\frac{d^2\Delta x}{dt^2} + 2\gamma\frac{d\Delta x}{dt} + \frac{1}{(1-x_{st})^2}\frac{dV}{dx_{st}}\Delta x = 0. \tag{7.103}$$

When $\frac{dV}{dx_{st}} > 0$ and $\frac{dV}{dx_{st}} < 0$, the equilibrium state is stable and unstable, respectively. The bifurcation diagram is described as Figure 7.21. In the range of $V < V_{sn}$, there are one stable and one unstable steady states. In the region of $V > V_{sn}$, there is no steady state and the unstable steady state in Figure 5.11, which was obtained by the analysis for the linearized system in Eq. (5.22) leading to an unacceptable phenomenon from a physical point of view, is disappeared. In the range of $V < V_{sn}$, the disturbance greater than the unstable equilibrium state is required to turn on the switch. In the range of $V > V_{sn}$, there is no equilibrium state and the displacement monotonically grows until the switch pulls in. To use it as switch, $V > V_{sn}$ is sufficient.

The phase portrait for $V < V_{sn}$ is described as Figure 7.22a, where point [F] is an equilibrium state classified as stable focus, which corresponds to the stable equilibrium state in Figure 7.21, and point [S] is another equilibrium state classified as saddle, which corresponds to the unstable equilibrium state in Figure 7.21. The phase portrait for $V > V_{sn}$ is described as Figure 7.22b. There are not any equilibrium states and the flow monotonically grows with time. In practice, for any initial condition or disturbance, the switch monotonically approaches to the electrode and turns on.

Next, we investigate the nonlinear dynamics near the bifurcation point at $V = V_{sn}, x = x_{sn}$. Let V and x be $V = V_{sn} + \Delta V$ and $x = x_{sn} + \Delta x$, respectively. Substituting these equation into Eq. (5.21) and neglecting third order higher terms with respect to Δx yields

$$\begin{aligned}\frac{d^2\Delta x}{dt^2} + 2\gamma\frac{d\Delta x}{dt} + x_{sn} + \Delta x = {} & \frac{V_{sn}}{(1-x_{sn})^2}\left\{1 + \frac{2\Delta x}{1-x_{sn}} + 3\left(\frac{\Delta x}{1-x_{sn}}\right)^2\right\} \\ & + \frac{\Delta V}{(1-x_{sn})^2}\left\{1 + \frac{2\Delta x}{1-x_{sn}} + 3\left(\frac{\Delta x}{1-x_{sn}}\right)^2\right\}.\end{aligned} \tag{7.104}$$

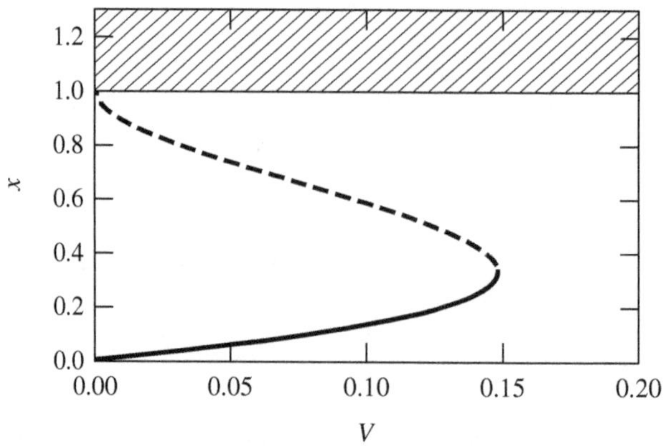

Figure 7.21 Bifurcation of MEMS switch.

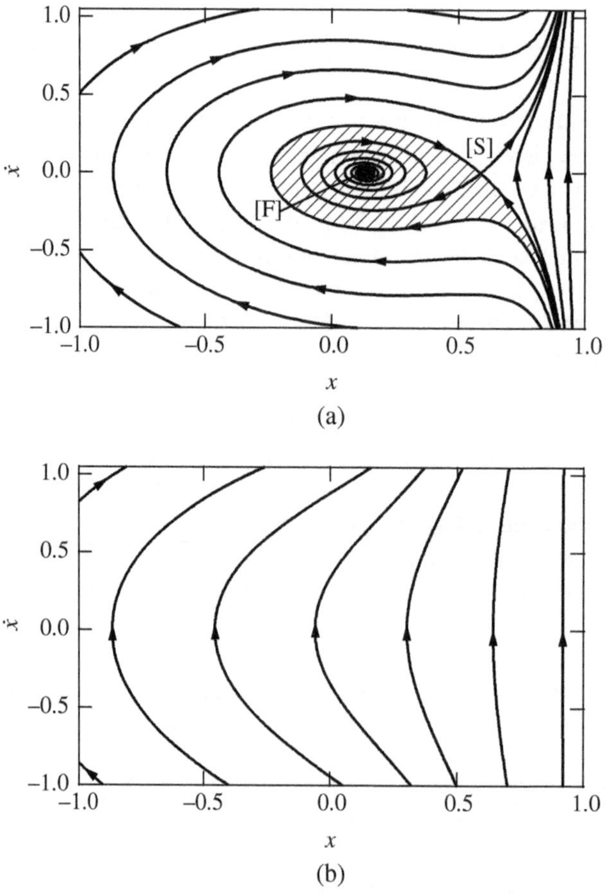

Figure 7.22 Phase space of MEMS swithc: (a) $V > V_{sn}$, (b) $V < V_{sn}$.

Taking into account Eq. (7.101) and $\left.\frac{dV}{dx_{st}}\right|_{x_{st}=x_{sn}} = 0$ yields

$$\frac{d^2\Delta x}{dt^2} + 2\gamma\frac{d\Delta x}{dt} = \frac{3V_{sn}}{(1-x_{sn})^3}\Delta x^2 + \frac{\Delta V}{(1-x_{sn})^2}\left\{1 + \frac{2\Delta x}{1-x_{sn}} + 3\left(\frac{\Delta x}{1-x_{sn}}\right)^2\right\}. \tag{7.105}$$

Neglecting the terms proportional to $\Delta V\Delta x$ and $\Delta V\Delta x^2$ due to their smallness, yields

$$\frac{d^2\Delta x}{dt^2} + 2\gamma\frac{d\Delta x}{dt} = \frac{3V_{sn}}{(1-x_{sn})^3}\Delta x^2 + \frac{1}{(1-x_{sn})^2}\Delta V. \tag{7.106}$$

Furthermore, as neglecting the inertia term in Section 7.2, we neglect the first term and obtain

$$\frac{d\Delta x}{dt} = \frac{1}{2\gamma(1-x_{sn})^2}\Delta V + \frac{3V_{sn}}{2\gamma(1-x_{sn})^3}\Delta x^2. \tag{7.107}$$

Considering ΔV as a control parameter, Eq. (7.107) is the normal form of saddle-node bifurcation of Eq. (7.44) and the bifurcation in Figure 7.21 is regarded as a saddle-node bifurcation.

References

Glocker, C. (2013). *Set-valued force laws: dynamics of non-smooth systems,* vol 1. Springer Science & Business Media.

Guckenheimer, J. and P. J. Holmes (1983). *Nonlinear oscillations, dynamical systems, and bifurcations of vector fields.* Springer.

Hetzler, H. (2014). Bifurcations in autonomous mechanical systems under the influence of joint damping. *Journal of Sound and Vibration* 333(23): 5953–5969.

Leine, R. I. and N. Van de Wouw (2007). *Stability and convergence of mechanical systems with unilateral constraints,* vol. 36. Springer Science & Business Media.

Nayfeh, A. H. (1993). *Method of Normal Forms.* Wiley.

Strogatz, S. (1994). *Nonlinear Dynamics and Chaos.* Westview.

Thomsen, J. J. (2003). *Vibrations and Stability.* Springer.

Yabuno, H. (2004). *Kougaku no tameno Hisenkei-kaiseki Nyuumon (The Elements of Nonlinear Analysis for Engineering).* Saiensu-sha.

Yabuno, H., R. Oowada, and N. Aoshima (1999). Effect of coulomb damping on buckling of a simply supported beam. *Proceedings of ASME Design Engineering Technical Conferences,* Number VIB–8056, pp. 1–12.

8

Reduction Methods of Nonlinear Dynamical Systems

In the neighborhood of bifurcation points, the phase space is spanned with the subspaces including a center subspace which is related to the zero real eigenvalue or pure imaginary eigenvalues. As mentioned in Section 2.5, the dynamics in the center subspace does not exponentially decay or grow, i.e. it is in a stationary state or vibrational one with a constant amplitude. On the other hand, the dynamics in the stable or unstable subspace is related to nonzero real eigenvalues and exponentially decays or grows (monotonically or oscillatory). By taking into account such a difference of time variation between the dynamics in the center subspace and that in the other subspaces, we can reduce the dimension of dynamics in the neighborhood of bifurcation points. In addition, we introduce a nonlinear coordinate transformations to extract the essential characteristics in nonlinear dynamics. The resulting simplified equations are called *normal forms* (Nayfeh, 1993).

8.1 Reduction of the Dimension of State Space by Center Manifold Theory

8.1.1 Nonlinear Stability Analysis at Pitchfork Bifurcation Point

We return to the following buckling problem discussed in Section 7.2 (see Eq. (7.33)):

$$
\begin{cases}
\dfrac{dy_1}{dt} = \dfrac{\epsilon}{2\gamma}y_1 - \dfrac{\alpha_3}{2\gamma}(y_1 + y_2)^3, & (8.1)\\[2ex]
\dfrac{dy_2}{dt} = -2\gamma y_2 + \dfrac{\alpha_3}{2\gamma}(y_1 + y_2)^3, & (8.2)
\end{cases}
$$

where $\gamma > 0$ and the terms with $O(\epsilon(y_1 + y_2)^3)$ are neglected. At the bifurcation point $\epsilon = 0$, the eigenvalues of matrix A are $\lambda_1 = 0$ and $\lambda_2 = -2\gamma$, thus Eqs. (8.1) and (8.2) lead to

$$
\begin{cases}
\dfrac{dy_1}{dt} = -\dfrac{\alpha_3}{2\gamma}(y_1 + y_2)^3, & (8.3)\\[2ex]
\dfrac{dy_2}{dt} = -2\gamma y_2 + \dfrac{\alpha_3}{2\gamma}(y_1 + y_2)^3. & (8.4)
\end{cases}
$$

Linear and Nonlinear Instabilities in Mechanical Systems: Analysis, Control and Application, First Edition. Hiroshi Yabuno.

Companion website: www.wiley.com/go/yabuno/instabilitiesinmechanicalsystems

There is no flow on the center subspace associated with the zero eigenvalue λ_1 if the nonlinear effects are neglected. The center subspace E^c expressed by a straight line should be curved by the nonlinear effects; the curved center subspace W^c is called a center manifold. When y_1 and y_2 approach to zero, since the nonlinear effects become much less than the linear term, they can be neglected. Therefore, in the neighborhood of the origin, the center manifold should be equivalent to the center subspace that is obtained by neglecting the nonlinear effects (see Figure 8.1). Hence, the center manifold $y_2 = h(y_1)$ passes the origin and is tangent to the center subspace E^c, and is assumed to be

$$y_2 = h(y_1) = \chi_1 y_1^2 + \chi_2 y_1^3 + \cdots . \tag{8.5}$$

We seek the coefficients, $\chi_1, \chi_2, \ldots$, to obtain the center manifold. By using Eq. (8.3) and the derivative of Eq. (8.5) with respect to y_1, the derivative of Eq. (8.5) with respect to t is expressed as

$$\begin{aligned}\frac{dy_2}{dt} &= \frac{\partial h}{\partial y_1}\frac{dy_1}{dt}\\ &= (2\chi_1 y_1 + 3\chi_2 y_1^2 + \cdots)\left[-\frac{\alpha_3}{2\gamma}\{y_1 + (\chi_1 y_1^2 + \chi_2 y_1^3 + \cdots)\}^3\right].\end{aligned} \tag{8.6}$$

On the other hand, by using Eq. (8.4), $\frac{dy_2}{dt}$ can be expressed also from as

$$\begin{aligned}\frac{dy_2}{dt} &= -2\gamma y_2 + \frac{\alpha_3}{2\gamma}(y_1 + y_2)^3\\ &= -2\gamma(\chi_1 y_1^2 + \chi_2 y_1^3) + \frac{\alpha_3}{2\gamma}\{y_1 + (\chi_1 y_1^2 + \chi_2 y_1^2 + \cdots)\}^3.\end{aligned} \tag{8.7}$$

Equating Eqs. (8.6) and (8.7) yields

$$-2\gamma\chi_1 y_1^2 + \left(-2\gamma\chi_2 + \frac{\alpha_3}{2\gamma}\right) y_1^3 + \cdots = 0. \tag{8.8}$$

We find

$$\chi_1 = 0, \quad \chi_2 = \frac{\alpha_3}{4\gamma^2} \tag{8.9}$$

and thus the center manifold is derived as

$$y_2 = \frac{\alpha_3}{4\gamma^2} y_1^3 + \cdots . \tag{8.10}$$

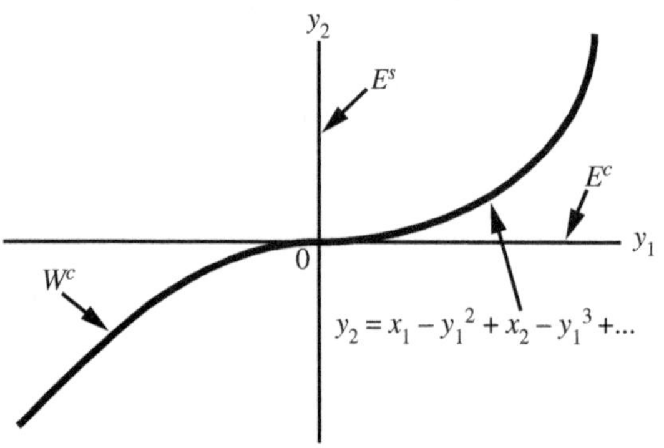

Figure 8.1 Center manifold tangent to center subspace.

Furthermore, substituting this result into Eq. (8.3) yields

$$\frac{dy_1}{dt} = -\frac{\alpha_3}{2\gamma} y_1^3 + O(y_1^5). \tag{8.11}$$

Because the effect of y_2 in Eq. (8.3) is $O(y_1^5)$, it is neglected.

The stable and unstable subspaces, E^s and E^u, expressed by a straight line should be curved by the nonlinear effects; the curved stable and unstable subspaces, W^s and W^u, are called a stable and unstable manifold, respectively. These manifolds can be obtained as the above method to obtain the center manifold.

In the linear analysis in which the nonlinear terms are neglected, there is no flow on the center subspace as mentioned before. Every flow rapidly approaches the center subspace, i.e. y_1-axis, but does not move in the direction parallel to the center subspace, because there is no direction of the flow on the center subspace. On the other hand, taking into account the nonlinear terms, we can detect the flow on the center manifold which is curved from the center subspace by the effect of the nonlinearity. Therefore, every flow rapidly approaches the center manifold and then moves along the center manifold in the manner described by the governing Eqs. (8.10) and (8.11). The speed of the flow on the center manifold is very slow because the velocity $\frac{dy_1}{dt}$ is $O(y_1^3)$, and the stability of the trivial equilibrium point $y_1 = y_2 = 0$ is determined by this slow dynamics. In this example, when α_3 is positive and negative, the system is stable and unstable, respectively (see Problem 8.1.1). The center manifold and phase portrait are described for $\alpha_3 > 0$ and $\alpha_3 < 0$ in Figure 8.2a,b, respectively. Thus, through the center manifold analysis, we can carry out the reduction of a stability problem to a lower dimensional problem at the critical state; in this example, the original two-dimensional dynamics is reduced to one-dimensional one.

Problem 8.1.1 We consider the dynamical system governed with Eq. (8.11). Determine the stability of the trivial equilibrium state $y_1 = 0$.

Ans: We consider the following equation:

$$\frac{dy}{dt} = \alpha_3' y^3 \tag{8.12}$$

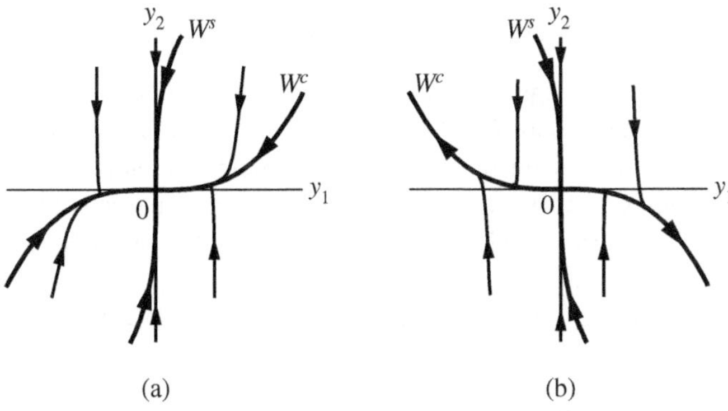

Figure 8.2 Center manifold and stability of pitchfork bifurcation point. (a) shows the stable and center manifolds for the supercritical pitchfork bifurcation point, which is stable. (b) shows the stable and center manifolds for the subcritical pitchfork bifurcation point, which is unstable.

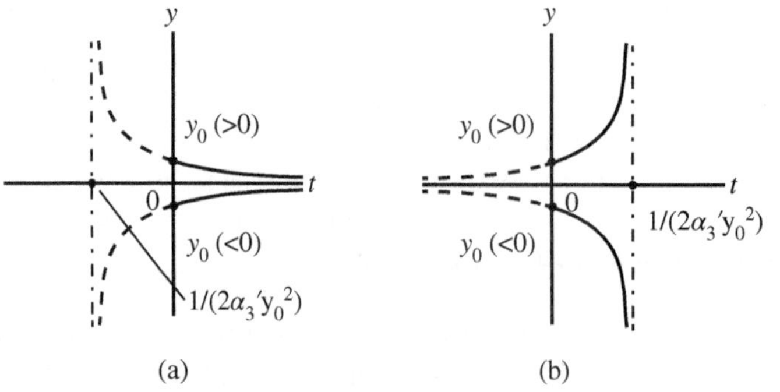

Figure 8.3 Graphs of the solutions of $\frac{dy}{dt} = \alpha_3' y^3$: (a) when $\alpha_3' < 0$, the solution is asymptotically converged to 0; (b) when $\alpha_3' > 0$, the solution is blown up at $t = \frac{1}{2\alpha_3' y(0)^2}$.

with the initial condition y_0 at $t = 0$. By the method of separation of variables, we have

$$\frac{dy}{y^3} = \alpha_3' \, dt. \tag{8.13}$$

Integrating both sides and considering the initial condition yields

$$t = -\frac{1}{2\alpha_3'}\left(\frac{1}{y^2} - \frac{1}{y(0)^2}\right). \tag{8.14}$$

The graphs of the solutions for $\alpha_3' > 0$ and $\alpha_3' < 0$ are depicted as Figure 8.3a,b, respectively. When $\alpha_3' > 0$, the solution is blown up at $t = \frac{1}{2\alpha_3' y(0)^2}$ as Figure 8.3a. On the other hand, when $\alpha_3' < 0$, the solution is not blown up and converges to 0. As a result, when $\alpha_3 = -2\gamma\alpha_3' > 0$ ($\alpha_3' < 0$), the trivial steady state of the system Eq. (8.11) is stable. When $\alpha_3 = -2\gamma\alpha_3' < 0$ ($\alpha_3' > 0$), the trivial steady state of the system Eq. (8.11) is unstable.

8.1.2 Reduction of Nonlinear Dynamics Near Bifurcation Point

In the case when a control parameter is slightly shifted from the critical value, the center manifold changes to a stable or unstable manifold. The dynamics on such a manifold still keeps much slower compared with those on the inherent stable manifold. Therefore, similar to the critical point, the reduction of the dynamics is expected also in a small neighborhood of critical point. We introduce the so called *suspension trick* method for systematically reducing the dimension of dynamics (Guckenheimer and Holmes 1983).

First, we regard the control parameter ϵ as an additional dependent variable as well as y_1 and y_2. Unlike Eq. (7.29), we express Eq. (7.13) as

$$\frac{d}{dt^*}\begin{bmatrix} x_1 \\ x_2 \\ \epsilon \end{bmatrix} = A'\begin{bmatrix} x_1 \\ x_2 \\ \epsilon \end{bmatrix} + (\epsilon x_1 - \alpha_3 x_1^3)\begin{bmatrix} 0 \\ 1 \\ 0 \end{bmatrix}, \tag{8.15}$$

where

$$A' = \begin{bmatrix} 0 & 1 & 0 \\ 0 & -2\gamma & 0 \\ 0 & 0 & 0 \end{bmatrix} \tag{8.16}$$

and the term including ϵx_1 is regarded as a nonlinear term. Matrix A' has zero multiple eigenvalues and a negative real eigenvalue -2γ and has three linear independent eigenvectors. We introduce the transformation as

$$\begin{bmatrix} x_1 \\ x_2 \\ \epsilon \end{bmatrix} = P \begin{bmatrix} y_1 \\ y_2 \\ \epsilon \end{bmatrix}, \tag{8.17}$$

where

$$P = \begin{bmatrix} 1 & 1 & 0 \\ 0 & -2\gamma & 0 \\ 0 & 0 & 1 \end{bmatrix}. \tag{8.18}$$

Substituting Eq. (8.17) into Eq. (8.15) and premultiplying the result by P^{-1}, we have the following matrix form whose linear part consists of a diagonal matrix:

$$\frac{d}{dt}\begin{bmatrix} y_1 \\ y_2 \\ \epsilon \end{bmatrix} = B' \begin{bmatrix} y_1 \\ y_2 \\ \epsilon \end{bmatrix} + \frac{1}{2\gamma}\{\epsilon(y_1 + y_2) - \alpha_3(y_1 + y_2)^3\} \begin{bmatrix} 1 \\ -1 \\ 0 \end{bmatrix}, \tag{8.19}$$

where

$$B' = \begin{bmatrix} 0 & 0 & 0 \\ 0 & -2\gamma & 0 \\ 0 & 0 & 0 \end{bmatrix}. \tag{8.20}$$

The y_2 axis is the stable subspace and the $y_1 - \epsilon$ plane-spanned by the eigenvectors corresponding to the zero eigenvalues of the matrix in the linear part is the center subspace. Because the center manifold is tangential to this plane, we assume it as

$$y_2 = h(y_1, \epsilon) = \chi_1 y_1^2 + \chi_2 y_1 \epsilon + \chi_3 \epsilon^2 + \chi_4 y_1^3 + \cdots . \tag{8.21}$$

Similar to the method in Section 8.1.1, the coefficients of $\chi_1, \chi_2, \chi_3, \ldots$ can be calculated by using

$$\begin{aligned} \frac{dy_2}{dt} &= \frac{\partial h}{\partial y_1}\frac{dy_1}{dt} + \frac{\partial h}{\partial \epsilon}\frac{d\epsilon}{dt} \\ &= (2\chi_1 y_1 + \chi_2 \epsilon + 3\chi_4 y_1^2 + \cdots)\left[\frac{1}{2\gamma}\{\epsilon(y_1 + h(y_1, \epsilon)) - \alpha_3(y_1 + h(y_1, \epsilon))^3\}\right]. \end{aligned} \tag{8.22}$$

Invoking that Eq. (8.22) is equivalent to the second row of Eq. (8.19), we can find $\chi_1 = 0, \chi_2 = -\frac{1}{4\gamma^2}, \chi_3 = 0, \chi_4 = \frac{\alpha_3}{4\gamma^2}$, and so on. Substituting Eq. (8.21) into the first row of Eq. (8.19) and neglecting fourth order terms leads to Eq. (7.35) again.

8.2 Reduction of Degree of Nonlinear Terms by the Method of Normal Forms

In Section 8.1, it is shown that the center manifold theory can reduce the dimension of dynamics near a critical point, where the dynamics in certain directions are much slower than the other directions. In this section, by using nonlinear coordinate transformations, we consider the reduction method of the number of nonlinear terms in nonlinear differential equations through some examples. The reduced systems may correspond to one of the normal forms introduced in Chapter 7. This is called *normal form method*, which categorizes the nonlinear phenomena expressed by nonlinear differential equations.

8.2.1 Reduction by Nonlinear Coordinate Transformation: Method of Normal Forms

By center manifold theory, we reduced the dimension of the systems in the neighborhood of a bifurcation point. In this section, we consider a method to transform the reduced systems to simpler forms, i.e. so called *normal forms*. We find a nonlinear coordinate transformation to eliminate as many nonlinear terms as possible. We consider a nonlinear system

$$\frac{d\boldsymbol{y}}{dt} = J\boldsymbol{y} + \boldsymbol{f}(\boldsymbol{y}) = J\boldsymbol{y} + \boldsymbol{f}_2(\boldsymbol{y}) + \boldsymbol{f}_3(\boldsymbol{y}) + \cdots, \tag{8.23}$$

where $\boldsymbol{f}_n(\boldsymbol{y})$ denotes nth order nonlinear terms with respect to $\boldsymbol{y}$. By a linear coordinate transformation, the linear part of Eq. (8.23) has already been changed into a diagonal matrix with real entries or a real Jordan form as matrix B in Section 2.4. By using a nonlinear transformation as

$$\boldsymbol{y} = \boldsymbol{z} + \boldsymbol{h}(\boldsymbol{z}) = \boldsymbol{z} + \boldsymbol{h}_2(\boldsymbol{z}) + \boldsymbol{h}_3(\boldsymbol{z}) + \cdots, \tag{8.24}$$

where $\boldsymbol{h}_n(\boldsymbol{z})$ denotes nth order nonlinear terms with respect to $\boldsymbol{z}$, let us consider to transform Eq. (8.23) into

$$\frac{d\boldsymbol{z}}{dt} = J\boldsymbol{z} + \boldsymbol{g}(\boldsymbol{z}) = J\boldsymbol{z} + \boldsymbol{g}_2(\boldsymbol{z}) + \boldsymbol{g}_3(\boldsymbol{z}) + \cdots, \tag{8.25}$$

where $\boldsymbol{g}_n(\boldsymbol{z})$ denotes nth order nonlinear terms with respect to $\boldsymbol{z}$. The objective is to choose the coefficients of Eq. (8.24) so that we can eliminate as many terms as possible in Eq. (8.25). First, we calculate $d\boldsymbol{y}/dt$ from Eq. (8.24) as

$$\frac{d\boldsymbol{y}}{dt} = \frac{d\boldsymbol{z}}{dt} + D_{\boldsymbol{z}}\boldsymbol{h}\frac{d\boldsymbol{z}}{dt}, \tag{8.26}$$

where $D_{\boldsymbol{z}}\boldsymbol{h}$ is Jacobian matrix of $\boldsymbol{h}$ with respect to $\boldsymbol{z}$. Substituting Eq. (8.25) into $d\boldsymbol{z}/dt$ in the above equation, we can rewrite $d\boldsymbol{y}/dt$ as

$$\frac{d\boldsymbol{y}}{dt} = J\boldsymbol{z} + \boldsymbol{g}(\boldsymbol{z}) + D_{\boldsymbol{z}}\boldsymbol{h}\{J\boldsymbol{z} + \boldsymbol{g}(\boldsymbol{z})\}. \tag{8.27}$$

On the other hand, substituting Eq. (8.24) into the right-hand side in Eq. (8.23), we obtain

$$\frac{d\boldsymbol{y}}{dt} = J(\boldsymbol{z} + \boldsymbol{h}(\boldsymbol{z})) + \boldsymbol{f}(\boldsymbol{z} + \boldsymbol{h}(\boldsymbol{z})). \tag{8.28}$$

The equality of Eqs. (8.27) and (8.28) leads to

$$J\boldsymbol{z} + \boldsymbol{g}(\boldsymbol{z}) + D_{\boldsymbol{z}}\boldsymbol{h}J\boldsymbol{z} + D_{\boldsymbol{z}}\boldsymbol{h}\boldsymbol{g}(\boldsymbol{z}) = J(\boldsymbol{z} + \boldsymbol{h}(\boldsymbol{z})) + \boldsymbol{f}(\boldsymbol{z} + \boldsymbol{h}(\boldsymbol{z})) \tag{8.29}$$

or

$$\boldsymbol{g}(\boldsymbol{z}) + D_{\boldsymbol{z}}\boldsymbol{h}J\boldsymbol{z} + D_{\boldsymbol{z}}\boldsymbol{h}\boldsymbol{g}(\boldsymbol{z}) = J\boldsymbol{h}(\boldsymbol{z}) + \boldsymbol{f}(\boldsymbol{z} + \boldsymbol{h}(\boldsymbol{z})). \tag{8.30}$$

Considering

$$\boldsymbol{f}(\boldsymbol{z} + \boldsymbol{h}(\boldsymbol{z})) \approx \boldsymbol{f}_2(\boldsymbol{z}) \tag{8.31}$$

and noticing that $D_{\boldsymbol{z}}\boldsymbol{h}_3J\boldsymbol{z}, \ldots$ is a term of higher order than third order, we can pick up the second order terms as

$$\boldsymbol{g}_2(\boldsymbol{z}) + D_{\boldsymbol{z}}\boldsymbol{h}_2J\boldsymbol{z} = J\boldsymbol{h}_2(\boldsymbol{z}) + \boldsymbol{f}_2(\boldsymbol{z}) \tag{8.32}$$

or

$$D_{\boldsymbol{z}}\boldsymbol{h}_2J\boldsymbol{z} - J\boldsymbol{h}_2(\boldsymbol{z}) = -\boldsymbol{g}_2(\boldsymbol{z}) + \boldsymbol{f}_2(\boldsymbol{z}). \tag{8.33}$$

While $\boldsymbol{f}_2(\boldsymbol{z})$ is given in the original equation (8.23), we can freely set $\boldsymbol{h}_2(\boldsymbol{z})$. By the suitable choice of $\boldsymbol{h}_2(\boldsymbol{z})$, i.e. the coefficients of the nonlinear coordinate, we eliminate as many terms as possible and we obtain reduced system of Eq. (8.25). This strategy is applicable for higher order nonlinear terms. In the next two sections, through examples, we understand the method of normal forms in more detail.

8.2.2 Case in which the Linear Part has Distinct Real Eigenvalues

We follow the above procedure to obtain a normal form of the following system:

$$\frac{d}{dt}\begin{bmatrix} y_1 \\ y_2 \end{bmatrix} = \begin{bmatrix} 2 & 0 \\ 0 & 1 \end{bmatrix}\begin{bmatrix} y_1 \\ y_2 \end{bmatrix} + \begin{bmatrix} \alpha_{20}y_1^2 + \alpha_{11}y_1y_2 + \alpha_{02}y_2^2 \\ \beta_{20}y_1^2 + \beta_{11}y_1y_2 + \beta_{02}y_2^2 \end{bmatrix}. \tag{8.34}$$

We set the nonlinear coordinate transformation of Eq. (8.24) as follows:

$$\boldsymbol{y} = \begin{bmatrix} y_1 \\ y_2 \end{bmatrix} = \begin{bmatrix} z_1 \\ z_2 \end{bmatrix} + \boldsymbol{h}_2(z_1, z_2), \tag{8.35}$$

where

$$\boldsymbol{h}_2(z_1, z_2) = \begin{bmatrix} h_{21}(z_1, z_2) \\ h_{22}(z_1, z_2) \end{bmatrix} = \begin{bmatrix} P_{20}z_1^2 + P_{11}z_1z_2 + P_{02}z_2^2 \\ Q_{20}z_1^2 + Q_{11}z_1z_2 + Q_{02}z_2^2 \end{bmatrix}. \tag{8.36}$$

Since Eq. (8.34) is a quadratic nonlinear system, we assume $\boldsymbol{g}_2(z_1, z_2)$ in Eq. (8.25) as

$$\boldsymbol{g}_2(z_1, z_2) = \begin{bmatrix} g_{21}(z_1, z_2) \\ g_{22}(z_1, z_2) \end{bmatrix} = \begin{bmatrix} R_{20}z_1^2 + R_{11}z_1z_2 + R_{02}z_2^2 \\ S_{20}z_1^2 + S_{11}z_1z_2 + S_{02}z_2^2 \end{bmatrix}. \tag{8.37}$$

Here, the objective is to eliminate as many terms in $R_{ij}(i,j = 0, 1, 2)$ and $S_{ij}(i,j = 0, 1, 2)$ by suitably choosing the coefficients $P_{ij}(i,j = 0, 1, 2)$ and $Q_{ij}(i,j = 0, 1, 2)$.

Thus, dy_1/dt and dy_2/dt can be expressed as

$$\frac{dy_1}{dt} = \frac{dz_1}{dt} + \frac{\partial h_{21}}{\partial z_1}\frac{dz_1}{dt} + \frac{\partial h_{21}}{\partial z_2}\frac{dz_2}{dt}$$

$$= \frac{dz_1}{dt} + (2P_{20}z_1 + P_{11}z_2)\frac{dz_1}{dt} + (P_{11}z_1 + 2P_{02}z_2)\frac{dz_2}{dt}, \tag{8.38}$$

$$\begin{aligned}\frac{dy_2}{dt} &= \frac{dz_2}{dt} + \frac{\partial h_{22}}{\partial z_1}\frac{dz_1}{dt} + \frac{\partial h_{22}}{\partial z_2}\frac{dz_2}{dt}\\ &= \frac{dz_2}{dt} + (2Q_{20}z_1 + Q_{11}z_2)\frac{dz_1}{dt} + (Q_{11}z_1 + 2Q_{02}z_2)\frac{dz_2}{dt}.\end{aligned} \tag{8.39}$$

Combining these equations yields

$$\begin{aligned}\frac{d}{dt}\begin{bmatrix} y_1 \\ y_2 \end{bmatrix} &= \frac{d}{dt}\begin{bmatrix} z_1 \\ z_2 \end{bmatrix} + \begin{bmatrix} \frac{\partial h_{21}}{\partial z_1} & \frac{\partial h_{21}}{\partial z_2} \\ \frac{\partial h_{22}}{\partial z_1} & \frac{\partial h_{22}}{\partial z_2} \end{bmatrix}\left(\frac{d}{dt}\begin{bmatrix} z_1 \\ z_2 \end{bmatrix}\right)\\ &= \frac{d}{dt}\begin{bmatrix} z_1 \\ z_2 \end{bmatrix} + \begin{bmatrix} 2P_{20}z_1 + P_{11}z_2 & P_{11}z_1 + 2P_{02}z_2 \\ 2Q_{20}z_1 + Q_{11}z_2 & Q_{11}z_1 + 2Q_{02}z_2 \end{bmatrix}\left(\frac{d}{dt}\begin{bmatrix} z_1 \\ z_2 \end{bmatrix}\right),\end{aligned} \tag{8.40}$$

where the second term in the right-hand side, that is the 2×2 matrix, corresponds to the-Jacobian matrix $D_z\boldsymbol{h}$ of Eq. (8.26).

Let us calculate both sides of Eq. (8.33). The left-hand side is

$$\begin{aligned} &D_z\boldsymbol{h}J\boldsymbol{z} - J\boldsymbol{h}(\boldsymbol{z})\\ &= \begin{bmatrix} 4P_{20}z_1^2 + 3P_{11}z_1z_2 + 2P_{02}z_2^2 \\ 4Q_{20}z_1^2 + 3Q_{11}z_1z_2 + 2Q_{02}z_2^2 \end{bmatrix} - \begin{bmatrix} 2P_{20}z_1^2 + 2P_{11}z_1z_2 + 2P_{02}z_2^2 \\ Q_{20}z_1^2 + Q_{11}z_1z_2 + Q_{02}z_2^2 \end{bmatrix}\\ &= \begin{bmatrix} 2P_{20}z_1^2 + P_{11}z_1z_2 + 0 \times P_{02}z_2^2 \\ 3Q_{20}z_1^2 + 2Q_{11}z_1z_2 + Q_{02}z_2^2 \end{bmatrix}.\end{aligned} \tag{8.41}$$

The right-hand side is

$$\begin{aligned} &-\boldsymbol{g}(\boldsymbol{z}) + \boldsymbol{f}(\boldsymbol{z})\\ &= -\begin{bmatrix} R_{20}z_1^2 + R_{11}z_1z_2 + R_{02}z_2^2 \\ S_{20}z_1^2 + S_{11}z_1z_2 + S_{02}z_2^2 \end{bmatrix} + \begin{bmatrix} \alpha_{20}z_1^2 + \alpha_{11}z_1z_2 + \alpha_{02}z_2^2 \\ \beta_{20}z_1^2 + \beta_{11}z_1z_2 + \beta_{02}z_2^2 \end{bmatrix}\\ &= \begin{bmatrix} (\alpha_{20} - R_{20})z_1^2 + (\alpha_{11} - R_{11})z_1z_2 + (\alpha_{02} - R_{02})z_2^2 \\ (\beta_{20} - S_{20})z_1^2 + (\beta_{11} - S_{11})z_1z_2 + (\beta_{02} - S_{02})z_2^2 \end{bmatrix}.\end{aligned} \tag{8.42}$$

Equating like powers of both sides yields
First row:

$$z_1^2 : 2P_{20} = \alpha_{20} - R_{20} \tag{8.43}$$

$$z_1z_2 : P_{11} = \alpha_{11} - R_{11} \tag{8.44}$$

$$z_2^2 : 0 \times P_{02} = \alpha_{02} - R_{02} \tag{8.45}$$

Second row:

$$z_1^2 : 3Q_{20} = \beta_{20} - S_{20} \tag{8.46}$$

$$z_1z_2 : 2Q_{11} = \beta_{11} - S_{11} \tag{8.47}$$

$$z_2^2 : Q_{02} = \beta_{02} - S_{02} \tag{8.48}$$

Here, because the left-hand side in Eq. (8.45) is zero independent of P_{02}, we cannot eliminate R_{02}, but R_{02} is determined to be $R_{02} = \alpha_{02}$. Then, we can set P_{02} to be 0. On the other hand, $R_{ij}(i,j = 0,1,2)$ other than R_{02} and all $S_{ij}(i,j = 0,1,2)$ can be eliminated by suitably setting P and Q as

$$P_{20} = \alpha_{20}/2,\ P_{11} = \alpha_{11},\ P_{02} = 0,\ Q_{20} = \beta_{20}/3,\ Q_{11} = \beta_{11}/2,\ Q_{02} = \beta_{02}. \tag{8.49}$$

As a result, by the nonlinear transformation

$$\mathbf{y} = \begin{bmatrix} y_1 \\ y_2 \end{bmatrix} = \begin{bmatrix} z_1 \\ z_2 \end{bmatrix} + \begin{bmatrix} \dfrac{\alpha_{20}}{2} z_1^2 + \alpha_{11} z_1 z_2 \\ \dfrac{\beta_{20}}{3} z_1^2 + \dfrac{\beta_{11}}{2} z_1 z_2 + \beta_{02} z_2^2 \end{bmatrix}, \tag{8.50}$$

Eq. (8.34) can be simplified as

$$\frac{d}{dt}\begin{bmatrix} z_1 \\ z_2 \end{bmatrix} = \begin{bmatrix} 2 & 0 \\ 0 & 1 \end{bmatrix}\begin{bmatrix} z_1 \\ z_2 \end{bmatrix} + \begin{bmatrix} \alpha_{02} z_2^2 \\ 0 \end{bmatrix}. \tag{8.51}$$

8.2.3 Nonlinear Term Remaining in Normal Form

We consider the nonlinear term which cannot be eliminated by the nonlinear transformation. We rewrite Eq. (8.34) as

$$\begin{cases} \dfrac{dy_1}{dt} - 2y_1 = \alpha_{20} y_1^2 + \alpha_{11} y_1 y_2 + \alpha_{02} y_2^2, & (8.52) \\ \dfrac{dy_2}{dt} - y_2 = \beta_{20} y_1^2 + \beta_{11} y_1 y_2 + \beta_{02} y_2^2, & (8.53) \end{cases}$$

by regarding the terms in the right-hand side as equivalent external excitations.

First, we consider only the linear terms as

$$\begin{cases} \dfrac{dy_1}{dt} - 2y_1 = 0, & (8.54) \\ \dfrac{dy_2}{dt} - y_2 = 0. & (8.55) \end{cases}$$

The solution is

$$\begin{cases} y_1 = c_1 e^{2t}, & (8.56) \\ y_2 = c_2 e^{t}, & (8.57) \end{cases}$$

where c_1 and c_2 are integral constants. Then, we substitute Eqs. (8.56) and (8.57) into the right-hand side in Eqs. (8.52) and (8.53) and evaluate the equivalent external excitations. Each term in the right-hand side is $\alpha_{20}y_1^2 = \alpha_{20}c_1^2e^{4t}$, $\alpha_{11}y_1y_2 = \alpha_{11}c_1c_2e^{3t}$, $\alpha_{02}y_2^2 = \alpha_{02}c_2^2e^{2t}$, $\beta_{20}y_1^2 = \beta_{20}c_1^2e^{4t}$, $\beta_{11}y_1y_2 = \beta_{11}c_1c_2e^{3t}$, and $\beta_{02}y_2^2 = \beta_{02}c_2^2e^{2t}$. In the linear part of Eq. (8.52), the eigenvalue is 2 as seen from Eq. (8.54). Since the coefficient of t in the exponent of the term $\alpha_{02}y_2^2 = \alpha_{02}c_2^2e^{2t}$ is equal to the eigenvalue 2, this term causes a resonance and is called a resonance term (see Appendix C). The coefficients of t in the exponents of the remaining terms on the right-hand side of Eq. (8.52), $\alpha_{20}y_1^2 = \alpha_{20}c_1^2e^{4t}$ and $\alpha_{11}y_1y_2 = \alpha_{11}c_1c_2e^{3t}$, are not equal to 2, therefore do not cause the resonance. Similarly, the coefficients of t in the exponents of all terms on the right-hand side of Eq. (8.53) are not equal to the eigenvalue 1, as seen from

Eq. (8.55), thus following that these terms are not resonance terms. Under the near identity transformation given in Eq. (8.35), in which z_1 and z_2 approximately correspond to y_1 and y_2, respectively, the resonance term is kept in the normal form, but *nonresonance terms* are eliminated. In other words, the normal form includes only the terms essentially governing the nonlinear dynamics. In fact, the particular solution of

$$\frac{dz_1}{dt} - 2z_1 = \alpha_{02} z_2^2 = \alpha_{02} c_2^2 e^{2t} \tag{8.58}$$

is

$$z_1 = \alpha_{02} c_2^2 t e^{2t} \tag{8.59}$$

and includes the *secular term* (see Section 9.1). Such a nonlinear term to produce the secular term cannot be eliminated even by the nonlinear transformation.

8.2.4 Reduction in the Neighborhood of Hopf Bifurcation Point

We consider the system

$$\frac{d}{dt}\begin{bmatrix} x_1 \\ x_2 \end{bmatrix} = \begin{bmatrix} \epsilon & \omega \\ -\omega & \epsilon \end{bmatrix}\begin{bmatrix} x_1 \\ x_2 \end{bmatrix} + \boldsymbol{f}_2(x_1, x_2) + \boldsymbol{f}_3(x_1, x_2), \tag{8.60}$$

where $\boldsymbol{f}_2(x_1, x_2)$ and $\boldsymbol{f}_3(x_1, x_2)$ are the quadratic and cubic nonlinear terms with respect to x_1 and x_2 as

$$\boldsymbol{f}_2(x_1, x_2) = \begin{bmatrix} \alpha_{20} x_1^2 + \alpha_{11} x_1 x_2 + \alpha_{02} x_2^2 \\ \beta_{20} x_1^2 + \beta_{11} x_1 x_2 + \beta_{02} x_2^2 \end{bmatrix}, \tag{8.61}$$

$$\boldsymbol{f}_3(x_1, x_2) = \begin{bmatrix} \alpha_{30} x_1^3 + \alpha_{21} x_1^2 x_2 + \alpha_{12} x_1 x_2^2 + \alpha_{03} x_2^3 \\ \beta_{30} x_1^3 + \beta_{21} x_1^2 x_2 + \beta_{12} x_1 x_2^2 + \beta_{03} x_2^3 \end{bmatrix}. \tag{8.62}$$

The linear part has the form of real Jordan canonical form as Eq. (2.86) in the case when the linear operator has complex eigenvalues $\epsilon \pm i\omega$, where ϵ and ω are real. Focusing on the dynamics at the critical point $\epsilon = 0$, where Hopf bifurcation occurs and the stability is changed (see Section 7.4.3), we derive a normal form.

We seek a nonlinear coordinate transformation to eliminate as many nonlinear terms as possible. Before applying the nonlinear coordinate transformation, to introduce the complex coordinate, we make the following linear transformation:

$$z = x_1 + ix_2, \tag{8.63}$$

i.e.

$$\begin{bmatrix} z \\ \bar{z} \end{bmatrix} = \begin{bmatrix} 1 & i \\ 1 & -i \end{bmatrix}\begin{bmatrix} x_1 \\ x_2 \end{bmatrix} \tag{8.64}$$

and

$$\begin{bmatrix} x_1 \\ x_2 \end{bmatrix} = \frac{1}{2}\begin{bmatrix} 1 & 1 \\ -i & i \end{bmatrix}\begin{bmatrix} z \\ \bar{z} \end{bmatrix}. \tag{8.65}$$

Then, without loss of generality, Eq. (8.60) is rewritten as

$$\frac{dz}{dt} = (\epsilon - i\omega)z + \gamma_{20}z^2 + \gamma_{11}|z|^2 + \gamma_{02}\bar{z}^2 + \gamma_{30}z^3 + \gamma_{21}|z|^2 z + \gamma_{12}|z|^2\bar{z} + \gamma_{03}\bar{z}^3 \tag{8.66}$$

or

$$\frac{d\bar{z}}{dt} = (\epsilon + i\omega)\bar{z} + \bar{\gamma}_{20}\bar{z}^2 + \bar{\gamma}_{11}|z|^2 + \bar{\gamma}_{02}z^2 + \bar{\gamma}_{30}\bar{z}^3 + \bar{\gamma}_{21}|z|^2\bar{z} + \bar{\gamma}_{12}|z|^2 z + \bar{\gamma}_{03}z^3, \tag{8.67}$$

where the complex constant coefficients, γ_{20}, γ_{11}, and γ_{02}, consist of the coefficients of the quadratic nonlinear terms of Eq. (8.61), and the complex constant coefficients, $\gamma_{30}, \gamma_{21}, \gamma_{12}$, and γ_{03}, consist of the coefficients of the cubic nonlinear terms of Eq. (8.62). First, by the nonlinear coordinate transformation

$$z = \xi + P_{20}\xi^2 + P_{11}\xi\bar{\xi} + P_{02}\bar{\xi}^2, \tag{8.68}$$

where P_{20}, P_{11}, and P_{02} are complex, we eliminate as many quadratic nonlinear terms as possible. In addition, we set the system obtained by the nonlinear transformation as

$$\frac{d\xi}{dt} = (\epsilon - i\omega)\xi + R_{20}\xi^2 + R_{11}\xi\bar{\xi} + R_{02}\bar{\xi}^2 + F_3(\xi, \bar{\xi}). \tag{8.69}$$

F_3 represents the cubic terms transformed from $\boldsymbol{f}_3(y_1, y_2)$ by Eqs. (8.63) and (8.68) and can be expressed as

$$F_3 = \gamma'_{30}\xi^3 + \gamma'_{21}\xi^2\bar{\xi} + \gamma'_{12}\xi\bar{\xi}^2 + \gamma'_{03}\bar{\xi}^3. \tag{8.70}$$

From Eqs. (8.68) and (8.69), we obtain

$$\begin{aligned}\frac{dz}{dt} &= \frac{\partial z}{\partial \xi}\frac{d\xi}{dt} + \frac{\partial z}{\partial \bar{\xi}}\frac{d\bar{\xi}}{dt}\\ &= (\epsilon - i\omega)\xi\\ &\quad + \{R_{20} + 2P_{20}(\epsilon - i\omega)\}\xi^2 + \{R_{11} + 2\epsilon P_{11}\}\xi\bar{\xi} + \{R_{02} + 2P_{02}(\epsilon + i\omega)\}\bar{\xi}^2 + O(3).\end{aligned} \tag{8.71}$$

On the other hand, from Eqs. (8.66) and (8.68)

$$\begin{aligned}\frac{dz}{dt} &= (\epsilon - i\omega)\xi\\ &\quad + \{P_{20}(\epsilon - i\omega) + \gamma_{20}\}\xi^2 + \{P_{11}(\epsilon - i\omega) + \gamma_{11}\}\xi\bar{\xi} + \{P_{02}(\epsilon - i\omega) + \gamma_{02}\}\bar{\xi}^2 + O(3).\end{aligned} \tag{8.72}$$

Equating like powers of ξ in the quadratic nonlinear terms of Eqs. (8.71) and (8.72) yields

$$R_{20} + P_{20}(\epsilon - i\omega) = \gamma_{20}, \tag{8.73}$$

$$R_{11} + P_{11}(\epsilon + i\omega) = \gamma_{11}, \tag{8.74}$$

$$R_{02} + P_{02}(\epsilon + 3i\omega) = \gamma_{02}. \tag{8.75}$$

Setting

$$P_{20} = \frac{\gamma_{20}}{\epsilon - i\omega}, \tag{8.76}$$

$$P_{11} = \frac{\gamma_{11}}{\epsilon + i\omega}, \tag{8.77}$$

$$P_{02} = \frac{\gamma_{02}}{\epsilon + 3i\omega}, \tag{8.78}$$

allows to eliminate the coefficients of all quadratic terms in Eq. (8.69), R_{20}, R_{11}, and R_{02}. As a result, under the nonlinear coordinate transformation of Eq. (8.68), Eq. (8.69) results in

$$\frac{d\xi}{dt} = (\epsilon - i\omega)\xi + \gamma'_{30}\xi^3 + \gamma'_{21}\xi^2\overline{\xi} + \gamma'_{12}\xi\overline{\xi}^2 + \gamma'_{03}\overline{\xi}^3. \tag{8.79}$$

As following step, we eliminate as many cubic nonlinear terms as possible in Eq. (8.79) by the nonlinear coordinate transformation:

$$\xi = \eta + P_{30}\eta^3 + P_{21}\eta^2\overline{\eta} + P_{12}\eta\overline{\eta}^2 + P_{03}\overline{\eta}^3 \tag{8.80}$$

and we set the system obtained by the nonlinear transformation as

$$\frac{d\eta}{dt} = (\epsilon - i\omega)\eta + R_{30}\eta^3 + R_{21}|\eta|^2\eta + R_{12}|\eta|^2\overline{\eta} + R_{03}\overline{\eta}^3. \tag{8.81}$$

Doing so, we obtain

$$\begin{aligned}\frac{d\xi}{dt} &= \frac{\partial\xi}{\partial\eta}\frac{d\eta}{dt} + \frac{\partial\xi}{\partial\overline{\eta}}\frac{d\overline{\eta}}{dt} \\ &= (\epsilon - i\omega)\eta + \{R_{30} + 3P_{30}(\epsilon - i\omega)\}\eta^3 + \{R_{21} + P_{21}(3\epsilon - i\omega)\}|\eta|^2\eta \\ &\quad + \{R_{12} + P_{12}(3\epsilon + i\omega)\}|\eta|^2\overline{\eta} + \{R_{03} + 3P_{03}(\epsilon + i\omega)\}\overline{\eta}^3.\end{aligned} \tag{8.82}$$

On the other hand, from Eqs. (8.79) and (8.80), we have

$$\begin{aligned}\frac{d\xi}{dt} &= (\epsilon - i\omega)\eta + \{P_{30}(\epsilon - i\omega) + \gamma'_{30}\}\eta^3 + \{P_{21}(\epsilon - i\omega) + \gamma'_{21}\}|\eta|^2\eta \\ &\quad + \{P_{12}(\epsilon - i\omega) + \gamma'_{12}\}|\eta|^2\overline{\eta} + \{P_{03}(\epsilon - i\omega) + \gamma'_{03}\}\overline{\eta}^3.\end{aligned} \tag{8.83}$$

Equating like powers of η in the cubic nonlinear terms of Eqs. (8.82) and (8.83) yields

$$R_{30} + 2P_{30}(\epsilon - i\omega) = \gamma'_{30}, \tag{8.84}$$

$$R_{21} + 2P_{21}\epsilon = \gamma'_{21}, \tag{8.85}$$

$$R_{12} + 2P_{12}(\epsilon + i\omega) = \gamma'_{12}, \tag{8.86}$$

$$R_{03} + 2P_{03}(\epsilon + 2i\omega) = \gamma'_{03}. \tag{8.87}$$

Setting

$$P_{30} = \frac{\gamma'_{30}}{2(\epsilon - i\omega)}, \tag{8.88}$$

$$P_{12} = \frac{\gamma'_{12}}{2(\epsilon + i\omega)}, \tag{8.89}$$

$$P_{03} = \frac{\gamma'_{03}}{2(\epsilon + 2i\omega)}, \tag{8.90}$$

eliminates R_{30}, R_{12}, and R_{03} in Eq. (8.81) . However, at the critical point $\epsilon = 0$, we cannot select P_{21} so that $R_{21} = 0$ because of Eq. (8.85). Therefore, the simplified equation in the neighborhood of $\epsilon = 0$ includes only the cubic nonlinear term proportional to $|\eta|^2\eta$ and is expressed as

$$\frac{d\eta}{dt} = (\epsilon - i\omega)\eta + R_{21}|\eta|^2\eta \tag{8.91}$$

or

$$\frac{d\eta}{dt} - (\epsilon - i\omega)\eta = R_{21}|\eta|^2\eta. \tag{8.92}$$

At the critical point $\epsilon = 0$, the linearized system is

$$\frac{d\eta}{dt} + i\omega\eta = 0 \tag{8.93}$$

and the solution is $\eta = Ae^{-i\omega t}$, where A is a complex number determined from the initial condition. Then, the right-hand side of Eq. (8.92) which is regarded as the external excitation to the linear system Eq. (8.93) is expressed as

$$R_{21}|\eta|^2\eta = R_{21}|A|^2Ae^{-i\omega t}. \tag{8.94}$$

The coefficient of t in the exponent of the exponential function is equal to the eigenvalue of Eq. (8.93), i.e. $-i\omega$. Therefore, $R_{21}|\eta|^2\eta$, which cannot be eliminated by the nonlinear coordinate transformation, is a resonance term. As well as the system in Section 8.2.2, the nonresonance nonlinear terms can be eliminated by the nonlinear transformation and the resonance term remains as an essential nonlinear term in the normal form.

We rewrite Eq. (8.91) into the real form by using the coordinate transformation

$$\eta = y_1 + iy_2, \tag{8.95}$$

where y_1 and y_2 are real. The complex coefficient R_{21} is expressed as

$$R_{21} = -\alpha_3 - i\beta_3. \tag{8.96}$$

Substituting Eqs. (8.95) and (8.96) into Eq. (8.91) and equating the real and imaginary parts on the left- and right-hand sides yields

$$\frac{dy_1}{dt} = \epsilon y_1 + \omega y_2 - \alpha_3(y_1^2 + y_2^2)y_1 + \beta_3(y_1^2 + y_2^2)y_2, \tag{8.97}$$

$$\frac{dy_2}{dt} = -\omega y_1 + \epsilon y_2 - \beta_3(y_1^2 + y_2^2)y_1 - \alpha_3(y_1^2 + y_2^2)y_2, \tag{8.98}$$

which correspond to Eq. (7.57). Therefore, Eq. (7.57) is a normal form of Hopf bifurcation, and the quadratic nonlinear terms do not affect the nonlinear dynamics.

References

Guckenheimer, J. and P. J. Holmes (1983). *Nonlinear oscillations, dynamical systems, and bifurcations of vector fields*. Springer.

Nayfeh, A. H. (1993). *Method of Normal Forms*. Wiley.

9

Method of Multiple Scales

Most differential equations governing the behaviors of various mechanical systems are nonlinear. If the behaviors are essentially related to the nonlinear components of systems, since the governing equations cannot be linearized, we have to employ numerical or analytically approximate approaches to solve the nonlinear equations. Numerical simulations, such Runge–Kutta method, finite element methods, to mention but a few, are widely applicable to multi-degree-of-freedom systems including infinite-degree-of-freedom systems with various types of nonlinearity. However, before performing the simulations, it is essential to numerically define the values of all system parameters as well as initial and boundary conditions when necessary, beforehand. Therefore, it is not straightforward to qualitatively grasp the features of the system dynamics.

On the other hand, the analytical approximate approaches are asymptotic methods which involve the use of small parameters in the governing equations, hence, such approaches may have a limited field of applications. However, these approaches can be performed by symbolic calculations and generally lead to solutions including symbolic expressions of the system parameters. Since the solutions are not limited to the specified values of all system parameters, initial and boundary conditions, they allow qualitative characterizations of the dynamics. In this chapter, the method of multiple time scales is introduced, which is well known as one of most effective analytical approximate approaches and is widely used for the analysis of nonlinear systems.

9.1 Spring-Mass System with Small Damping

The multiple scales method may be explained and fully understood through the application to a simple system which the exact solution is known. Let us consider the dimensionless equation of motion of a spring-mass system with small damping, whose exact solution is obtained in Chapter 2:

$$\frac{d^2x}{dt^2} + 2\gamma\frac{dx}{dt} + x = 0, \tag{9.1}$$

Linear and Nonlinear Instabilities in Mechanical Systems: Analysis, Control and Application, First Edition. Hiroshi Yabuno.

Companion website: www.wiley.com/go/yabuno/instabilitiesinmechanicalsystems

where $0 < \gamma \ll 1$. First, introducing an order parameter of $0 < \epsilon \ll 1$, we quantitatively express the smallness of the damping ratio γ as

$$\gamma = \epsilon\hat{\gamma} \quad (\hat{\gamma} = O(1)). \tag{9.2}$$

Then, Eq. (9.1) is rewritten as

$$\frac{d^2x}{dt^2} + 2\epsilon\hat{\gamma}\frac{dx}{dt} + x = 0. \tag{9.3}$$

By using ϵ, a uniform asymptotic expansion of the solution is sought in the form

$$x = x_0 + \epsilon x_1 + \epsilon^2 x_2 + \cdots . \tag{9.4}$$

The first term intends to capture the leading characteristic of the solution and has the leading order. The second one corrects the first term. After substituting Eq. (9.4) into Eq. (9.1) and equating coefficients of like powers of ϵ yields

$$O(\epsilon^0) : \frac{d^2x_0}{dt^2} + x_0 = 0, \tag{9.5}$$

$$O(\epsilon) : \frac{d^2x_1}{dt^2} + x_1 = -2\hat{\gamma}\frac{dx_0}{dt}. \tag{9.6}$$

Equation (9.5) expresses a harmonic oscillator system without damping and its natural frequency is 1. From Eq. (9.5), the solution is

$$x_0 = Ae^{it} + \bar{A}e^{-it}. \tag{9.7}$$

The complex amplitude A is constant and is expressed as

$$A = \frac{1}{2}ae^{i\phi}. \tag{9.8}$$

Then, Eq. (9.7) is

$$x_0 = a\cos(t + \phi). \tag{9.9}$$

Substituting Eq. (9.9) into the right-hand side of Eq. (9.6) yields

$$\frac{d^2x_1}{dt^2} + x_1 = 2\hat{\gamma}a\sin(t + \phi), \tag{9.10}$$

where a and ϕ are the amplitude and the initial phase, respectively, and determined from the initial displacement $x|_{t=0} = x(0)$ and velocity $\left.\frac{dx}{dt}\right|_{t=0} = v(0)$. This can be regarded as the equation of the harmonic oscillator with the unit natural frequency under the harmonic excitations expressed by the right-hand side. Because the excitation frequency is equal to the unit natural frequency, the resonance occurs as shown in Section C.1. In fact, a particular solution of Eq. (9.10) is

$$x_1 = -\hat{\gamma}at\cos(t + \phi), \tag{9.11}$$

where the homogeneous solution is not generally taken into account because it has the same form as Eq. (9.9). As a result, Eq. (9.4) is expressed as

$$\begin{aligned} x &= a\cos(t + \phi) - \epsilon\hat{\gamma}at\cos(t + \phi) + O(\epsilon^2) \\ &= a\cos(t + \phi) - \gamma at\cos(t + \phi) + O(\epsilon^2). \end{aligned} \tag{9.12}$$

Problem 9.1 Express x using the initial conditions $x|_{t=0} = x(0)$, $\left.\frac{dx}{dt}\right|_{t=0} = v(0)$.

Ans: We have the following equations about $a \cos\phi$ and $a \sin\phi$:

$$\begin{cases} a\cos\phi = x(0), & (9.13) \\ a\sin\phi = -v(0) - \gamma x(0). & (9.14) \end{cases}$$

Therefore, a and ϕ are expressed as

$$\begin{cases} a = \sqrt{x(0)^2 + \{v(0) + \gamma x(0)\}^2}, & (9.15) \\ \phi = -\arctan\dfrac{v(0) + \gamma x(0)}{x(0)}, & (9.16) \end{cases}$$

where the signs of $x(0)$ and $v(0) + \gamma x(0)$ determine the quadrant in which ϕ is located. By using Eqs. (9.13) and (9.14), Eq. (9.12) is also expressed as

$$x = (1 - \gamma t)\{x(0)\cos t + (v(0) + \gamma x(0))\sin t\} + O(\epsilon^2). \tag{9.17}$$

Since x_1 is proportional to time t, it grows unbounded with time. At $t = O(1/\epsilon)$, the second term of Eq. (9.4) becomes $O(1)$, which is the order of the first term of Eq. (9.4). Spending more time, at $t = O(1/\epsilon^2)$, the second term becomes $O(1/\epsilon)$, which exceeds the order of the first term. Hence, after $t = O(1/\epsilon)$, the second term is no longer of the order $O(\epsilon)$ and does not play the role to correct the first term. The assumption that the expansion of Eq. (9.4) should be uniformly valid is violated. Thus, the second term x_1, which is proportional to t and enforces the regular perturbation expansion of Eq. (9.4) to break down, is called a *secular term* (for example, (Nayfeh 1981,2008; Witelski and Bowen 2015)).

For example, in case of $\gamma = 0.1$, and initial conditions $x|_{t=0} = 1$and $\left.\frac{dx}{dt}\right|_{t=0} = 0$, we compare the exact solution and the approximate one obtained from Eq. (9.12). Their time histories are described as Figure 9.1, where the solid and dashed lines denote the exact and approximate solutions, respectively. The approximate solution deviates from the exact solution with time.

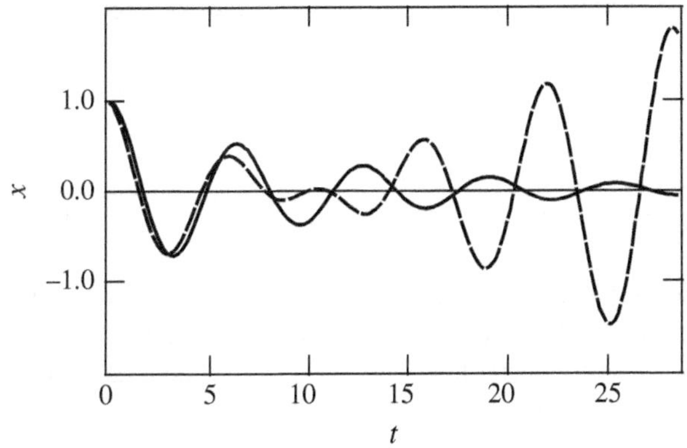

Figure 9.1 Time histories; the exact solution and the approximate solution obtained by the multiple time scales are described by solid and dashed lines, respectively.

9.2 Introduction of Multiple Time Scales

Let us further consider the characteristics of the dynamics governed by Eq. (9.3) to seek a method which overcomes the issue of the breakdown in the regular perturbation due to the secular term. If $\epsilon = 0$, Eq. (9.3) is reduced to

$$\frac{d^2x_0}{dt^2} + x_0 = 0, \tag{9.18}$$

where x_0 denotes x in case of $\epsilon = 0$. The solution of x_0 exhibits an undamped harmonic oscillation. Hence, x_0 is an element of the dynamics of x, which is determined by the balance of the first and the third terms in Eq. (9.3), i.e. the inertia and spring forces. The remaining term $2\epsilon\hat{\gamma}\frac{dx}{dt}$ seems to have no terms to be balanced in Eq. (9.3).

Let us seek the effect that can balance this small term. In addition to the original time $t \overset{\text{def}}{=} t_0$, we introduce another time scale $\epsilon t \overset{\text{def}}{=} t_1$. Because ϵ is any small value, t_1 is measured by the long unit time $1/\epsilon$. For example, setting $\epsilon = 1/60$, we can regard t_0 and t_1 as the time scales by the minute and hour hands in a clock, respectively (Nayfeh 1981). In this case, 60 (minutes) measured by t_0 is equal to one hour measured by t_1. By the chain rule, the first and second derivatives with respect to time t are

$$\left\{\begin{aligned}
\frac{d}{dt} &= \frac{\partial}{\partial t_0}\frac{dt_0}{dt} + \frac{\partial}{\partial t_1}\frac{dt_1}{dt} = \frac{\partial}{\partial t_0} + \epsilon\frac{\partial}{\partial t_1}, && (9.19)\\
\frac{d^2}{dt^2} &= \frac{d}{dt}\left(\frac{d}{dt}\right)\\
&= \frac{\partial}{\partial t_0}\left(\frac{\partial}{\partial t_0} + \epsilon\frac{\partial}{\partial t_1}\right)\frac{dt_0}{dt} + \frac{\partial}{\partial t_1}\left(\frac{\partial}{\partial t_0} + \epsilon\frac{\partial}{\partial t_1}\right)\frac{dt_1}{dt}\\
&= \frac{\partial^2}{\partial t_0^2} + 2\epsilon\frac{\partial^2}{\partial t_0\,\partial t_1} + \epsilon^2\frac{\partial^2}{\partial t_1^2}. && (9.20)
\end{aligned}\right.$$

Then, Eq (9.3) is rewritten by neglecting the term of $O(\epsilon^2)$ as

$$\frac{\partial^2 x}{\partial t_0^2} + 2\epsilon\frac{\partial^2 x}{\partial t_0\,\partial t_1} + 2\epsilon\hat{\gamma}\frac{\partial x}{\partial t_0} + x = 0, \tag{9.21}$$

where the first and second terms express the inertia force, and the third and fourth terms the damping and spring forces, respectively. Equating coefficients of like powers of ϵ, we have

$$\left\{\begin{aligned}
&O(\epsilon^0): \quad \frac{\partial^2 x}{\partial t_0^2} + x = 0, && (9.22)\\
&O(\epsilon): \quad \frac{\partial}{\partial t_1}\left(\frac{\partial x}{\partial t_0}\right) + \hat{\gamma}\frac{\partial x}{\partial t_0} = 0. && (9.23)
\end{aligned}\right.$$

These can be regarded as two "hierarchical levels" in the scale (Yoshida 2010). Equation (9.22) is equivalent to Eq. (9.18). At the first hierarchical level of $O(\epsilon^0)$, that is the *dominant level*, the balance of the spring force to the inertia force yields the undamped oscillation, which can be described with the original time scale $t = t_0$. On the other hand, Eq. (9.23) at the second hierarchical level of $O(\epsilon)$, *that is subdominant level*, indicates the balance of the damping force to the inertia one $2\epsilon\frac{\partial^2 x}{\partial t_0\,\partial t_1}$ and presents the characteristic of a slowly varying component which can be observed by using the slow time scale t_1. Hence, the change ratio of the velocity $\frac{dx}{dt}\left(\approx \frac{\partial x}{\partial t_0}\right)$ with respect to the slow time scale t_1, i.e. $\frac{\partial}{\partial t_1}\left(\frac{\partial x}{\partial t_0}\right)$, is proportional to the velocity itself with proportional constant $-\hat{\gamma}$. Therefore, it may be regarded as a method that avoids the secular term to introduce such an additional time scale.

The method based on the introduction of multiple time scales is well known as *the method of multiple scales*, which is an analytical way to take into account both the displacement-scale hierarchies as x_0 and ϵx_1 in the expanded solution and the time-scale hierarchies as $t_0 = t$ and $t_1 = \epsilon t$.

9.3 Method of Multiple Scales

We recall the problem of Eq. (9.1) and assume Eq. (9.4) as a uniform expansion of the solution. Introducing the fast and slow time scales, $t_0 = t$ and $t_1 = \epsilon t$, and substituting Eq. (9.4) into Eq. (9.1) yields

$$\epsilon^0\left(\frac{\partial^2 x_0}{\partial t_0^2} + x_0\right) + \epsilon\left\{\frac{\partial^2 x_1}{\partial t_0^2} + x_1 + 2\frac{\partial}{\partial t_1}\left(\frac{\partial x_0}{\partial t_0}\right) + 2\hat{\gamma}\frac{\partial x_0}{\partial t_0}\right\} = 0, \tag{9.24}$$

or

$$\begin{cases} O(\epsilon^0): & \dfrac{\partial^2 x_0}{\partial t_0^2} + x_0 = 0, & (9.25) \\ O(\epsilon): & \dfrac{\partial^2 x_1}{\partial t_0^2} + x_1 = -2\dfrac{\partial}{\partial t_1}\left(\dfrac{\partial x_0}{\partial t_0}\right) - 2\hat{\gamma}\dfrac{\partial x_0}{\partial t_0}, & (9.26) \end{cases}$$

where the first equation with respect to $O(\epsilon^0)$ in the first hierarchical level and the second equation with respect to $O(\epsilon)$ in the second hierarchical level denote the dominant and subdominant balances, respectively.

First, we solve Eq. (9.25) easily as

$$x_0 = A(t_1)e^{it_0} + \mathrm{CC}, \tag{9.27}$$

where CC denotes the complex conjugate of the preceding term and the complex amplitude A can be a function of t_1 because Eq. (9.25) is a partial differential equation with respect to time t_0 different from Eq. (9.5). Furthermore, A is expressed in the polar form as

$$A(t_1) = \frac{1}{2}a(t_1)e^{i\phi(t_1)}, \tag{9.28}$$

where a and ϕ are amplitude and phase angle, respectively, and can be a function of t_1. Then, Eq. (9.27) is

$$x_0 = a(t_1)\cos\{t_0 + \phi(t_1)\}. \tag{9.29}$$

Substituting x_0 into Eq. (9.26) yields

$$\frac{\partial^2 x_1}{\partial t_0^2} + x_1 = 2\left(\frac{\partial a}{\partial t_1} + \hat{\gamma}a\right)\sin(t_0 + \phi) + 2a\frac{\partial \phi}{\partial t_1}\cos(t_0 + \phi). \tag{9.30}$$

Again, we have a particular solution expressed by the secular term proportional to t_0 as

$$x_1 = -\left(\frac{\partial a}{\partial t_1} + \hat{\gamma}a\right)t_0\cos(t_0 + \phi) - a\frac{\partial \phi}{\partial t_1}t_0\sin(t_0 + \phi). \tag{9.31}$$

However, conversely to the case of the regular perturbation in Section 9.1, we can avoid the divergence of x_1 due to the secular term by demanding

$$\begin{cases} \dfrac{\partial a}{\partial t_1} + \hat{\gamma} a = 0, & (9.32) \\ a\dfrac{\partial \phi}{\partial t_1} = 0. & (9.33) \end{cases}$$

The condition to avoid the breakdown in the perturbation method is called a *solvability condition* for x_1 (see Appendix G). Their solutions are

$$\begin{cases} a = a_0 e^{-\hat{\gamma} t_1} = a_0 e^{-\gamma t}, & (9.34) \\ \phi = \phi_0, & (9.35) \end{cases}$$

where $t_1 = \epsilon t$ and $\hat{\gamma} = \gamma/\epsilon$ are used. Therefore, we obtain the approximate solution of Eq. (9.1) as

$$x = x_0 + O(\epsilon) = a_0 e^{-\gamma t} \cos(t + \phi_0) + O(\epsilon). \tag{9.36}$$

The time variation of the amplitude is detected with the slow time scale t_1. On the other hand, the phase angle is regarded as a constant and the time variation may be detected with a much slower time scale as $\epsilon^2 t \overset{\text{def}}{=} t_2$.

Equations (9.32) and (9.33), which are described at the second hierarchical level, clarify the characteristics of the solution x_0 at the first hierarchical level with respect to the slow time scales. Hence, the fast time variation of the dynamics is described at the first hierarchical level and the slow time variation cannot be clarified at this level, but at the higher level. Such a situation causes a kind of "resonance" and the break down in the regular perturbation. A method to overcome this issue is the method of multiple scales, which is one of the singular perturbation methods (for example, (Kevorkian and Cole 2012; Nayfeh 1981)).

9.4 Slow Time Scale Variation of Amplitude and Stability of Periodic Solutions

From Eq. (9.34), if γ is zero (no damping), because the amplitude is not varied even with slow time scale, the systems exhibits oscillation with constant amplitude. In case of $\gamma > 0$ (positive damping), since the amplitude is decreased according to Eq. (9.34), this periodic behavior decays with time as shown in Figure 9.2a, where the dashed line is the envelope expressed by Eq. (9.34). On the other hand, in case of $\gamma < 0$ (negative damping), since the amplitude is increased according to Eq. (9.34), the solution oscillatory grows without bound as shown in Figure 9.2b. From the perspective of determining the stability of periodic motion, the amplitude equation takes an important role, which is obtained by extracting the slow time dynamics from the system behavior.

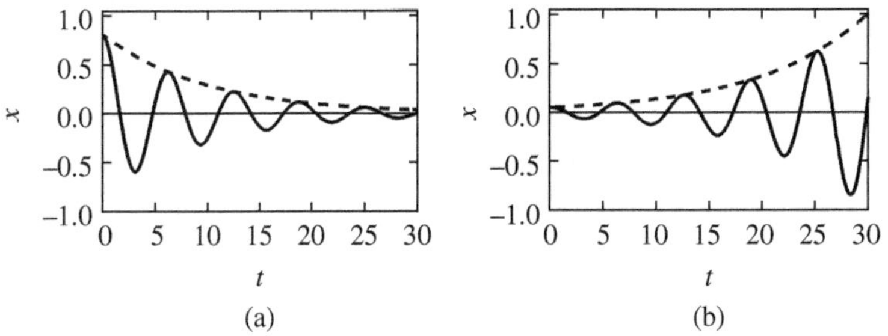

Figure 9.2 Time histories in cases of (a) positive damping $\gamma = 0.1$ and (b) negative damping $\gamma = -0.1$. The dashed line denotes the envelope denoted by slow time scale t_1 for the oscillation denoted by fast time scale t_0.

References

Kevorkian, J. K. and J. D. Cole (2012). *Multiple scale and singular perturbation methods*, Volume 114. Springer Science & Business Media.

Nayfeh, A. H. (1981). *Introduction to Perturbation Technique*. Wiley.

Nayfeh, A. H. (2008). *Perturbation methods*. John Wiley & Sons.

Witelski, T. and M. Bowen (2015). *Methods of mathematical modelling: Continuous systems and differential equations*. Springer.

Yoshida, Z. (2010). *Nonlinear science: the challenge of complex systems*. Springer Science & Business Media.

10

Nonlinear Characteristics of Dynamic Instability

The dynamic destabilization based on Hopf bifurcation by the change of the c' parameter value along arrow (2) in Figure 2.7 causes the amplitude growth, i.e. self-excited oscillation due to a negative damping effect. Also, dynamic destabilization based on Hopf bifurcation or Hamiltonian Hopf bifurcation by circulatory forces causes the growth as predicted in Section 3.2.4. The linear analyses in Sections 2.4.4 and 4.4.2 for these behaviors could not determine if the growth continues infinitely or not. In this chapter, taking into account the nonlinear characteristics of the system neglected in the linear analyses, we determine if the amplitude continues the growth according to the linear analytical results or different from that, becomes constant, i.e. steady state. For the systems whose qualitative characteristics are first determined by considering nonlinearity as introduced in literature (for example, Lacarbonara 2013; Nayfeh and Mook 2008; Shaw and Balachandran 2008; Thompson and Stewart 2002; Thomsen 2003; Troger and Steindl 2012), the nonlinear analysis is essential. Such systems include self-excited systems discussed in this chapter, parametrically excited systems in Chapter 11, and so on.

In this chapter, we perform nonlinear analysis for some self-excited systems by applying the method of multiple scales or seeking solvability condition and obtain the amplitude equations expressing the time variation of the complex amplitude. From the result, we obtain the magnitude of the steady-state amplitude and determine its stability.

Furthermore, we introduce an amplitude control method for self-excited systems by applying nonlinear feedback. This method can be applicable to the amplitude reduction control in which self-excitation is positively utilized, in particular, for measurement systems as the atomic force microscope (AFM) and mass sensors dealt in Chapters 13 and 14, respectively.

10.1 Effect of Nonlinearity on Dynamic Instability Due to Negative Damping Force

10.1.1 Cubic Nonlinear Damping (Rayleigh Type and van der Pol Type)

As seen from the investigation in Sections 2.6 and 9.4, a self-excited oscillation occurs in the linear system with negative damping. The linear system Eq. (9.1) analyzed in Sections 2.4.4

Linear and Nonlinear Instabilities in Mechanical Systems: Analysis, Control and Application, First Edition. Hiroshi Yabuno.

Companion website: www.wiley.com/go/yabuno/instabilitiesinmechanicalsystems

and 9.3 can be regarded as the linearized equation of the following rather general oscillator through Taylor expansion:

$$\frac{d^2x}{dt^2}+f_d\left(x,\frac{dx}{dt}\right)+f_s(x)=0, \tag{10.1}$$

where f_d is a nonlinear function with respect to x and $\frac{dx}{dt}$, and f_s is a nonlinear function with respect to x. In the case when the infinite growth is predicted in the linear theory, i.e. $\gamma < 0$, we have to recalculate the equation including some nonlinear terms truncated in the linear analysis but even in the case when the decrease of amplitude is predicted in the linear theory, i.e. $\gamma > 0$, we have to do that. The second and third terms show the damping and stiffness, respectively, and can be usually expressed by polynomials with the linear and symmetric cubic nonlinear terms through Taylor expansion. Here, we consider the following terms:

$$\begin{cases} f_d = 2\gamma\dfrac{dx}{dt}+\gamma_3\left(\dfrac{dx}{dt}\right)^3+\beta_3x^2\dfrac{dx}{dt}, & (10.2)\\ f_s = x+\alpha_3x^3, & (10.3)\end{cases}$$

where the first terms in Eqs. (10.2) and (10.3) are the linear damping and stiffness, respectively, which are already considered in Eq. (9.1). The second and third terms in Eq. (10.2) are the cubic nonlinearity of the damping and the amplitude dependent damping effect, respectively (for example, Polunin et al. 2016). The second term in Eq. (10.3) is the cubic stiffness term. The nonlinear terms, $\gamma_3\left(\frac{dx}{dt}\right)^3$, $\beta_3x^2\frac{dx}{dt}$, and α_3x^3, correspond to the nonlinearities included in Rayleigh's equation, van der Pol's equation, and Duffing's equation, respectively (for example, see Kovacic and Brennan 2011; Nayfeh and Mook 2008). These equations are as follows:

$$\text{Rayleigh's equation:}\quad \frac{d^2x}{dt^2}+x+2\gamma\frac{dx}{dt}+\gamma_3\left(\frac{dx}{dt}\right)^3=0, \tag{10.4}$$

$$\text{van der Pol's equation:}\quad \frac{d^2x}{dt^2}+x+2\gamma\frac{dx}{dt}+\beta_3x^2\frac{dx}{dt}=0, \tag{10.5}$$

$$\text{Duffing's equation:}\quad \frac{d^2x}{dt^2}+x+\alpha_3x^3=0. \tag{10.6}$$

10.1.2 Self-Excited Oscillation Produced Through Hopf Bifurcation

We analyze the following equation:

$$\frac{d^2x}{dt^2}+\left\{2\gamma+\gamma_3\left(\frac{dx}{dt}\right)^2+\beta_3x^2\right\}\frac{dx}{dt}+x+\alpha_3x^3=0. \tag{10.7}$$

First, focusing on the behavior in the neighborhood of the critical point of stability, $\gamma = 0$, which are obtained by the linear analysis, we assume the linear damping coefficient γ to be small and express γ as

$$\gamma = \epsilon\hat{\gamma}, \tag{10.8}$$

where $\hat{\gamma} = O(1)$. Considering that the balance between the linear and cubic nonlinear damping effects can produce the oscillation with steady-state amplitude, we assume the following uniform expansion of the solution (see Appendix D for detail):

$$x = \epsilon^{1/2}x_1 + \epsilon^{3/2}x_3 + \cdots . \tag{10.9}$$

Introducing the fast and slow time scales, $t_0 = t$ and $t_1 = \epsilon t$, and substituting Eq. (10.9) into Eq. (10.7) yields

$$\begin{cases} O(\epsilon^{1/2}):\ D_0^2x_1 + x_1 = 0, & (10.10) \\ O(\epsilon^{3/2}):\ D_0^2x_3 + x_3 = -2D_0D_1x_1 - 2\hat{\gamma}D_0x_1 - \gamma_3(D_0x_1)^3 - \alpha_3x_1^3 - \beta_3x_1^2D_0x_1, & (10.11) \end{cases}$$

where $D_0 = \frac{\partial}{\partial t_0}$, $D_0^2 = \frac{\partial^2}{\partial t_0^2}$, and $D_0D_1 = \frac{\partial^2}{\partial t_0 \partial t_1}$. The solution of Eq. (10.10) is

$$x_1 = A(t_1)e^{it_0} + \text{CC}, \tag{10.12}$$

where CC denotes the complex conjugate of the preceding term. Furthermore, without loss of generality, the complex amplitude A is expressed in the polar form as

$$A(t_1) = \frac{1}{2}\hat{a}(t_1)e^{i\phi(t_1)}. \tag{10.13}$$

Then, Eq. (10.12) becomes

$$x_1 = \hat{a}(t_1)\cos\{t_0 + \phi(t_1)\}, \tag{10.14}$$

and Eq. (10.9) becomes

$$x = a(t_1)\cos\{t_0 + \phi(t_1)\} + O(\epsilon^{3/2}), \tag{10.15}$$

where $a \overset{\text{def}}{=} \epsilon^{1/2}\hat{a}$. Because of $t_1 = \epsilon t$, Eq. (10.15) can be also expressed as

$$x = a(t)\cos\{t + \phi(t)\} + O(\epsilon^{3/2}). \tag{10.16}$$

Substituting Eq. (10.12) into the right-hand side of Eq. (10.11) yields

$$D_0^2x_3 + x_3 = -\{i(2D_1A + 2\hat{\gamma}A + \Gamma|A|^2A) + 3\alpha_3|A|^2A\}e^{it_0} + \text{CC} + \text{NST}, \tag{10.17}$$

where $\Gamma = 3\gamma_3 + \beta_3$, and CC and NST denote the complex conjugate of the preceding terms and the terms not to produce the secular term in x_3, that is, not proportional to e^{it_0} or e^{-it_0}. Therefore, the condition not to produce the secular term in x_3 is

$$D_1A = -\hat{\gamma}A - \frac{\Gamma}{2}|A|^2A + i\frac{3}{2}\alpha_3|A|^2A. \tag{10.18}$$

We want to know if the nontrivial steady state exists or not in the negative damping case ($\hat{\gamma} < 0$) and when there exists, to obtain the magnitude of the steady-state amplitude. The condition of the steady-state oscillation is that there is no time variation of a and ϕ, i.e. $D_1A = 0$. Under this condition, Eq. (10.18) leads to

$$0 = \left(\hat{\gamma} + \frac{\Gamma}{2}|A|^2\right)A + \frac{3}{2}\alpha_3|A|^2Ae^{-i\pi/2}. \tag{10.19}$$

From this equation, it is noticed that the term including Γ related to the nonlinear damping effects expressed by β_3 and γ_3 can balance the linear damping term related to $\hat{\gamma}$, and the nonlinear damping can play a role in suppressing the infinite growth of amplitude due to the negative linear damping effect. On the other hand, the cubic stiffness expressed by α_3

cannot balance the linear damping term, since the phase of this effect is different from that of the negative damping effect with $-\pi/2$ as seen from Eq. (10.19).

Next, we determine the stability of the steady-state amplitude using Eq. (10.18). Substituting Eq. (10.13) into Eq. (10.18) yields

$$\begin{cases} D_1\hat{a} = -\hat{\gamma}\hat{a} - \dfrac{\Gamma}{8}\hat{a}^3, & (10.20) \\ \hat{a}D_1\phi = \dfrac{3}{8}\alpha_3\hat{a}^3. & (10.21) \end{cases}$$

These are the equations governing the slow time variation of the amplitude and phase, respectively, which are expressed by the slow time scale t_1. Equivalently, these are rewritten as

$$\begin{cases} \dfrac{da}{dt} = -\gamma a - \dfrac{\Gamma}{8}a^3, & (10.22) \\ a\dfrac{d\phi}{dt} = \dfrac{3}{8}\alpha_3 a^3. & (10.23) \end{cases}$$

Letting $\frac{da}{dt} = 0$, i.e. giving the condition that the amplitude is not varied with time, yields the equilibrium equation for the steady-state amplitude as

$$a_{st}\left(\gamma + \frac{\Gamma}{8}a_{st}^2\right) = 0. \tag{10.24}$$

As predicted from Eq. (10.19), it is seen that the response amplitude is governed by the linear and nonlinear effects, γ and Γ, independent of the nonlinear stiffness α_3. The steady-state amplitudes are

$$\begin{cases} a_{st} = 0, & (10.25) \\ a_{st} = \sqrt{-\dfrac{8\gamma}{\Gamma}}. & (10.26) \end{cases}$$

In case when we do not consider the nonlinear damping terms in Eq. (10.7) ($\beta_3 = \gamma_3 = 0$, i.e. $\Gamma = 0$), from Eq. (10.24), we have only the trivial steady-state amplitude $a_{st} = 0$ and cannot find the nontrivial steady-state amplitude of Eq. (10.26). The trivial steady-state amplitude of Eq. (10.25) exists for any γ, but the nontrivial steady-state amplitude exits only in the parameter range of $\gamma\Gamma < 0$.

Next, let us determine the stability of the steady-state amplitudes. Substituting

$$a(t) = a_{st} + \Delta a(t) \tag{10.27}$$

into Eq. (10.22) and taking the linear term with respect to Δa yields

$$\frac{d\Delta a}{dt} = -\left(\gamma + \frac{3\Gamma}{8}a_{st}^2\right)\Delta a. \tag{10.28}$$

First, substituting the trivial steady-state amplitude $a_{st} = 0$ into Eq. (10.28) yields

$$\frac{d\Delta a}{dt} = -\gamma\Delta a. \tag{10.29}$$

Therefore, as predicted in the linear theory, the trivial steady-state amplitude $a_{st} = 0$ is stable and unstable when $\gamma > 0$ (positive damping) and $\gamma < 0$ (negative damping), respectively.

On the other hand, substituting $a_{st} = \sqrt{-\frac{8\gamma}{\Gamma}}$ into Eq. (10.28) yields

$$\frac{d\Delta a}{dt} = 2\gamma\Delta a. \tag{10.30}$$

Therefore, the nontrivial steady-state amplitudes existing in $\gamma > 0$ and $\gamma < 0$ are unstable and stable, respectively.

Using the above result, we can describe the bifurcation diagrams for the cases of $\Gamma > 0$ and $\Gamma < 0$ as Figure 10.1a,b, respectively. In both cases, like the linear analytical result, the trivial steady-state amplitude is stable and unstable, in the positive damping $\gamma > 0$ and in the negative damping $\gamma < 0$, respectively. The unstable trivial steady-state amplitude indicates the production of the self-excited oscillation due to the negative damping mentioned in Section 2.6.

On the other hand, there is qualitatively difference between the nontrivial steady states depending on the sign of Γ. In the case of $\Gamma > 0$, the stable nontrivial steady-state amplitude exists in the region of $\gamma < 0$ where the trivial steady state is unstable due to the self-excited oscillation, but the response amplitude does not grow infinitely and converges to the constant stable steady-state amplitude as Figure 10.2a. Even if the initial amplitude is larger than the stable nontrivial steady-state amplitude, the amplitude converges to the steady-state amplitude as Figure 10.2b. For the stable steady-state amplitude, ϕ is expressed from Eq. (10.23) as

$$\phi = \frac{3}{8}\alpha_3 a_{st}^2 t + \phi_0, \tag{10.31}$$

where ϕ_0 is an integral constant and corresponds to the phase angle. Therefore, the stable steady-state oscillation with the stable steady-state amplitude is expressed from Eq. (10.16) as

$$x = a_{st}\cos(\omega_n t + \phi_0) + O(\epsilon^{3/2}), \tag{10.32}$$

where

$$\omega_n = 1 + \frac{3}{8}\alpha_3 a_{st}^2. \tag{10.33}$$

The response frequency ω_n is deviated from the linear natural frequency 1 by the effect of nonlinear stiffness α_3 and depends on the response amplitude a_{st}: in case of $\alpha_3 > 0$ (hard

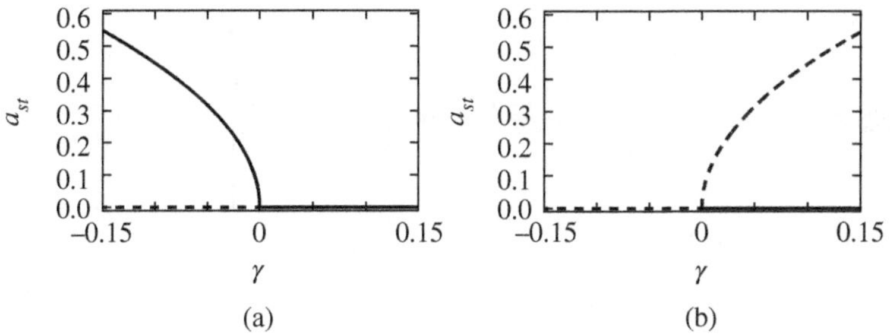

Figure 10.1 Bifurcation diagrams of nonlinear self-excited oscillation produced through Hopf bifurcation, where the solid and dashed lines denote stable and unstable steady-state amplitudes, respectively: (a) supercritical Hopf bifurcation ($\Gamma = 4 > 0$); (b) subcritical Hopf bifurcation ($\Gamma = -4 < 0$).

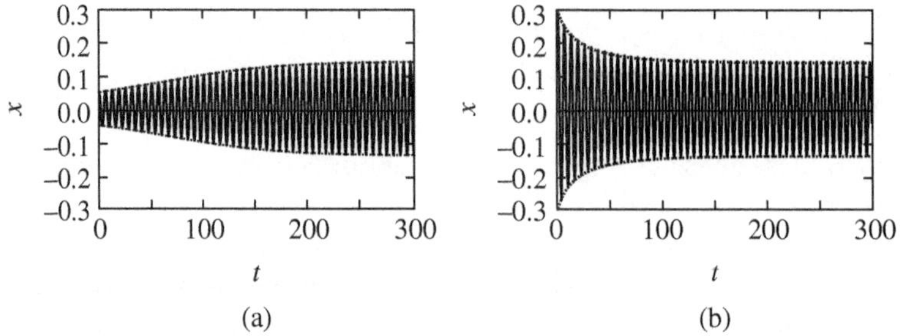

Figure 10.2 Time histories in the case when the nontrivial stable steady-state amplitude exists due to supercritical Hopf bifurcation: $\gamma = -0.01, \Gamma = 4, \alpha_3 = 1$. (a) $x(0) = 0.05$ and $\frac{dx}{dt} = 0$, (b) $x(0) = 0.3$ and $\frac{dx}{dt} = 0$. The time history is obtained directly from Eq. (10.7) by numerical integration and the envelope is obtained from Eq. (10.22).

spring), the response frequency increases as the response amplitude becomes larger; in case of $\alpha_3 < 0$ (soft spring), the response frequency increases as the response amplitude becomes smaller. This agrees with the nonlinear characteristic of the natural frequency of the Duffing oscillator expressed by Eq. (G.45).

In the case of $\Gamma < 0$, the bifurcation diagram is Figure 10.1b and the unstable nontrivial steady-state amplitude exists in the range of $\gamma > 0$ where the trivial steady state is stable in a linear sense. If the disturbance is smaller than the unstable steady-state amplitude, since the disturbance decays as Figure 10.3a, the behavior results in that obtained by the linear analysis. On the other hand, if the disturbance is larger than the unstable steady-state amplitude, the response grows with time as Figure 10.3b. In the case of $\gamma < 0$, the response grows with time as the linear system with negative damping independent of the magnitude of disturbance.

By the way, Eq. (10.16) and its time derivative obtained from Eq. (10.15) by neglecting the derivative the slow time scale t_1, i.e. $\frac{dx}{dt} = D_0x + \epsilon D_1 + \cdots \approx D_0x$, are expressed as

$$\begin{cases} x = a(t)\cos\Phi(t) + O(\epsilon^{3/2}), & (10.34) \\ \dfrac{dx}{dt} = a(t)\sin\Phi(t) + O(\epsilon^{3/2}), & (10.35) \end{cases}$$

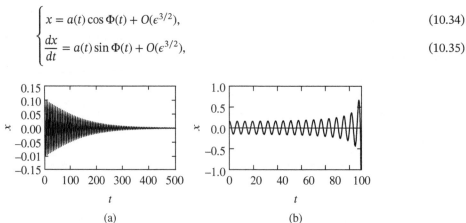

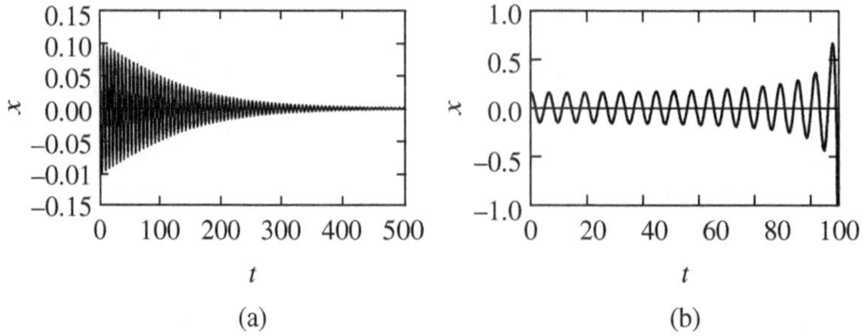

Figure 10.3 Time histories in the case when the nontrivial unstable steady-state amplitude exists due to subcritical Hopf bifurcation: $\gamma = 0.01, \Gamma = -4, \alpha_3 = 1$. (a) $x(0) = 0.1$ and $\frac{dx}{dt} = 0$, (b) $x(0) = 0.16$ and $\frac{dx}{dt} = 0$. The time history is obtained directly from Eq. (10.7).

where $\Phi(t) = -t - \phi(t)$ and from Eqs. (10.22) and (10.23), the time variations of $a(t)(\neq 0)$ and Φ are governed respectively by

$$\begin{cases} \dfrac{da}{dt} = -\gamma a - \dfrac{\Gamma}{8} a^3, & (10.36) \\ \dfrac{d\Phi}{dt} = -\left(1 + \dfrac{3}{8}\alpha_3 a^2\right). & (10.37) \end{cases}$$

It is noticed that Eqs. (10.34) and (10.35) is equivalent to Eqs. (7.58) and (7.59), respectively, and Eqs. (10.36) and (10.37) are equivalent to Eqs.(7.62) and (7.63), respectively. In the case of $\Gamma > 0$ (supercritical Hopf bifurcation), the phase portraits in cases of $\gamma > 0$ and $\gamma < 0$ are Figure 7.6a,b, respectively. Even if the damping is negative, the response amplitude does not grow infinitely, but converges to the stable steady-state amplitude corresponding to the radius of the stable limit cycle in Figure 7.6b.

In the case of $\Gamma < 0$ (subcritical Hopf bifurcation), the phase portraits in cases of $\gamma > 0$ and $\gamma < 0$ are reported in Figure 7.7a,b, respectively. Even if the damping is positive, the response amplitude grows infinitely in the case when the disturbance is so large that the initial amplitude is greater than the unstable steady-state amplitude corresponding to the radius of the unstable limit cycle in Figure 7.7a.

10.1.3 Self-Excited Oscillation by Linear Feedback and Its Amplitude Control by Nonlinear Feedback

The nonlinear dynamics of van der Pol oscillator can be utilized to the amplitude control of self-excited resonators in an AFM mentioned in Chapter 13 and in a mass sensor mentioned in Chapter 14. In this section, we propose a linear-plus-nonlinear feedback control to change a linear damped oscillator to van der Pol oscillator. We consider a damped cantilever-type resonator shown in Figure 10.4a, which is usually utilized for micro electro mechanical systems (MEMS) sensors, such as AFM, mass sensor, viscometer, and so on. In the case when only the first mode is self-excited, the behavior corresponds to the single-degree-of-freedom damped oscillator shown in Figure 10.4b. We apply the displacement input Δx for the control at the supporting point as shown in Figure 10.5a, which corresponds to the system schematized in Figure 10.5b. The equation of motion of the damped oscillator shown in Figure 10.4b is

$$m\frac{d^2x}{dt^2} + c\frac{dx}{dt} + kx = 0. \tag{10.38}$$

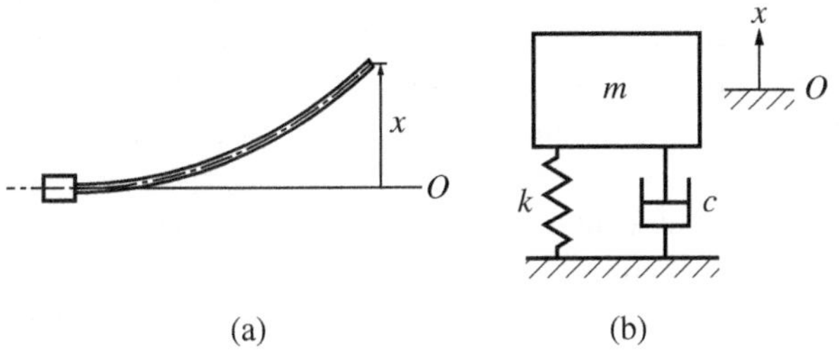

Figure 10.4 Cantilever and single-degree-of-freedom model.

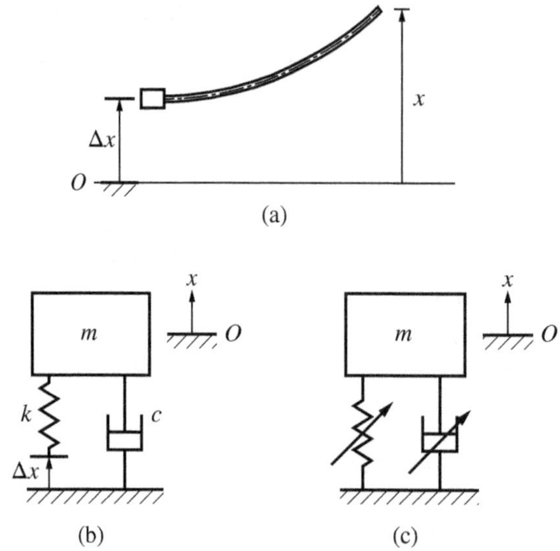

Figure 10.5 Feedback Control system to realize van der Pol oscillator.

Under the control input Δx as Figure 10.5b, this equation of motion is changed to

$$m\frac{d^2x}{dt^2} + c\frac{dx}{dt} + k(x - \Delta x) = 0. \tag{10.39}$$

Setting Δx by the following linear-plus-nonlinear feedback:

$$\Delta x = \frac{c_{lin}}{k}\frac{dx}{dt} + \frac{c_{non}}{k}x^2\frac{dx}{dt}, \tag{10.40}$$

we have the modified equation of motion as

$$m\frac{d^2x}{dt^2} + kx + (c - c_{lin})\frac{dx}{dt} - c_{non}x^2\frac{dx}{dt} = 0. \tag{10.41}$$

By the following transformation:

$$x^* = \frac{x}{L}, \quad t^* = \sqrt{\frac{k}{m}}t, \quad 2\gamma = \frac{c - c_{lin}}{\sqrt{mk}}, \quad \beta_3 = -\frac{c_{non}x_{st}^2}{\sqrt{mk}}, \tag{10.42}$$

where L is a representative length, this system is written as

$$\frac{d^2x^*}{dt^{*2}} + x^* + 2\gamma\frac{dx^*}{dt^*} + \beta_3 x^{*2}\frac{dx^*}{dt^*} = 0, \tag{10.43}$$

which is equivalent to van der Pol's equation (10.5). Therefore, we can produce the dynamics of van der Pol oscillator for the damped oscillator system by linear plus nonlinear feedback. In particular, as seen from Section 10.1.2, when we set the linear feedback gain c_{lin} so that γ is negative, the self-excited oscillation is produced. Furthermore, by setting the positive nonlinear feedback gain c_{non}, i.e. $\beta_3 > 0$ so that the nonlinear characteristic becomes supercritical, the self-excited oscillation shows the nontrivial steady-state amplitude and the high gain nonlinear feedback realizes small steady-state amplitude; small steady-state amplitude can be widely utilized for the measurement methods based on self-excited oscillation as in the case of noncontact AFM in Chapter 13, mass sensing in Chapter 14, and so on.

10.2 Effect of Nonlinearity on Dynamic Instability Due to Circulatory Force

In this section, we analyze the nonlinear effect on the self-excited oscillation due to circulatory force. The linear characteristics have been investigated in Chapter 3. Unlike the self-excited oscillation due to the negative damping, the resonance occurs only in multi-degree-of-freedom systems. The nonsymmetric characteristic of matrix K' expressing the positional force in Eq. (3.1) is necessary for the self-excited oscillation. In this section, we show how to apply the method of multiple scales on the nonlinear analysis for a two-degree-of-freedom system with nonsymmetric K'. To obtain the dependence of the complex amplitude on the slow time scale, we seek a solvability condition for the higher-order problem with nonlinear effects. Appendix G summarizes an elementary way to solve such problem highlighting the fundamental concepts.

10.2.1 Derivation of Amplitude Equations by Solvability Condition

We consider the following equations of a railway vehicle wheelset:

$$\begin{cases} \ddot{y} + \dfrac{d_{11}}{v^*}\dot{y} + k_{11}y + k_{12}\psi + \alpha y^3 = 0, & (10.44) \\ \ddot{\psi} + \dfrac{d_{22}}{v^*}\dot{\psi} + k_{21}y + \psi = 0, & (10.45) \end{cases}$$

where y and ψ describe the lateral and yaw motions, respectively. In contrast with Eqs. (6.40) and (6.41), we take into account only the cubic nonlinearity of the restoring force in the lateral direction as a representative nonlinear force. We set

$$v^* = v^*_{cr}(1 + \Delta v), \tag{10.46}$$

where v^*_{cr} is the critical speed indicated by point (5) in Figure 6.12, at which two of the four eigenvalues are a complex conjugate pair of pure imaginary and Δv is a deviation of velocity from the critical running speed.

Assuming $|\Delta v| \ll 1$, we express $\frac{d_{11}}{v^*}$ and $\frac{d_{22}}{v^*}$ as

$$\frac{d_{11}}{v^*} = \frac{d_{11}}{v^*_{cr}}(1 - \Delta v) + O(\Delta v^2), \quad \frac{d_{22}}{v^*} = \frac{d_{22}}{v^*_{cr}}(1 - \Delta v) + O(\Delta v^2). \tag{10.47}$$

Using an order parameter $\epsilon (0 < \epsilon \ll 1)$, we set y and ψ as

$$\begin{cases} y = y_0 + \epsilon y_1 + O(\epsilon^2), & (10.48) \\ \psi = \psi_0 + \epsilon\psi_1 + O(\epsilon^2). & (10.49) \end{cases}$$

Also, we set the order of Δv and α as

$$\Delta v = \epsilon \Delta\hat{v}, \quad \alpha = \epsilon\hat{\alpha}. \tag{10.50}$$

Introducing multiple time scales:

$$t_0 = t, \quad t_1 = \epsilon t, \tag{10.51}$$

we can express the time derivative as follows:

$$\frac{d}{dt} = D_0 + \epsilon D_0, \tag{10.52}$$

$$\frac{d^2}{dt^2} = D_0^2 + 2\epsilon D_1 D_0 + O(\epsilon^2), \tag{10.53}$$

where $D^n = \frac{\partial^n}{\partial t^n}$ and the term of higher order $\epsilon^2 D_1^2$ is neglected. Here, we focus on the mode corresponding to the eigenvalue $i\omega$ at the critical running speed v_{cr}. Then, we set y_0 and ψ_0 as

$$\begin{bmatrix} y_0 \\ \psi_0 \end{bmatrix} = A(t_1) e^{i\omega t_0} \mathbf{\Phi}_0 + \text{CC}. \tag{10.54}$$

As Eq. (6.49), $\mathbf{\Phi}_0$ is the same reported in Section 6.3, which is expressed as

$$\mathbf{\Phi}_0 = \begin{bmatrix} k_{12} \\ \omega^2 - k_{11} - i\dfrac{d_{11}}{v_{cr}^*}\omega \end{bmatrix} \tag{10.55}$$

and satisfies

$$L(v_{cr}^*)\mathbf{\Phi}_0 = \mathbf{0}, \tag{10.56}$$

where $L(v_{cr}^*)$ is expressed by Eq. (6.48), i.e.

$$L(v_{cr}^*) = \begin{bmatrix} k_{11} - \omega^2 + i\dfrac{d_{11}}{v_{cr}^*}\omega & k_{12} \\ k_{21} & 1 - \omega^2 + i\dfrac{d_{22}}{v_{cr}^*}\omega \end{bmatrix}, \tag{10.57}$$

and ω is given by Eq. (6.54). In addition, we set y_1 and ψ_1 as

$$\begin{bmatrix} y_1 \\ \psi_1 \end{bmatrix} = e^{i\omega t_0} \mathbf{\Phi}_1(t_1) + \text{CC}. \tag{10.58}$$

We substitute Eqs. (10.48) and (10.49) into Eqs. (10.44) and (10.45) and pick up all terms including $e^{i\omega t_0}$. Considering Eqs. (10.50)–(10.53) and equating the coefficients of like powers of ϵ to zero yields the following equations for $O(\epsilon^0)$ and $O(\epsilon)$:

$$O(\epsilon^0) : L(v_{cr}^*)\mathbf{\Phi}_0 A = \mathbf{0}, \tag{10.59}$$

$$O(\epsilon) : L(v_{cr}^*)\mathbf{\Phi}_1 = -(2i\omega I + C_c)\mathbf{\Phi}_0 D_1 A + i\omega \Delta\hat{v} C_c \mathbf{\Phi}_0 A - 3\hat{\alpha} k_{12}^3 |A|^2 A \begin{bmatrix} 1 \\ 0 \end{bmatrix}, \tag{10.60}$$

where C_c is

$$C_c = \begin{bmatrix} \dfrac{d_{11}}{v_{cr}^*} & 0 \\ 0 & \dfrac{d_{22}}{v_{cr}^*} \end{bmatrix}. \tag{10.61}$$

First, we analyze Eq. (10.59). By expressing $A(t_1)$ in the polar form as

$$A(t_1) = \frac{1}{2}a(t_1)e^{i\phi(t_1)}, \tag{10.62}$$

$[y_0 \;\; \psi_0]^T$ is expressed as

$$\begin{bmatrix} y_0 \\ \psi_0 \end{bmatrix} = a(t_1)\cos\{\omega t_0 + \phi(t_1)\}\mathbf{\Phi}_0. \tag{10.63}$$

Next, we derive the equations governing the slow time variation of complex amplitude A or $a(t_1)$ and $\phi(t_1)$ from the solvability condition of Φ_1 (see Appendixes F and G). Since $L(v_{cr}^*)$ is nonself-adjoint, i.e. $L(v_{cr}^*) \neq \overline{L}^T(v_{cr}^*)$, we introduce an adjoint linear operator $\overline{L}^T(v_{cr}^*)$ of $L(v_{cr}^*)$

$$\overline{L}^T(v_{cr}^*) = \begin{bmatrix} k_{11} - \omega^2 - i\dfrac{d_{11}}{v_{cr}^*}\omega & k_{21} \\ k_{12} & 1 - \omega^2 - i\dfrac{d_{22}}{v_{cr}^*}\omega \end{bmatrix} \tag{10.64}$$

and define the adjoint vector $\mathbf{\Xi}_0$ associated with $\mathbf{\Phi}_0$ so as to satisfy

$$\overline{L}^T(v_{cr}^*)\mathbf{\Xi}_0 = \mathbf{0} \quad \text{or} \quad \overline{\mathbf{\Xi}}_0^T L(v_{cr}^*) = \mathbf{0}, \tag{10.65}$$

i.e.

$$\mathbf{\Xi}_0 = \begin{bmatrix} -1 + \omega^2 + i\dfrac{d_{22}}{v_{cr}^*}\omega \\ k_{12} \end{bmatrix}. \tag{10.66}$$

We take the inner product of Eq. (10.60) and $\mathbf{\Xi}_0$ or multiply the both sides of Eq. (10.60) by $\overline{\mathbf{\Xi}}_0^T$ from the left. Then, the left-hand side of the result becomes zero, and each term on the right-hand side is calculated as

$$\begin{aligned}
-\overline{\mathbf{\Xi}}_0^T(2i\omega I + C_c)\mathbf{\Phi}_0 D_1 A &= \Bigg[\frac{-3\omega^2 k_{12}(d_{11}+d_{22}) + k_{12}(d_{11}+d_{22}k_{11})}{v_{cr}^*} \\
&\quad -2\omega k_{12}\left(2\omega^2 - k_{11} - 1 - \frac{d_{11}d_{22}}{v_{cr}^{*2}}\right)i\Bigg] D_1 A \\
&= \Bigg[-\frac{2(d_{11}+d_{22}k_{11})k_{12}}{v_{cr}^*} + \frac{2k_{12}\sqrt{d_{11}+d_{22}k_{11}}}{(d_{11}+d_{22})^{3/2}v_{cr}^{*2}} \\
&\quad \times \{d_{11}^2 d_{22} - d_{22}(k_{11}-1)v_{cr}^{*2} + d_{11}(d_{22}^2 + (k_{11}-1)v_{cr}^{*2})\}i\Bigg] D_1 A,
\end{aligned}$$

$$\begin{aligned}
i\omega\hat{v}\overline{\mathbf{\Xi}}_0^T C_c \mathbf{\Phi}_0 A &= \Bigg[\frac{2\omega^2 k_{12} d_{11} d_{22}}{v_{cr}^{*2}} \\
&\quad + \frac{\omega k_{12}\{\omega^2(d_{11}+d_{22}) - d_{11} - d_{22}k_{11}\}}{v_{cr}^*}i\Bigg] \Delta\hat{v}A \\
&= \frac{2d_{11}d_{22}(d_{11}+d_{22}k_{11})k_{12}}{(d_{11}+d_{22})v_{cr}^{*2}}\Delta\hat{v}A,
\end{aligned}$$

$$-3\hat{\alpha}k_{12}^3|A|^2A\boldsymbol{\Xi}_0^T\begin{bmatrix}1\\0\end{bmatrix} = 3k_{12}^3\left(1-\omega^2+i\frac{\omega d_{22}}{v_{cr}^*}\right)\hat{\alpha}|A|^2A$$

$$= \left\{-\frac{3d_{22}(k_{11}-1)k_{12}^3}{d_{11}+d_{22}} + \frac{3d_{22}k_{12}^3}{v_{cr}^*}\sqrt{\frac{d_{11}+d_{22}k_{11}}{d_{11}+d_{22}}}i\right\}\hat{\alpha}|A|^2A.$$

As a result, the solvability condition of $\boldsymbol{\Phi}_1$ is

$$PD_1A + Q\Delta\hat{v}A + R\hat{\alpha}|A|^2A = 0, \tag{10.67}$$

or

$$D_1A + \frac{Q\overline{P}}{|P|^2}\Delta\hat{v}A + \frac{R\overline{P}}{|P|^2}\hat{\alpha}|A|^2A = 0, \tag{10.68}$$

where

$$P = -\frac{2(d_{11}+d_{22}k_{11})k_{12}}{v_{cr}^*} + \frac{2k_{12}\sqrt{d_{11}+d_{22}k_{11}}}{(d_{11}+d_{22})^{3/2}v_{cr}^{*2}}$$
$$\times[d_{11}^2d_{22} - d_{22}(k_{11}-1)v_{cr}^{*2} + d_{11}\{d_{22}^2 + (k_{11}-1)v_{cr}^{*2}\}]i, \tag{10.69}$$

$$Q = \frac{2d_{11}d_{22}(d_{11}+d_{22}k_{11})k_{12}}{(d_{11}+d_{22})v_{cr}^{*2}}, \tag{10.70}$$

$$R = -\frac{3d_{22}(k_{11}-1)k_{12}^3}{d_{11}+d_{22}} + \frac{3d_{22}k_{12}^3}{v_{cr}^*}\sqrt{\frac{d_{11}+d_{22}k_{11}}{d_{11}+d_{22}}}i. \tag{10.71}$$

Substituting Eq. (10.62) into Eq. (10.68) and separating the real and imaginary parts yields

$$\begin{cases} D_1a = -\dfrac{\text{Re}\,(Q\overline{P})}{|P|^2}\Delta\hat{v}a - \dfrac{\text{Re}\,(R\overline{P})}{4|P|^2}\hat{\alpha}a^3, & (10.72)\\[2ex] aD_1\phi = -\dfrac{\text{Im}(Q\overline{P})}{|P|^2}\Delta\hat{v}a - \dfrac{\text{Im}(R\overline{P})}{4|P|^2}\hat{\alpha}a^3, & (10.73)\end{cases}$$

or equivalently

$$\begin{cases} \dfrac{da}{dt} = -\dfrac{\text{Re}\,(Q\overline{P})}{|P|^2}\Delta va - \dfrac{\text{Re}\,(R\overline{P})}{4|P|^2}\alpha a^3, & (10.74)\\[2ex] a\dfrac{d\phi}{dt} = -\dfrac{\text{Im}(Q\overline{P})}{|P|^2}\Delta va - \dfrac{\text{Im}(R\overline{P})}{4|P|^2}\alpha a^3, & (10.75)\end{cases}$$

where

$$\text{Re}(Q\overline{P}) = -\frac{4d_{11}d_{22}(d_{11}+d_{22}k_{11})^2k_{12}^2}{(d_{11}+d_{22})v_{cr}^3} < 0, \tag{10.76}$$

$$\text{Re}(R\overline{P}) = \frac{6d_{11}d_{22}k_{12}^4(d_{11}+d_{22}k_{11})\{d_{11}d_{22}+d_{22}^2+2(k_{11}-1)v_{cr}^2\}}{(d_{11}+d_{22})^2v_{cr}^3}, \tag{10.77}$$

$$\text{Im}(Q\overline{P}) = -\frac{4d_{11}d_{22}k_{12}^2}{(d_{11}+d_{22})v_{cr}^4}\left(\frac{d_{11}+d_{22}k_{11}}{d_{11}+d_{22}}\right)^{3/2}$$

$$\times [d_{11}^2 d_{22} - d_{22}(k_{11} - 1)v_{cr}^2 + d_{11}\{d_{22}^2 + (k_{11} - 1)v_{cr}^2\}], \tag{10.78}$$

$$\begin{aligned}\mathrm{Im}(R\overline{P}) = &-\frac{6d_{22}k_{12}^4}{(d_{11}+d_{22})^2 v_{cr}^2}\sqrt{\frac{d_{11}+d_{22}k_{11}}{d_{11}+d_{22}}}\\ &\times \left[(d_{11}+d_{22}k_{11})(d_{11}+d_{22})^2 - d_{11}^2 d_{22}\right.\\ &\left.+d_{22}(k_{11}-1)v_{cr}^2 - d_{11}\{d_{22}^2 + (k_{11}-1)v_{cr}^2\}\right].\end{aligned} \tag{10.79}$$

10.2.2 Effect of Cubic Nonlinear Stiffness on Steady-State Response

Using a and ϕ, whose time variation is governed with Eqs. (10.74) and (10.75), y and ψ are expressed as

$$\begin{aligned}\begin{bmatrix} y \\ \psi \end{bmatrix} &= \begin{bmatrix} y_0 \\ \psi_0 \end{bmatrix} + O(\epsilon) = \frac{a(t)}{2}\mathrm{e}^{i(\omega t+\phi(t))}\Phi_0 + \mathrm{CC} + O(\epsilon)\\ &= \begin{bmatrix} a(t)k_{12}\cos\{\omega t + \phi(t)\} \\ a(t)(\omega^2 - k_{11})\cos\{\omega t + \phi(t)\} + a(t)\dfrac{d_{11}\omega}{v_{cr}}\sin\{\omega t + \phi(t)\} \end{bmatrix} + O(\epsilon).\end{aligned} \tag{10.80}$$

Let the steady-state amplitudes under the condition $\frac{da}{dt} = 0$ for Eq. (10.74) be $a \stackrel{\text{def}}{=} a_{st}$, which satisfies the following equilibrium equation:

$$0 = \{4\mathrm{Re}\,(Q\overline{P})\Delta v + \mathrm{Re}\,(R\overline{P})\alpha a_{st}^2\}a_{st}. \tag{10.81}$$

There are the following trivial and nontrivial steady states:

$$a_{st} = 0, \tag{10.82}$$

$$a_{st} = 2\sqrt{-\frac{\mathrm{Re}(Q\overline{P})\Delta v}{\mathrm{Re}(R\overline{P})\alpha}}. \tag{10.83}$$

In order to determine the stability of the steady-state amplitude, substituting

$$a(t) = a_{st} + \Delta a(t) \tag{10.84}$$

into Eq. (10.74) and taking only the linear term with respect to Δa, we obtain

$$\frac{d\Delta a}{dt} = -\left\{\frac{\mathrm{Re}(Q\overline{P})}{|P|^2}\Delta v + \frac{3\mathrm{Re}(R\overline{P})}{4|P|^2}\alpha a_{st}^2\right\}\Delta a. \tag{10.85}$$

For the trivial steady state Eq. (10.82), this equation leads to

$$\frac{d\Delta a}{dt} = -\frac{\mathrm{Re}(Q\overline{P})}{|P|^2}\Delta v \Delta a. \tag{10.86}$$

Then, the solution Δa is expressed as

$$\Delta a = c\,\exp\left(-\frac{\mathrm{Re}(Q\overline{P})\Delta v}{|P|^2}t\right), \tag{10.87}$$

where c is the integral constant that is determined by the initial condition. Therefore, when Δv is negative or positive, the trivial steady-state amplitude is stable or unstable, respectively, because of $\mathrm{Re}(Q\overline{P}) < 0$.

For the nontrivial steady state Eq. (10.83), Eq. (10.85) leads to

$$\frac{d\Delta a}{dt} = \frac{2\mathrm{Re}(Q\overline{P})}{|P|^2}\Delta v \Delta a. \tag{10.88}$$

Then, the solution Δa is expressed as

$$\Delta a = c \exp\left(\frac{2\mathrm{Re}(Q\overline{P})\Delta v}{|P|^2}t\right), \tag{10.89}$$

where c is the integral constant that is determined by the initial condition. Since $\mathrm{Re}(Q\overline{P})$ is positive, the nontrivial steady-state amplitude existing in the range of $\Delta v < 0$ and $\Delta v > 0$ is unstable and stable, respectively.

Thus, when the coefficient of Δv in Eq. (10.83), $-\dfrac{\mathrm{Re}(Q\overline{P})}{\mathrm{Re}(R\overline{P})\alpha}$, is positive (negative), the bifurcation at the critical speed, i.e. $\Delta v = 0$, is a supercritical (subcritical) Hopf bifurcation.

Since $\mathrm{Re}(Q\overline{P}) < 0$, it is determined by the sign of $\mathrm{Re}(R\overline{P})\alpha$ whether the nonlinear characteristic of the nontrivial steady-state amplitude is supercritical or subcritical. In other words, the nonlinear characteristic is determined not only by the sign of the coefficient of the nonlinear term α but also by the sign of P and R related to the coefficients of the linear term. As seen from Eq. (10.22), the cubic nonlinear stiffness cannot suppress the growth of response amplitude of the self-excitation due to negative damping, but the nonlinear damping contribution can ensure the finite response amplitude. On the other hand, it is seen from Eq. (10.74) that the growth of the response amplitude of self-excited oscillation due to the circulatory force can be suppressed by the cubic nonlinear stiffness.

In the nontrivial steady state ($a_{st} \neq 0$), ϕ is expressed from Eq. (10.75) as

$$\phi = -\left\{\frac{\mathrm{Im}(Q\overline{P})}{|P|^2}\Delta v + \frac{\mathrm{Im}(R\overline{P})}{4|P|^2}\alpha a_{st}^2\right\} t + \phi_0, \tag{10.90}$$

where ϕ_0 is the integral constant which is determined by the initial condition. Therefore, the steady-state response is expressed from Eq. (10.80) as

$$y = a_{st}k_{12}\cos\left[\left\{\omega - \frac{\mathrm{Im}(Q\overline{P})}{|P|^2}\Delta v - \frac{\mathrm{Im}(R\overline{P})}{4|P|^2}\alpha a_{st}^2\right\} t + \phi_0\right] + O(\epsilon), \tag{10.91}$$

$$\begin{aligned} \psi = {} & a_{st}(\omega^2 - k_{11})\cos\left[\left\{\omega - \frac{\mathrm{Im}(Q\overline{P})}{|P|^2}\Delta v - \frac{\mathrm{Im}(R\overline{P})}{4|P|^2}\alpha a_{st}^2\right\} t + \phi_0\right] \\ & + a_{st}\frac{d_{11}\omega}{v_{cr}}\sin\left[\left\{\omega - \frac{\mathrm{Im}(Q\overline{P})}{|P|^2}\Delta v - \frac{\mathrm{Im}(R\overline{P})}{4|P|^2}\alpha a_{st}^2\right\} t + \phi_0\right] + O(\epsilon). \end{aligned} \tag{10.92}$$

References

Kovacic, I. and M. J. Brennan (2011). *The Duffing equation: nonlinear oscillators and their behaviour*. John Wiley & Sons.

Lacarbonara, W. (2013). *Nonlinear structural mechanics: theory, dynamical phenomena and modeling*. Springer Science & Business Media.

Nayfeh, A. H. and D. T. Mook (2008). *Nonlinear oscillations*. John Wiley & Sons.

Polunin, P. M., Y. Yang, M. I. Dykman, T. W. Kenny, and S. W. Shaw (2016). Characterization of mems resonator nonlinearities using the ringdown response. *Journal of Microelectromechanical Systems* 25(2), 297–303.

Shaw, S. W. and B. Balachandran (2008). A review of nonlinear dynamics of mechanical systems in year 2008. *Journal of System Design and Dynamics* 2(3), 611–640.

Thompson, J. and H. Stewart (2002). *Nonlinear Dynamics and Chaos*. Wiley.

Thomsen, J. J. (2003). *Vibrations and Stability*. Springer.

Troger, H. and A. Steindl (2012). *Nonlinear stability and bifurcation theory: an introduction for engineers and applied scientists*. Springer Science & Business Media.

11

Parametric Resonance and Pitchfork Bifurcation

In this chapter, we consider the resonance induced in linear oscillatory systems with time periodic coefficient, which is expressed by the Mathieu equation (Nayfeh and Mook 2008). Such a periodic excitation is called *parametric excitation* and the resonance is called parametric resonance. The system is nonautonomous as that subject to the external or forced excitation, but homogeneous. Therefore, there exists the zero amplitude response. In specific frequency ranges of the parametric excitation, the stable zero amplitude response is destabilized through pitchfork bifurcation. Similar to the self-excited oscillation discussed in Chapter 10, when the zero amplitude response is destabilized, the nonlinear analysis is essential to determine if the growth of amplitude continues infinitely or not. Applying the method of multiple scales on the nonlinear governing equation, i.e. nonlinear Mathieu equation, we clarify the linear and nonlinear characteristics of parametric resonance. The nonlinear dynamics under the parametric excitation with high-frequency excitation will be discussed in Chapter 12.

The parametric resonance has been regarded as an undesired phenomenon for a long time. On the other hand, the unique linear and nonlinear characteristics with respect to the steady states and their stability become to be positively exploited in rich new fields, in particular, microelectromechanical systems (For example, Moran et al. 2013; Rhoads et al. 2010). Application of the parametric excitation can increase the response amplitude in the resonance produced under the external excitation. For example, it enhances the sensitivity of resonator-type microsensors in cases where noise dominates. Such an amplification method using parametric resonance is known as *parametric amplification* (for example, Carr et al. 2000; Rugar and Grütter 1991; Zhang and Turner 2005). The parametric resonance is also utilized as an excitation method in energy harvesters (for example, Daqaq et al. 2009; Jia et al. 2016). In addition, there are applications to mass sensors (for example, Zhang et al. 2002), to mechanical filters (for example, Zhang et al. 2003), and so on.

11.1 Parametric Resonance of Vertically Excited Inverted Pendulum

11.1.1 Equation of Motion

We consider an inverted pendulum supported by a hinge with an elastic restoring moment as shown in Figure 11.1a.

Linear and Nonlinear Instabilities in Mechanical Systems: Analysis, Control and Application,
First Edition. Hiroshi Yabuno.

Companion website: www.wiley.com/go/yabuno/instabilitiesinmechanicalsystems

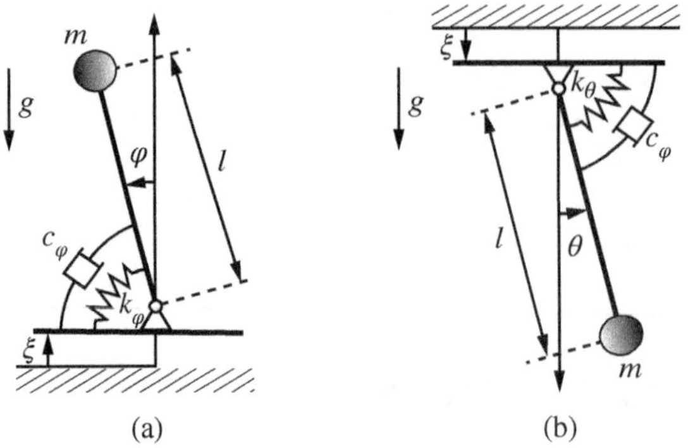

Figure 11.1 Vertically excited pendulum: (a) inverted pendulum supported by a hinge with an elastic restoring moment; (b) pendulum supported by a hinge.

The vertical excitation is

$$\xi = a_e \cos \nu t, \tag{11.1}$$

where a_e and ν are the excitation amplitude and frequency, respectively. The hinge is subject to moment stiffness and damping, $M = k_\varphi \varphi$ and $c_\varphi \frac{d\varphi}{dt}$, respectively. The equation of motion is

$$ml\frac{d^2\varphi}{dt^2} = -\frac{c_\varphi}{l}\frac{d\varphi}{dt} + m\left(g - a_e\nu^2 \cos \nu t\right)\sin\varphi - \frac{k_\varphi}{l}\varphi. \tag{11.2}$$

For reference, we show the equation of motion for Figure 11.1b.

$$ml\frac{d^2\theta}{dt^2} = -\frac{c_\theta}{l}\frac{d\theta}{dt} - m\left(g + a_e\nu^2 \cos \nu t\right)\sin\theta - \frac{k_\theta}{l}\theta, \tag{11.3}$$

where the moment stiffness and damping at the hinged are assumed as $k_\theta \theta$ and $c_\theta \frac{d\theta}{dt}$, respectively.

11.2 Dynamics in Case Without Excitation

In case without excitation ($a_e = 0$), approximating $\sin\varphi$ as $\sin\varphi \approx \varphi - \frac{1}{6}\varphi^3$ yields

$$ml\frac{d^2\varphi}{dt^2} + \frac{c_\varphi}{l}\frac{d\varphi}{dt} + \left(\frac{k_\varphi}{l} - mg\right)\varphi + \frac{1}{6}mg\varphi^3 = 0. \tag{11.4}$$

The equivalent linear stiffness is

$$k_{equiv} = \frac{k_\varphi}{l} - mg, \tag{11.5}$$

and when the mass equals the critical value

$$m_{cr} = \frac{k_\varphi}{lg}, \tag{11.6}$$

the system is buckled; this state corresponds to the case with $k' = 0$ and $c' > 0$ in Figure 2.7. The linear stiffness is positive or negative for $m < m_{cr}$ or $m > m_{cr}$, respectively, while the cubic nonlinearity due to the gravity is hardening and independent of m. The linear and nonlinear characteristics for $m < m_{cr}$ and $m > m_{cr}$ are qualitatively same as those in

Figures A.2a and A.3a, respectively. An increase of the mass produces a variation in the dynamics as for the two-link system subject to compressive forces with $\alpha_3 > 0$ in Chapter 7.

In the system of Figure 11.1b, the equation of motion corresponding to Eq. (11.4) that considers the cubic nonlinearity is expressed as

$$ml\frac{d^2\theta}{dt^2} + \frac{c_\theta}{l}\frac{d\theta}{dt} + \left(\frac{k_\varphi}{l} + mg\right)\theta - \frac{1}{6}mg\theta^3 = 0. \tag{11.7}$$

Then, independent of the mass, the linear stiffness is positive and the cubic nonlinearity is softening. The nonlinear characteristic is shown as Figure A.2b.

11.2.1 Dimensionless Equation of Motion Subject to Vertical Excitation

In this chapter, below the critical mass of $m < m_{cr}$, we consider the dynamics of the pendulum under the vertical excitation. Setting the representative time to the inverse of natural frequency as $T_r = \sqrt{\frac{ml^2}{k_\varphi - mgl}}$ and approximating $\sin\varphi$ as $\sin\varphi \approx \varphi - \frac{1}{6}\varphi^3$ yields the dimensionless form of Eq. (11.2) as follows:

$$\frac{d^2\varphi}{dt^{*2}} + 2\gamma\frac{d\varphi}{dt^*} + (1 + a_e^* v^{*2}\cos v^* t^*)\varphi + \left(\alpha_3 - \frac{1}{6}a_e^* v^{*2}\cos v^* t^*\right)\varphi^3 = 0, \tag{11.8}$$

where the dimensionless parameters are

$$\gamma = \frac{c_\varphi}{2}\sqrt{\frac{1}{ml^2(k_\varphi - mgl)}}, \quad a_e^* = \frac{a_e}{l}, \quad v^* = T_r v, \quad \alpha_3 = \frac{mgl}{6(k_\varphi - mgl)}. \tag{11.9}$$

By making the assumptions of weak damping and excitation amplitude, γ and a_e^* are described, respectively, as $\gamma = \epsilon\hat{\gamma}$ and $a_e^* = \epsilon\hat{a}_e^*$, where $0 < \epsilon \ll 1$, and $\hat{\gamma}$ and $\hat{a}_e^*$ are $O(1)$. Then, the system takes the following form:

$$\frac{d^2\varphi}{dt^{*2}} + 2\epsilon\hat{\gamma}\frac{d\varphi}{dt^*} + (1 + \epsilon\hat{a}_e^* v^{*2}\cos v^* t^*)\varphi + \left(\alpha_3 - \frac{\epsilon}{6}\hat{a}_e^* v^{*2}\cos v^* t^*\right)\varphi^3 = 0. \tag{11.10}$$

Hereafter, the asterisk is omitted. We analyze the primary parametric resonance that occurs in the case when the excitation frequency is in the neighborhood of twice the natural frequency, i.e. $v \approx 2$. The proximity is expressed as

$$v^* = 2 + \sigma \quad (\sigma = \epsilon\hat{\sigma}), \tag{11.11}$$

where σ is a detuning parameter and $\hat{\sigma} = O(1)$. Using the method of multiple scales, we obtain an analytical approximate solution. We consider that the steady-state response is realized by the balance between the parametric excitation and the cubic nonlinear effect (see Section D.7), i.e.

$$\epsilon\hat{a}_e v^2 \cos vt\varphi \sim \alpha_3\varphi^3 \Leftrightarrow \epsilon\varphi \sim \varphi^3 \Leftrightarrow \varphi \sim \epsilon^{1/2}. \tag{11.12}$$

Then, because the leading order of φ is $O(\epsilon^{1/2})$, the uniform expansion of the solution is sought in the form

$$\varphi(t;\epsilon) = \epsilon^{1/2}\varphi_1(t_0, t_1) + \epsilon^{3/2}\varphi_3(t_0, t_1) + \cdots, \tag{11.13}$$

where the fast time scale t_0 and the slow time scale t_1 are given by

$$t_0 = t, \quad t_1 = \epsilon t. \tag{11.14}$$

It is noted from Eq. (11.11) that

$$\frac{\nu}{2}t_0 = \left(1 + \frac{\epsilon\hat{\sigma}}{2}\right)t_0 = t_0 + \frac{\hat{\sigma}}{2}t_1. \tag{11.15}$$

The time derivative with respect to t and the second time derivative are transformed as

$$\frac{d}{dt} = \frac{\partial}{\partial t_0} + \epsilon\frac{\partial}{\partial t_1} = D_0 + \epsilon D_1, \tag{11.16}$$

$$\frac{d^2}{dt^2} = \frac{\partial^2}{\partial t_0^2} + 2\epsilon\frac{\partial^2}{\partial t_0 \partial t_1} + O(\epsilon^2) = D_0^2 + 2\epsilon D_0 D_1 + O(\epsilon^2), \tag{11.17}$$

where $D_0 = \frac{\partial}{\partial t_0}, D_1 = \frac{\partial}{\partial t_1}$, $D_0^2 = \frac{\partial^2}{\partial t_0^2}$, and $D_0 D_1 = \frac{\partial^2}{\partial t_0 \partial t_1}$. Substituting Eqs. (11.13), (11.16), and (11.17) into Eq. (11.10), we have the hierarchy of equations for $O(\epsilon^{1/2})$ and $O(\epsilon^{3/2})$, respectively, as

$$D_0^2\varphi_1 + \varphi_1 = 0, \tag{11.18}$$

$$D_0^2\varphi_3 + \varphi_3 = -2D_0D_1\varphi_1 - 2\hat{\gamma}D_0\varphi_1 - 4\hat{a}_e \cos \nu t\varphi_1 - \alpha_3\varphi_1^3. \tag{11.19}$$

Then, the solution of Eq. (11.18) is

$$\varphi_1(t_0, t_1) = A(t_1)e^{it_0} + \mathrm{CC}, \tag{11.20}$$

where A is a complex amplitude and CC denotes the complex conjugate of the preceding term. Substituting this equation into Eq. (11.19) yields

$$\begin{aligned} D_0^2\varphi_3 + \varphi_3 = &-(2iD_1A + 2i\hat{\gamma}A + 2\hat{a}_e e^{i\hat{\sigma}t_1}\bar{A} + 3\alpha_3|A|^2A)e^{it_0} \\ &-(2\hat{a}_e A e^{i\hat{\sigma}t_1} + \alpha_3 A^3)e^{3it_0} + \mathrm{CC} + \mathrm{NST}. \end{aligned} \tag{11.21}$$

A condition not to produce the secular terms in φ_3 is

$$2iD_1A + 2i\hat{\gamma}A + 2\hat{a}_e e^{i\hat{\sigma}t_1}\bar{A} + 3\alpha_3|A|^2A = 0. \tag{11.22}$$

Introducing the transformation

$$A = Be^{i\hat{\sigma}t_1/2} \tag{11.23}$$

leads to the autonomous equation

$$2iD_1B - \hat{\sigma}B + 2i\hat{\gamma}B + 2\hat{a}_e\overline{B} + 3\alpha_3|B|^2B = 0. \tag{11.24}$$

By expressing the complex value B in the Cartesian form as

$$B(t_1) = \frac{\hat{a}_c(t_1)}{2} + i\frac{\hat{a}_s(t_1)}{2}, \tag{11.25}$$

and using Eq. (11.15), the solution of φ in Eq. (11.13) is expressed as

$$\begin{aligned} \varphi(t;\epsilon) &= \epsilon^{1/2}\left\{A(t_1)e^{it_0} + \mathrm{CC}\right\} + O(\epsilon^{3/2}) = \epsilon^{1/2}\left\{B(t_1)e^{i(t_0 + \hat{\sigma}t_1/2)} + \mathrm{CC}\right\} + O(\epsilon^{3/2}) \\ &= \epsilon^{1/2}\left\{B(t_1)e^{i\frac{\nu}{2}t_0} + \mathrm{CC}\right\} + O(\epsilon^{3/2}) \\ &= a_c \cos\frac{\nu}{2}t - a_s \sin\frac{\nu}{2}t + O(\epsilon^{3/2}), \end{aligned} \tag{11.26}$$

where $a_c = \epsilon^{1/2}\hat{a}_c$ and $a_s = \epsilon^{1/2}\hat{a}_s$.

On the other hand, by introducing the polar form of the complex value B as

$$B(t_1) = \frac{\hat{a}(t_1)}{2} e^{i\phi(t_1)}, \tag{11.27}$$

the solution of φ in Eq. (11.13) is expressed as

$$\begin{aligned} \varphi(t;\epsilon) &= \epsilon^{1/2}\left\{B(t_1)e^{i\frac{\nu}{2}t_0} + \mathrm{CC}\right\} + O(\epsilon^{3/2}) \\ &= a\cos\left(\frac{\nu}{2}t + \phi\right) + O(\epsilon^{3/2}), \end{aligned} \tag{11.28}$$

where $a = \epsilon^{1/2}\hat{a}$.

11.2.2 Trivial Equilibrium State and Its Stability

It is obvious that Eq. (11.10) has the trivial equilibrium state $\varphi = 0$, which corresponds to the trivial steady-state amplitude $a_c = a_s = 0$ in Eq. (11.26), i.e. $\hat{a}_c = \hat{a}_s = 0$. Referring to the method in Section 2.3, let us determine the stability of $\hat{a}_c = \hat{a}_s = 0$. Since $\boldsymbol{x} = [\hat{a}_c, \hat{a}_e]^T$, $\Delta\boldsymbol{x} = [\Delta\hat{a}_c, \Delta\hat{a}_e]^T$, and $\boldsymbol{x}_{st} = [0, 0]^T$ in Eq. (2.25), substituting

$$B(t_1) = \frac{\Delta\hat{a}_c(t_1)}{2} + i\frac{\Delta\hat{a}_s(t_1)}{2} \tag{11.29}$$

into Eq. (11.24) and separating the result into the real and imaginary parts yields

$$\begin{cases} D_1\Delta\hat{a}_c = -\hat{\gamma}\Delta\hat{a}_c + \left(\frac{\hat{\sigma}}{2} + \hat{a}_e\right)\Delta\hat{a}_s + O(2), & (11.30) \\ D_1\Delta\hat{a}_s = \left(\hat{a}_e - \frac{\hat{\sigma}}{2}\right)\Delta\hat{a}_c - \hat{\gamma}\Delta\hat{a}_s + O(2), & (11.31) \end{cases}$$

or equivalently,

$$\frac{\partial}{\partial t_1}\begin{bmatrix}\Delta\hat{a}_c \\ \Delta\hat{a}_s\end{bmatrix} = \begin{bmatrix} -\hat{\gamma} & \frac{\hat{\sigma}}{2} + \hat{a}_e \\ \hat{a}_e - \frac{\hat{\sigma}}{2} & -\hat{\gamma} \end{bmatrix}\begin{bmatrix}\Delta\hat{a}_c \\ \Delta\hat{a}_s\end{bmatrix} + O(2), \tag{11.32}$$

where $O(2)$ denotes higher order terms than the second order of $\Delta\hat{a}_c$ and $\Delta\hat{a}_s$. The characteristic equation of the matrix is

$$\lambda^2 + 2\hat{\gamma}\lambda + \hat{\gamma}^2 + \frac{\hat{\sigma}^2}{4} - \hat{a}_e^2 = 0. \tag{11.33}$$

Because of $c' = 2\hat{\gamma} > 0$ and $k' = \hat{\gamma}^2 + \frac{\hat{\sigma}^2}{4} - \hat{a}_e^2$ in Eq. (2.113), the stability boundary of the trivial steady state is expressed as

$$\hat{\gamma}^2 + \frac{\hat{\sigma}^2}{4} - \hat{a}_e^2 = 0 \tag{11.34}$$

or

$$\gamma^2 + \frac{\sigma^2}{4} - a_e^2 = 0. \tag{11.35}$$

The trivial steady state is stable and unstable in the cases of $k' > 0$ and $k' < 0$, respectively, therefore the trivial steady-state amplitude is unstable in the region

$$\sigma_- < \sigma < \sigma_+, \tag{11.36}$$

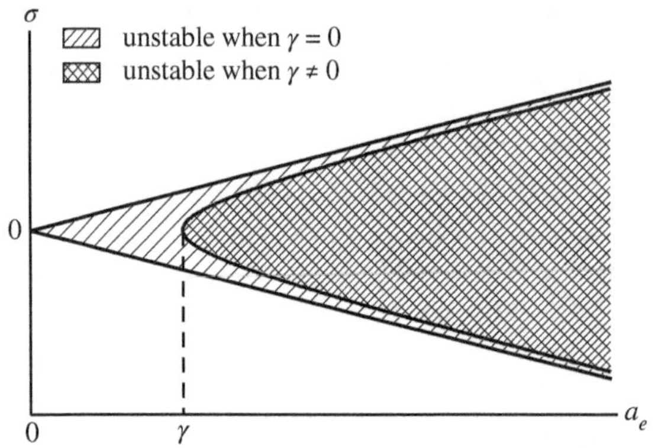

Figure 11.2 Unstable region for parametric resonance. The boundary is described by Eq. (11.35). When the combination of the detuning parameter of the excitation frequency σ and the excitation amplitude a_e is located in the hatched region, the trivial steady-state amplitude is unstable.

and stable in the other region, where $\sigma_- = -2\sqrt{a_e^2 - \gamma^2}$ and $\sigma_+ = 2\sqrt{a_e^2 - \gamma^2}$. The unstable region is depicted as Figure 11.2. Increasing the effect of damping decreases the unstable region of the trivial equilibrium state.

11.2.3 Nontrivial Steady-State Amplitude and Its Stability

There are also nontrivial steady states of complex amplitude in this system. Using the polar form, we investigate the steady states and their stability. Substituting Eq. (11.27) into Eq. (11.24) and separating the result into real and imaginary parts yields

$$\begin{cases} D_1\hat{a} = -\hat{\gamma}\hat{a} + \hat{a}_e \sin 2\phi\hat{a}, & (11.37) \\ D_1\phi = -\frac{\hat{\sigma}}{2} + \hat{a}_e \cos 2\phi + \frac{3}{8}\alpha_3\hat{a}^2. & (11.38) \end{cases}$$

Let the equilibrium states under the condition $D_1\hat{a} = D_1\phi = 0$ be $\hat{a} \overset{\text{def}}{=} \hat{a}_{st}$ and $\phi \overset{\text{def}}{=} \phi_{st}$; $a_{st} = \epsilon^{1/2}\hat{a}_{st}$ and ϕ_{st} are the amplitude and phase angle in the steady state, respectively. Hence, $\hat{a}_{st}$ and ϕ_{st} satisfy the equilibrium equations

$$\begin{cases} 0 = -\hat{\gamma} + \hat{a}_e \sin 2\phi_{st}, & (11.39) \\ 0 = -\frac{\hat{\sigma}}{2} + \hat{a}_e \cos 2\phi_{st} + \frac{3}{8}\alpha_3\hat{a}_{st}^2, & (11.40) \end{cases}$$

or equivalently,

$$\begin{cases} \hat{a}_e \sin 2\phi_{st} = \hat{\gamma}, & (11.41) \\ \hat{a}_e \cos 2\phi_{st} = \frac{\hat{\sigma}}{2} - \frac{3}{8}\alpha_3\hat{a}_{st}^2. & (11.42) \end{cases}$$

Squaring Eqs. (11.41) and (11.42), and adding the results, we obtain the relationship between the detuning parameter of excitation frequency $\sigma(=\epsilon\hat{\sigma})$ and the steady-state amplitude $a_{st}(=\epsilon^{1/2}\hat{a}_{st})$ as follows:

$$\hat{a}_{st} = \sqrt{\frac{8}{3\alpha_3}\left(\frac{\hat{\sigma}}{2} \pm \sqrt{\hat{a}_e^2 - \hat{\gamma}^2}\right)} \tag{11.43}$$

or equivalently,

$$a_{st} = \epsilon^{1/2}\hat{a}_{st} = \sqrt{\frac{8}{3\alpha_3}\left(\frac{\sigma}{2} \pm \sqrt{a_e^2 - \gamma^2}\right)}. \tag{11.44}$$

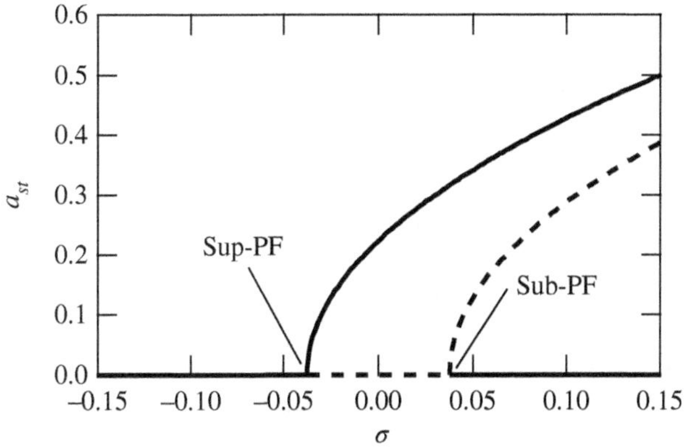

Figure 11.3 Frequency response curve for parametrically excited inverted pendulum subject to restoring moment: Sup-PF is a supercritical pitchfork bifurcation point; Sub-PF is a subcritical pitchfork bifurcation point. $a_e = 0.02, \gamma = 0.01, \alpha_3 = 1$.

We express these two steady states as

$$a_{st-} = \sqrt{\frac{4}{3\alpha_3}\left(\sigma - \sigma_-\right)}, \tag{11.45}$$

$$a_{st+} = \sqrt{\frac{4}{3\alpha_3}\left(\sigma - \sigma_+\right)}. \tag{11.46}$$

Then, Eq. (11.42) can be written as

$$\hat{a}_e \cos 2\phi_{st} = \mp\sqrt{\hat{a}_e^2 - \hat{\gamma}^2}, \tag{11.47}$$

where the double sign corresponds to that of Eq. (11.43), and the phase angles ϕ_{st+} and ϕ_{st-} associated with a_{st+} and a_{st-}, respectively, satisfy

$$\hat{a}_e \cos 2\phi_{st-} = \frac{\sigma_-}{2} < 0, \tag{11.48}$$

$$\hat{a}_e \cos 2\phi_{st+} = \frac{\sigma_+}{2} > 0. \tag{11.49}$$

The relationship between the detuning parameter σ and the steady-state amplitude a_{st}, i.e. frequency response curve, is described as Figure 11.3. In this system, because the cubic nonlinearity is hardening $\alpha_3 > 0$, both the response curves lean toward the right, i.e. the higher frequency, resulting in a hardening response.

Next, we determine the stability of the steady-state amplitudes. Substituting $\hat{a} = \hat{a}_{st} + \Delta\hat{a}$ and $\phi = \phi_{st} + \Delta\phi$ into Eqs. (11.37) and (11.38) and neglecting the nonlinear terms with respect to $\Delta\hat{a}$ and $\Delta\phi$ yields

$$\frac{\partial}{\partial t_1}\begin{bmatrix}\Delta\hat{a}\\ \Delta\phi\end{bmatrix} = \begin{bmatrix}-\hat{\gamma} + \hat{a}_e \sin 2\phi_{st} & 2\hat{a}_e \cos 2\phi_{st}\hat{a}_{st}\\ \frac{3}{4}\alpha_3\hat{a}_{st} & -2\hat{a}_e \sin 2\phi_{st}\end{bmatrix}\begin{bmatrix}\Delta\hat{a}\\ \Delta\phi\end{bmatrix}. \tag{11.50}$$

The characteristic equation is

$$\lambda^2 + 2\hat{\gamma}\lambda - \frac{3}{2}\alpha_3\hat{a}_e \cos 2\phi_{st}\hat{a}_{st}^2 = 0. \tag{11.51}$$

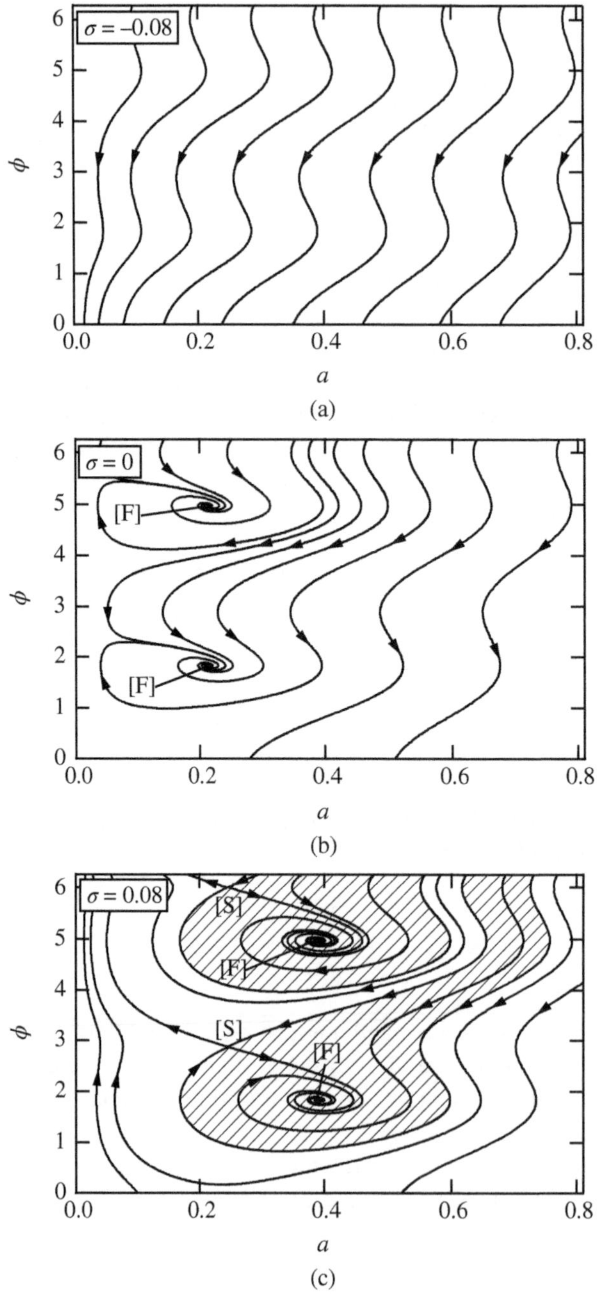

Figure 11.4 Phase space for parametrically excited inverted pendulum subject to restoring moment: (a) $\sigma < \sigma_-$; (b) $\sigma_- < \sigma < \sigma_+$; (c) $\sigma_+ < \sigma$. $a_e = 0.02, \gamma = 0.01, \alpha_3 = 1, \sigma_+ = 0.035, \sigma_- = -0.035$.

Because of $c' = 2\hat{\gamma} > 0$ and $k' = -\frac{3}{2}\alpha_3\hat{a}_e \cos 2\phi_{st}\hat{a}_{st}^2$ in Eq. (2.113), we can determine the stability of the steady state from Figure 2.7. Considering Eqs. (11.48) and (11.49) and $\alpha_3 > 0$, the steady-state amplitudes, a_{st-} and a_{st+}, are stable and unstable, respectively. In Figure 11.3, the stable and unstable nontrivial steady-state amplitudes are bifurcated from the trivial equilibrium state through a supercritical pitchfork bifurcation at $\sigma = \sigma_-$ and a subcritical one at $\sigma = \sigma_+$, respectively. We show phase portraits for the cases of (a) $\sigma < \sigma_-$, (b) $\sigma_- < \sigma < \sigma_+$, and (c) $\sigma_+ < \sigma$ in Figure 11.4a–c, respectively. In case (a), only one stable steady state is trivial, i.e. zero amplitude as seen from the frequency response curve Figure 11.3, and we do not see any nontrivial steady-state amplitude in Figure 11.4a. In case (b), there are a stable nontrivial steady-state amplitude and the unstable trivial steady state as seen from the frequency response curve in Figure 11.3, and we can observe the stable nontrivial steady-state amplitude at point [F] in Figure 11.4b that is a stable focus. In case (c), there are stable and unstable nontrivial steady-state amplitudes and the stable trivial steady state as seen from the frequency response curve of Figure 11.3. We can observe the stable and unstable nontrivial steady-state amplitudes, at points [F] and [S] in Figure 11.4c, respectively. Point [F] is a stable focus as (C) in Figure 2.7 and point [S] is a saddle as (A) in Figure 2.7 generating separatrix (Thompson and Stewart 2002).

References

Carr, D. W., S. Evoy, L. Sekaric, H. Craighead, and J. Parpia (2000). Parametric amplification in a torsional microresonator. *Applied Physics Letters* 77(10), 1545–1547.

Daqaq, M. F., C. Stabler, Y. Qaroush, and T. Seuaciuc-Osório (2009). Investigation of power harvesting via parametric excitations. *Journal of Intelligent Material Systems and Structures* 20(5), 545–557.

Jia, Y., S. Du, and A. A. Seshia (2016). Twenty-eight orders of parametric resonance in a microelectromechanical device for multi-band vibration energy harvesting. *Scientific reports* 6, 30167.

Moran, K., C. Burgner, S. Shaw, and K. Turner (2013). A review of parametric resonance in microelectromechanical systems. *Nonlinear Theory and Its Applications, IEICE* 4(3), 198–224.

Nayfeh, A. H. and D. T. Mook (2008). *Nonlinear oscillations*. John Wiley & Sons.

Rhoads, J. F., S. W. Shaw, and K. L. Turner (2010). Nonlinear dynamics and its applications in micro-and nanoresonators. *Journal of dynamic systems, measurement, and control* 132(3), 034001.

Rugar, D. and P. Grütter (1991). Mechanical parametric amplification and thermomechanical noise squeezing. *Physical Review Letters* 67(6), 699.

Thompson, J. and H. Stewart (2002). *Nonlinear Dynamics and Chaos*. Wiley.

Zhang, W., R. Baskaran, and K. Turner (2003). Tuning the dynamic behavior of parametric resonance in a micromechanical oscillator. *Applied physics letters* 82(1), 130–132.

Zhang, W., R. Baskaran, and K. L. Turner (2002). Effect of cubic nonlinearity on auto-parametrically amplified resonant mems mass sensor. *Sensors and Actuators A: Physical* 102(1), 139–150.

Zhang, W. and K. L. Turner (2005). Application of parametric resonance amplification in a single-crystal silicon micro-oscillator based mass sensor. *Sensors and Actuators A: Physical* 122(1), 23–30.

12

Stabilization of Inverted Pendulum Under High-Frequency Excitation

As shown in Section 11.2, the upright position in the inverted pendulum is statically unstable in the case without restoring moment. The stabilization for such an inverted pendulum is dealt with as a good example of the application of modern control theory (for example, Dorf and Bishop 2011). In this chapter, we consider another stabilization method with no feedback control which is based on the high frequency excitation parallel to the gravity direction. The stabilization of statically unstable equilibrium states is called *dynamic stabilization phenomenon* and has been recognized in the field of mechanics for a considerably long time (Stephenson 1908a,b). The stabilization mechanism is theoretically explained by Kapitza and such a stabilized pendulum is called Kapitza pendulum (Kapitza 1965; Landau and Lifshitz 1960). The method is based on an asymptotic separation of fast and slow variables yielding a renormalized potential, i.e. effective potential. The increase of the stability of elastic systems as seen in Videoclips 16.8(2) and 16.8(3) in Section 16.8 is an application of dynamic stabilization phenomenon. Applications in microscopic fields are summarized in Richards et al. (2018). In this chapter, after the asymptotic separation by the method of multiple scales, the stabilization mechanism is analytically clarified by bifurcation analysis.

12.1 Equation of Motion

As shown in Figure 12.1, we consider the inverted pendulum subject to the vertical excitation, where a_e and ν are the excitation amplitude and frequency, respectively.

The equation of motion is expressed as

$$ml\frac{d^2\varphi}{dt^2} = -\frac{c}{l}\frac{d\varphi}{dt} + m(g - a_e\nu^2 \cos \nu t)\sin \varphi, \tag{12.1}$$

where $c\frac{d\varphi}{dt}$ is the moment damping at the pivot. Unlike the representative time in Chapter 11, since we consider the case when the excitation frequency is much higher than the natural frequency, i.e. $\nu >> \sqrt{\frac{g}{l}}$ of high-frequency excitation, we set the representative time as $1/\nu$ for the nondimensionalization of equation of motion. Then, the nondimensional natural frequency appears as a small parameter in the nondimensional equation of motion and by the smallness, we can apply the method of multiple scales to obtain the analytical

Linear and Nonlinear Instabilities in Mechanical Systems: Analysis, Control and Application, First Edition. Hiroshi Yabuno.

Companion website: www.wiley.com/go/yabuno/instabilitiesinmechanicalsystems

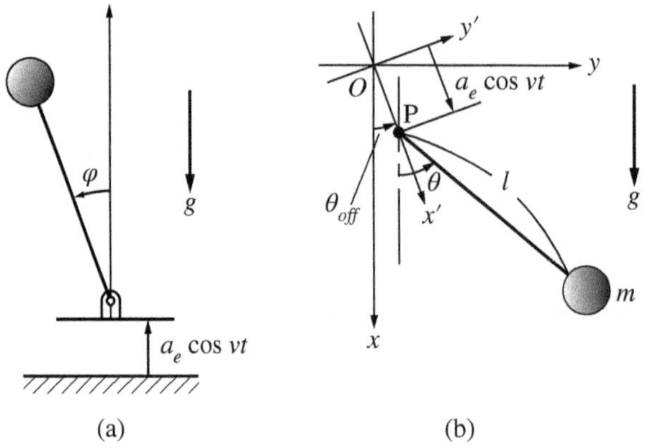

Figure 12.1 Harmonically excited pendulum in case without restoring moment at the pivot: (a) vertical excitation, (b) excitation along x'-direction inclined from the vertical direction of θ_{off}.

solution. The nondimensional equation of motion is

$$\frac{d^2\varphi}{dt^{*2}} + \mu\frac{d\varphi}{dt^*} + (-\sigma + a_e^* \cos t^*)\sin\varphi = 0, \tag{12.2}$$

or equivalently,

$$\frac{d^2\varphi}{dt^{*2}} + \mu\frac{d\varphi}{dt^*} - \sigma\sin\varphi + a_e^* \cos t^* \sin\varphi = 0, \tag{12.3}$$

where $a_e^* = a_e/l$ and $\sigma = (g/l)/\nu^2$ are the nondimensional excitation amplitude and the natural frequency squared, respectively. Small σ indicates the high-frequency excitation. The nondimensional damping effect is $\mu = c/(ml^2\nu)$.

12.2 Analysis by the Method of Multiple Scales

12.2.1 Scaling of Some Parameters

Before the application of the method of multiple scales, we perform the scaling of dimensionless parameters. Considering the case when the excitation amplitude is sufficiently smaller than the length of the pendulum, i.e. $0 < a_e^* \ll 1$, we set a_e^* as

$$a_e^* = \epsilon\hat{a}_e. \tag{12.4}$$

First, we set the order of the high frequency, i.e. the smallness of the natural frequency relative to the excitation frequency as $O(\epsilon^2)$:

$$\sigma = \epsilon^2\hat{\sigma}. \tag{12.5}$$

In addition, μ is set as $\mu = \epsilon\hat{\mu}$. We assume the uniform expansion to be

$$\varphi = \varphi_0 + \epsilon\varphi_1 + \epsilon^2\varphi_2 + \cdots \tag{12.6}$$

and introduce the multiple time scales

$$t_0 = t^*, \quad t_1 = \epsilon t^*, \quad t_2 = \epsilon^2 t^*. \tag{12.7}$$

The initial angular velocity is expressed as follows:

$$\left.\frac{d\varphi}{dt^*}\right|_{t^*=0} = (D_0 + \epsilon D_1 + \cdots)(\varphi_0 + \epsilon\varphi_1 + \cdots)|_{t^*=0}$$
$$= \{D_0\varphi_0 + O(\epsilon)\}|_{t^*=0}. \tag{12.8}$$

We consider the case when the initial angular velocity is small, i.e. $D_0\varphi_0 = 0$. Substituting Eq. (12.6) into Eq. (12.2) and equating coefficients of like powers of ϵ yields the following equations of the orders $O(\epsilon^0)$, $O(\epsilon)$, and $O(\epsilon^2)$:

$$O(\epsilon^0) : D_0^2\varphi_0 = 0, \tag{12.9}$$

$$O(\epsilon) : D_0^2\varphi_1 = -2D_0D_1\varphi_0 - \hat{\mu}D_0\varphi_0 - \hat{a}_e \sin\varphi_0 \cos t_0, \tag{12.10}$$

$$O(\epsilon^2) : D_0^2\varphi_2 = -2D_0D_1\varphi_1 - 2D_0D_2\varphi_0 - D_1^2\varphi_0 - \hat{\mu}(D_0\varphi_1 + D_1\varphi_0)$$
$$+ \hat{\sigma}\sin\varphi_0 - \hat{a}_e\varphi_1\cos\varphi_0\cos t_0. \tag{12.11}$$

12.2.2 Averaging by the Method of Multiple Scales

Taking into account $D_0\varphi_0|_{t^*=0} = 0$, we have from Eq. (12.9)

$$\varphi_0 = c_0(t_1, t_2, \ldots). \tag{12.12}$$

Then, Eq. (12.10) leads to

$$D_0^2\varphi_1 = -\hat{a}_e \sin\varphi_0 \cos t_0. \tag{12.13}$$

A particular solution is

$$\varphi_1 = \hat{a}_e \sin\varphi_0 \cos t_0. \tag{12.14}$$

Substituting Eqs. (12.12) and (12.14) into Eq. (12.11) yields

$$D_0^2\varphi_2 = q_0(t_1, t_2) + q_1(t_1, t_2)\sin t_0 + q_2(t_1, t_2)\cos 2t_0, \tag{12.15}$$

where

$$q_0(t_1, t_2) = -D_1^2\varphi_0 - \hat{\mu}D_1\varphi_0 + \hat{\sigma}\sin\varphi_0 - \frac{1}{4}\hat{a}_e^2\sin 2\varphi_0, \tag{12.16}$$

$$q_1(t_1, t_2) = 2\hat{a}_e\cos\varphi_0 D_1\varphi_0 + \hat{\mu}\hat{a}_e\sin\varphi_0, \tag{12.17}$$

$$q_2(t_1, t_2) = -\frac{1}{4}\hat{a}_e^2\sin 2\varphi_0. \tag{12.18}$$

A particular solution of Eq. (12.15) is

$$\varphi_2 = \frac{1}{2}q_0(t_1, t_2)t_0^2 - q_1(t_1, t_2)\sin t_0 - \frac{1}{4}q_2(t_1, t_2)\cos 2t_0. \tag{12.19}$$

Since the first term is a secular term, we set $q_0(t_1, t_2)$ to be zero, i.e.

$$D_1^2\varphi_0 + \hat{\mu}D_1\varphi_0 - \hat{\sigma}\sin\varphi_0 + \frac{1}{4}\hat{a}_e^2\sin 2\varphi_0 = 0. \tag{12.20}$$

Multiplying both sides by ϵ^2 yields

$$\epsilon^2D_1^2\varphi_0 + \epsilon^2\hat{\mu}D_1\varphi_0 - \epsilon^2\hat{\sigma}\sin\varphi_0 + \frac{1}{4}\epsilon^2\hat{a}_e^2\sin 2\varphi_0 = 0. \tag{12.21}$$

Taking into account that φ_0 is independent of t_0 and $\varphi = \varphi_0 + O(\epsilon)$, we have the following equations:

$$\frac{d\varphi}{dt^*} = \epsilon D_1\varphi_0 + O(\epsilon^2) = \epsilon D_1\varphi + O(\epsilon^2), \tag{12.22}$$

$$\frac{d^2\varphi}{dt^{*2}} = \epsilon^2 D_1^2\varphi_0 + O(\epsilon^3) = \epsilon^2 D_1^2\varphi + O(\epsilon^3). \tag{12.23}$$

Furthermore, recalling $\mu = \epsilon\hat{\mu}$, $\sigma = \epsilon^2\hat{\sigma}$, and $a_e^* = \epsilon\hat{a}_e$, we can rewrite Eq. (12.21) in the accuracy of $O(\epsilon^2)$ as

$$\frac{d^2\varphi}{dt^{*2}} + \mu\frac{d\varphi}{dt^*} - \sigma\sin\varphi + \frac{1}{4}a_e^{*2}\sin 2\varphi = 0. \tag{12.24}$$

As a result, the nonautonomous original equation (12.2) is transformed into the autonomous Eq. (12.24) by using the method of multiple scales. This transformation allows us to perform the linear and nonlinear analyses in Chapters 5 and 7.

It should be also noticed that the averaged equation (12.24) is applicable to the analysis of the global dynamics since we do not assume that φ is small in the analytical process. Generally, the method of multiple scales is a method for weakly nonlinear dynamics. However, in the case when the first order solution is independent of the fastest time scale t_0, the averaged equation to approximately describe the global dynamics can be obtained. Such a global dynamics in an underactuated manipulator is obtained in Chapter 15 by the method of multiple scales.

12.3 Bifurcation Analysis of Inverted Pendulum Under High-Frequency Excitation

12.3.1 Subcritical Pitchfork Bifurcation and Stabilization of Inverted Pendulum

Let us consider the case of $|\varphi| \ll 1$. Equation (12.24) is approximately expressed in the accuracy of $O(\varphi^3)$ as

$$\frac{d^2\varphi}{dt^{*2}} + \mu\frac{d\varphi}{dt^*} + \left(\frac{a_e^{*2}}{2} - \sigma\right)\varphi - \left(\frac{a_e^{*2}}{3} - \frac{\sigma}{6}\right)\varphi^3 + O(\varphi^5) = 0. \tag{12.25}$$

Since $F_{equiv}^{\#*}$ corresponding to Eq. (7.8) is

$$F_{equiv}^{\#*} = \left(\frac{a_e^{*2}}{2} - \sigma\right)\varphi - \left(\frac{a_e^{*2}}{3} - \frac{\sigma}{6}\right)\varphi^3 + O(\varphi^5), \tag{12.26}$$

Equation (12.24) shows the same dynamics as that in the two-link system subject to the compressive forces introduced in Section 5.1. We focus on the parameter σ as a control parameter; σ plays the same role as p in Eq. (7.8), i.e. the compressive force in the two-link system. Since σ is inversely proportional to the excitation frequency ν^2 of the supporting point of the inverted pendulum, the increase of the excitation frequency corresponds to the decrease of the compressive force in the two-link system.

We investigate the stability of the trivial equilibrium state $\varphi = 0$, i.e. the upright position. The linearized equation of Eq. (12.25) is

$$\frac{d^2\varphi}{dt^{*2}} + \mu \frac{d\varphi}{dt^*} + \left(\frac{a_e^{*2}}{2} - \sigma \right) \varphi = 0. \tag{12.27}$$

This system is equivalent to Eq. (2.115). Increasing the excitation frequency, i.e. decreasing σ, the phase portrait is changed as (A)→(J)→(E)→(F)→(G) in Figure 2.7. The critical point of stability is at (J), i.e. $\frac{a_e^{*2}}{2} - \sigma = 0$. If $\sigma > \frac{a_e^{*2}}{2}$, the equivalent stiffness is negative and the inverted pendulum is statically unstable as the case without vertical excitation. If $\sigma < \frac{a_e^{*2}}{2}$, the equivalent stiffness is changed to be positive and the inverted pendulum is stabilized. The stabilization phenomenon is called *dynamic stabilization*. The stabilization condition is expressed in dimensional form as

$$a_e \nu > \sqrt{2lg}. \tag{12.28}$$

When the excitation amplitude and frequency, a_e and ν, satisfy the above conditions, the inverted pendulum is stable without feedback control.

Next, we consider the cubic nonlinear term in Eqs. (12.25) and (12.26). In order to analyze the nonlinear dynamics in the neighborhood of the critical point of linear stability, $\sigma = a_e^{*2}/2$, we set σ to be

$$\sigma = \frac{a_e^{*2}}{2} + \epsilon \quad (|\epsilon| \ll 1). \tag{12.29}$$

Substituting this equation into Eq. (12.25) yields

$$\frac{d^2\varphi}{dt^{*2}} + \mu \frac{d\varphi}{dt^*} - \epsilon\varphi - \frac{a_e^{*2}}{4}\varphi^3 = 0, \tag{12.30}$$

where $\epsilon\varphi^3/6$ is neglected due to the higher order smallness. Then, Eq. (12.26) becomes

$$F^{\#*}_{equiv} = -\epsilon\varphi - \frac{a_e^{*2}}{4}\varphi^3 + O(\varphi^5). \tag{12.31}$$

Because of the negative coefficient of the cubic nonlinear term, i.e. the softening cubic nonlinearity, from the discussion in Section 7.1.2, we find that the subcritical pitchfork bifurcation occurs at the critical point $\epsilon = 0$, i.e. $\sigma = a_e^{*2}/2$.

The bifurcation phenomenon is shown in Figure 12.2a. The subcritical characteristic indicates that the higher excitation frequency can stabilize the upright position of the pendulum even against larger disturbances to the pendulum.

12.3.2 Global Stability of Equilibrium States

From Eq. (12.24), the equilibrium equation is

$$\sigma \sin\varphi_{st} - \frac{1}{4}a_e^{*2} \sin 2\varphi_{st} = \sin\varphi_{st}\left(\sigma - \frac{1}{2}a_e^{*2}\cos\varphi_{st}\right) = 0. \tag{12.32}$$

In addition to the trivial equilibrium state $\varphi_{st} = 0$, i.e. the upright position, there are nontrivial equilibrium states in the region of $\sigma < \frac{a_e^{*2}}{2}$ as

$$\varphi_{st} = \pm\cos^{-1}\frac{2\sigma}{a_e^{*2}}. \tag{12.33}$$

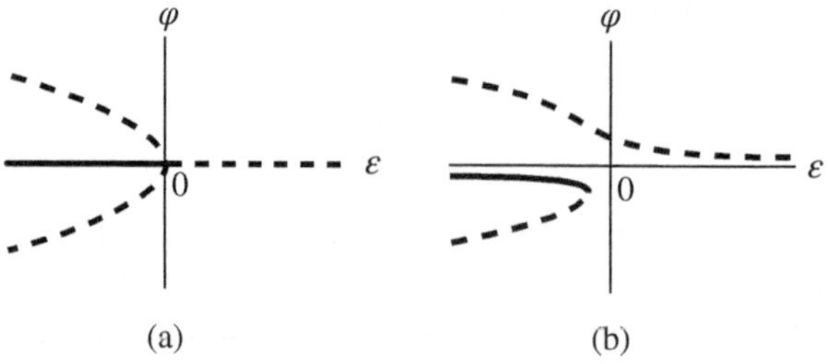

Figure 12.2 Dynamic stabilization by high-frequency vertical excitation and subcritical pitchfork bifurcation. (a) is a subcritical pitchfork bifurcation in the case when the excitation direction is exactly parallel to the gravity direction. (b) is a perturbed subcritical pitchfork bifurcation in the case when the excitation direction is not exactly parallel to the gravity direction, i.e. in the case of $\Delta\theta_{off} < 0$ as will be shown in Section 12.5.2.

Let us determine the stability of the equilibrium states. Substituting $\varphi = \varphi_{st} + \Delta\varphi$ into Eq. (12.24) and considering Eq. (12.32) yields the linearized equation for $\Delta\varphi$

$$\frac{d^2\Delta\varphi}{dt^{*2}} + \mu\frac{d\Delta\varphi}{dt^*} + \left\{-\sigma\cos\varphi_{st} + \frac{a_e^{*2}}{2}(2\cos^2\varphi_{st} - 1)\right\}\Delta\varphi = 0. \tag{12.34}$$

For the trivial equilibrium state $\varphi_{st} = 0$, this equation leads to

$$\frac{d^2\Delta\varphi}{dt^{*2}} + \mu\frac{d\Delta\varphi}{dt^*} + \left(-\sigma + \frac{a_e^{*2}}{2}\right)\Delta\varphi = 0, \tag{12.35}$$

which is equivalent to Eq. (12.27). The stability of the trivial equilibrium state is changed at $\sigma = \frac{a_e^{*2}}{2}$ as the result in Section 12.3.1. For the nontrivial equilibrium states expressed by Eqs. (12.33) and (12.34) leads to

$$\frac{d^2\Delta\varphi}{dt^{*2}} + \mu\frac{d\Delta\varphi}{dt^*} - \frac{a_e^{*2}}{2}(1 - \cos^2\varphi_{st})\Delta\varphi = 0. \tag{12.36}$$

Since this equation corresponds to Eq. (2.115) with negative k', the nontrivial equilibrium states are unstable. As a result, the bifurcation diagram is described as in Figure 12.3.

The equivalent external and conservative force in Eq. (12.24) is

$$F^*_{equiv} = -F^{\#*}_{equiv} = \sigma\sin\varphi - \frac{1}{4}a_e^{*2}\sin 2\varphi = \sin\varphi\left(\sigma - \frac{1}{2}a_e^{*2}\cos\varphi\right). \tag{12.37}$$

The nondimensional potential energy U^* is

$$U^* = -\int F^*_{equiv}d\varphi = -\sigma\cos\varphi + \frac{a_e^{*2}}{8}\cos 2\varphi, \tag{12.38}$$

which is described for some values of σ in Figure 12.3. Because higher excitation frequency increases the absolute value of two unstable equilibrium states, i.e. the distance between the maximum states in the potential energy curve increases, the robustness of the stability against the disturbance is enhanced. If we would apply the excitation with infinite-high frequency, i.e. $\sigma = 0$, the upright position is stabilized even against large disturbance as almost $\varphi = \frac{\pi}{2}$.

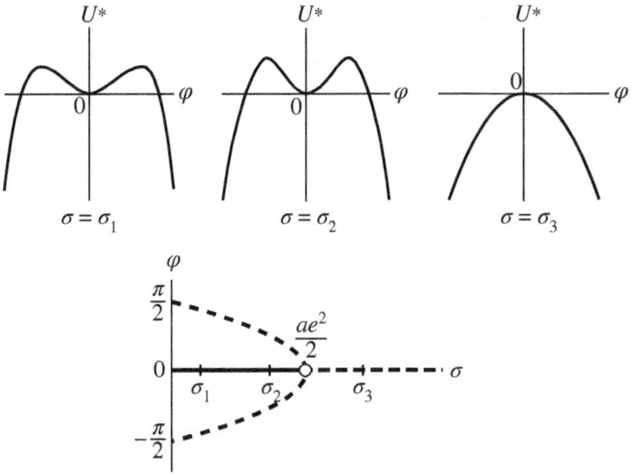

Figure 12.3 Global bifurcation diagram of inverted pendulum under high-frequency excitation and potential energy curves. In case with not enough excitation frequency $\sigma = \sigma_3$, the potential energy curve has only one maximum at the trivial steady state. When the trivial steady state is stabilized ($\sigma = \sigma_1$ and σ_2), the maximum at the trivial steady state becomes a minimum and two maxima appear. Increasing the excitation frequency leads to an increase in the distance of these maxima corresponding to unstable steady states and enhances the stability against disturbance.

12.4 Experiments

Figure 12.4 shows a picture of stabilization of an inverted pendulum subject to the high-frequency excitation, where the length and tip mass of the pendulum are 5.37×10^{-2} m and 2.35×10^{-2} kg, respectively. The excitation amplitude and frequency are 4.5×10^{-3} m and 35 Hz, respectively; see (Yabuno et al. 2004) for more detail.

We observed the phenomenon in Videoclip 16.8(1) as mentioned in Section 16.8. The high-frequency vertical excitation is applied to the supporting point. The stabilized state is not at the complete upright position because the excitation direction cannot be set completely coincidental with the gravity direction in experiment. The inclination is investigated in the subsequent section.

12.5 Effects of the Excitation Direction on the Bifurcation

We investigate that the excitation direction changes qualitatively the nonlinear characteristics of the bifurcation produced in the pendulum (Ciezkowski 2011; VanDalen 2004; Weibel and Baillieul 1998). We consider the dynamics of the pendulum with length l in Figure 12.1b, where the periodic excitation $a_e \cos \nu t$ at the supporting point P is along x'-direction inclined from the vertical direction of θ_{off}. The equation of motion of the pendulum is

$$\frac{d^2\theta}{dt^2} + \frac{g}{l}\sin\theta + \frac{a_e}{l}\nu^2 \cos \nu t \sin(\theta - \theta_{off}) = 0, \tag{12.39}$$

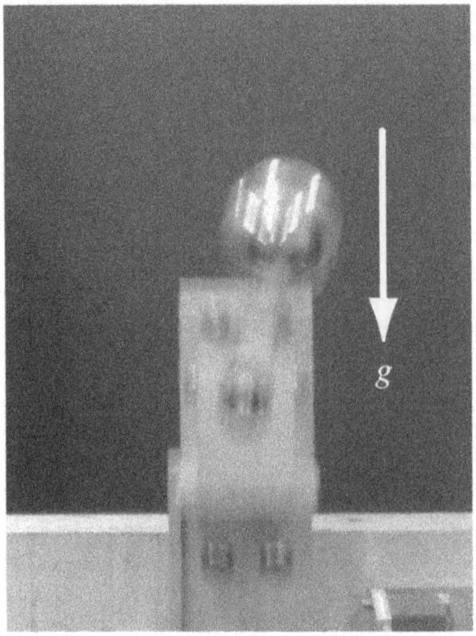

Figure 12.4 Stabilization of inverted pendulum by high-frequency vertical excitation; see Videoclip 16.8(1) as mentioned in Section 16.8. Source: Reprinted with permission (Yabuno 2004).

where g is the gravity acceleration and l is the length of a pendulum. The dimensionless equation of motion is

$$\frac{d^2\theta}{dt^{*2}} + \mu\frac{d\theta}{dt^*} + \sigma\sin\theta + a_e^*\cos t^*\sin(\theta - \theta_{off}) = 0, \tag{12.40}$$

where $a_e^* = a_e/l$ and $\sigma = (g/l)/\nu^2$ are the nondimensional excitation amplitude and the natural frequency squared, respectively. The viscous damping is assumed as the second term on the left-hand side.

When the excitation is applied exactly along the horizontal direction ($\theta_{off} = \frac{\pi}{2}$), Eq. (12.40) leads to

$$\frac{d^2\theta}{dt^{*2}} + \mu\frac{d\theta}{dt^*} + \sigma\sin\theta - a_e^*\cos t^*\cos\theta = 0. \tag{12.41}$$

In the case with the inclination $\Delta\theta_{off}$ from the horizontal direction, by substituting $\theta_{off} = \frac{\pi}{2} + \Delta\theta_{off}$ into Eq. (12.40), the nondimensional equation of motion is rewritten as

$$\frac{d^2\theta}{dt^{*2}} + \mu\frac{d\theta}{dt^*} + \sigma\sin\theta - a_e^*\cos t^*\cos(\theta - \Delta\theta_{off}) = 0, \tag{12.42}$$

which turns into

$$\frac{d^2\theta}{dt^{*2}} + \mu\frac{d\theta}{dt^*} - \sigma\sin(\theta + \pi) + a_e^*\cos t^*\sin\left(\theta - \frac{\pi}{2} - \Delta\theta_{off}\right) = 0. \tag{12.43}$$

This equation will be used in Section 12.5.1.

In the case with the inclination $\Delta\theta_{off}$ from the vertical direction, by substituting $\theta_{off} = \pi + \Delta\theta_{off}$ into Eq. (12.40), the dimensionless equation of motion is rewritten as

$$\frac{d^2\varphi}{dt^{*2}} + \mu\frac{d\varphi}{dt^*} - \sigma\sin\varphi + a_e^*\cos t^*\sin(\varphi - \Delta\theta_{off}) = 0, \tag{12.44}$$

where $\varphi = \theta - \pi$. If $\Delta\theta_{off}$ is set to be zero, Eq. (12.44) leads to Eq. (12.3).

12.5.1 Averaging by the Method of Multiple Scales

As in Section 12.2, we set the orders of a_e^*, σ, and μ as $a_e^* = \epsilon\hat{a}_e$ $\sigma = \epsilon^2\hat{\sigma}$, $\mu = \epsilon\hat{\mu}$, respectively, where $|\epsilon| \ll 1$. For Eqs. (12.43) and (12.44), we assume uniform expansions as

$$\theta(t;\epsilon) = \theta_0(t_0, t_1, t_2) + \epsilon\theta_1(t_0, t_1, t_2) + \epsilon^2\theta_2(t_0, t_1, t_2) + \cdots, \tag{12.45}$$

$$\varphi(t;\epsilon) = \varphi_0(t_0, t_1, t_2) + \epsilon\varphi_1(t_0, t_1, t_2) + \epsilon^2\varphi_2(t_0, t_1, t_2) + \cdots, \tag{12.46}$$

where the multiple time scales are $t_0 = t$, $t_1 = \epsilon t$, and $t_2 = \epsilon^2 t$. According to the same method described in Section 12.2, we can have the solvability conditions of θ_2 and φ_2 for Eqs. (12.43) and (12.44), respectively. From the relationship of the nonautonomous original equation (12.3) and the resulting autonomous averaged equation (12.24), we can easily obtain the averaged equations associated to Eqs. (12.43) and (12.44) as

$$\frac{d^2\theta}{dt^{*2}} + \mu\frac{d\theta}{dt^*} + \sigma\sin\theta - \frac{1}{4}a_e^{*2}\sin 2(\theta - \Delta\theta_{off}) = 0 \tag{12.47}$$

and

$$\frac{d^2\varphi}{dt^{*2}} + \mu\frac{d\varphi}{dt^*} - \sigma\sin\varphi + \frac{1}{4}a_e^{*2}\sin 2(\varphi - \Delta\theta_{off}) = 0, \tag{12.48}$$

respectively. Since these are autonomous, it is possible to directly apply the bifurcation analysis method introduced in Section 7.5 as mentioned in Section 12.5.2.

12.5.2 Excitation Inclined from the Vertical Direction and Perturbed Subcritical Pitchfork Bifurcation

We consider the case of $|\varphi| \ll 1$ and $|\Delta\theta_{off}| \ll 1$. Neglecting the terms with $O(\varphi^5)$ and $O(\Delta\theta_{off}\varphi)$ yields from Eq. (12.48)

$$\frac{d^2\varphi}{dt^{*2}} + \mu\frac{d\varphi}{dt^*} + \left(\frac{a_e^{*2}}{2} - \sigma\right)\varphi - \left(\frac{a_e^{*2}}{3} - \frac{\sigma}{6}\right)\varphi^3 - \frac{a_e^{*2}}{2}\Delta\theta_{off} = 0. \tag{12.49}$$

We investigate the bifurcation near the bifurcation point in case of the exactly vertical excitation $\Delta\theta_{off} = 0$. The bifurcation point is $\sigma = \frac{a_e^{*2}}{2}$ as derived in Section 12.3. Substituting Eq. (12.29) into Eq. (12.49), we have

$$\frac{d^2\varphi}{dt^{*2}} + \mu\frac{d\varphi}{dt^*} - \epsilon\varphi - \frac{a_e^{*2}}{4}\varphi^3 - \frac{a_e^{*2}}{2}\Delta\theta_{off} = 0, \tag{12.50}$$

where $\frac{\epsilon\varphi^3}{6}$ is neglected due to higher order smallness as in Section 12.3. By regarding the constant $\frac{a_e^{*2}}{2}\Delta\theta_{off}$ as σ in Eq. (7.71), Eq. (12.50) corresponds to Eq. (7.71) with negative α_3, i.e. the softening cubic nonlinearity. Therefore, the inclination of the excitation from the vertical direction perturbs the subcritical pitchfork bifurcation as in Figure 12.2b.

12.5.3 Supercritical Pitchfork Bifurcation in Horizontal Excitation and Its Perturbation Due to Inclination of the Excitation Direction

As in Section 12.5.2, we consider the case of $|\theta| \ll 1$ and $|\Delta\theta_{off}| \ll 1$. Neglecting the terms with $O(\theta^5)$ and $O(\Delta\theta_{off}\theta)$ yields from Eq. (12.47)

$$\frac{d^2\theta}{dt^{*2}} + \frac{d\theta}{dt^*} + \left(\sigma - \frac{a_e^{*2}}{2}\right)\theta + \left(\frac{a_e^{*2}}{3} - \frac{\sigma}{6}\right)\theta^3 + \frac{a_e^2}{2}\Delta\theta_{off} = 0. \tag{12.51}$$

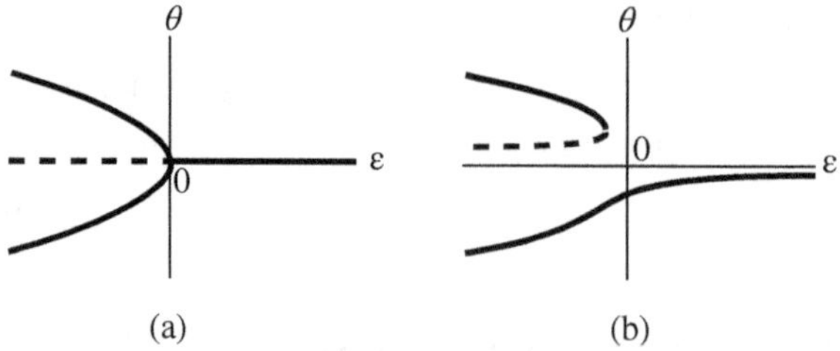

Figure 12.5 Destabilization of the trivial equilibrium state through supercritical pitchfork by high-frequency horizontal excitation. (a) is a supercritical pitchfork bifurcation in the case when the excitation direction is exactly horizontal. (b) is a perturbed supercritical pitchfork bifurcation in the case when the excitation direction is not exactly horizontal, i.e. in case of $\Delta\theta_{off} > 0$.

In the case of $\Delta\theta_{off} = 0$, the stability of the trivial equilibrium state changes at $\sigma = \frac{a_e^{*2}}{2}$, therefore we investigate the bifurcation near this critical point. Substituting Eq. (12.29) into Eq. (12.51), we obtain

$$\frac{d^2\theta}{dt^{*2}} + \mu\frac{d\theta}{dt^*} + \epsilon\theta + \frac{a_e^{*2}}{4}\theta^3 + \frac{a_e^{*2}}{2}\Delta\theta_{off} = 0 \tag{12.52}$$

where $\epsilon\varphi^3/6$ is neglected due to higher order smallness as in Section 12.3. By regarding the constant $\frac{a_e^{*2}}{2}\Delta\theta_{off}$ as σ in Eq. (7.71), Eq. (12.52) corresponds to Eq. (7.71) with positive α_3, i.e. the hardening cubic nonlinearity. Therefore, in the case without inclination, i.e. $\Delta\theta_{off} = 0$, a supercritical pitchfork bifurcation occurs as Figure 12.5a. The inclination of the excitation from the horizontal direction perturbs the supercritical pitchfork bifurcation as in Figure 12.5b.

As seen from the above results, the variations of the excitation frequency and the direction against the gravity cause the changes of the stable equilibrium states. In Chapter 15, exploiting such phenomena based on bifurcations, we realize the motion control of an underactuated manipulator without state feedback control.

12.6 Stabilization of Statically Unstable Equilibrium States by High-Frequency Excitation

We have shown the stabilization of the statically unstable equilibrium state of inverted pendulum by high-frequency excitation. From the stabilization mechanism, it is noticed that high-frequency excitation can stabilize a wide range of statically unstable equilibrium states. Here, we introduce the stabilization of the buckling phenomenon, which is statically destabilized as shown in Chapter 5, by high-frequency excitation.

The axially high-frequency excitation can stabilize the destabilized trivial equilibrium state of a clamped–clamped beam buckled by compressive force (Jensen 2000; Jensen et al. 2000; Yabuno and Tsumoto 2007) and of a cantilever buckled by its own weight (Champneys and Fraser 2000), as experimentally shown in Videoclips 16.8(2) and 16.8(3) in Section 16.8.

References

Champneys, A. R. and W. B. Fraser (2000). The 'indian rope trick' for a parametrically excited flexible rod: linearized analysis. In *Proceedings of the Royal Society of London A: Mathematical, Physical and Engineering Sciences*, Volume 456, pp. 553–570. The Royal Society.

Ciezkowski, M. (2011). Stabilization of pendulum in various inclinations using open-loop control. *Acta Mechanica et Automatica* 5(4), 22–28.

Dorf, R. C. and R. H. Bishop (2011). *Modern control systems*. Pearson.

Jensen, J. S. (2000). Buckling of an elastic beam with added high-frequency excitation. *International Journal of Non-Linear Mechanics* 35(2), 217–227.

Jensen, J. S., D. Tcherniak, and J. J. Thomsen (2000). Stiffening effects of high-frequency excitation: experiments for an axially loaded beam. *Journal of applied mechanics* 67(2), 397–402.

Kapitza, P. L. (1965). Dynamical stability of a pendulum when its point of suspension vibrates, and pendulum with a vibrating suspension. *Collected papers of PL Kapitza* 2, 714–737.

Landau, L. and E. Lifshitz (1960). Mechanics pergamon. *New York*, 87.

Richards, C. J., T. J. Smart, P. H. Jones, and D. Cubero (2018). A microscopic kapitza pendulum. *Scientific reports* 8(1), 1–10.

Stephenson, A. (1908a). L. on periodic nongenerating force of high frequency. *The London, Edinburgh, and Dublin Philosophical Magazine and Journal of Science* 16(94), 616–621.

Stephenson, A. (1908b). Xx. on induced stability. *The London, Edinburgh, and Dublin Philosophical Magazine and Journal of Science* 15(86), 233–236.

VanDalen, G. J. (2004). The driven pendulum at arbitrary drive angles. *American Journal of Physics* 72(4), 484–491.

Weibel, S. P. and J. Baillieul (1998). Open-loop oscillatory stabilization of an n-pendulum. *International Journal of Control* 71(5), 931–957.

Yabuno, H. (2004). *Kougaku no tameno Hisenkei-kaiseki Nyuumon (The Elements of Nonlinear Analysis for Engineering)*. Saiensu-sha.

Yabuno, H., M. Miura, and N. Aoshima (2004). Bifurcation in an inverted pendulum with tilted high-frequency excitation: analytical and experimental investigations on the symmetry-breaking of the bifurcation. *Journal of sound and vibration* 273(3), 493–513.

Yabuno, H. and K. Tsumoto (2007). Experimental investigation of a buckled beam under high-frequency excitation. *Archive of Applied Mechanics* 77(5), 339–351.

13

Self-Excited Resonator in Atomic Force Microscopy (Utilization of Dynamic Instability)

Atomic force microscopy (AFM) is an instrument to measure the profile of sample surfaces with atomic scale by detecting the variation in the dynamics of a force sensing cantilever probe. There are some imaging modes such as contact mode, amplitude modulation (AM) mode, frequency modulation (FM) mode, and so on (for example, Israelachvili (1991) and Morita et al. (2012)). We here focus on FM–AFM, which has received considerable attention as a method to obtain AFM images without contact of the probe to a sample. We start with a brief review of the measurement methods by FM–AFM, which are based on a periodically external excitation and a self-excited oscillation.

13.1 Principle of Frequency Modulation Atomic Force Microscope (FM–AFM)

Figure 13.1 schematically describes the configuration of the microcantilever probe and the sample surface at time t. We introduce a $z' - x$ coordinate system, whose origin O' is located at the supporting point of the probe at the start of scanning. The sample surface is measured by scanning the microcantilever probe in x-direction. The vertical position of the tip of the probe is expressed by z' and the vertical distance between the supporting point of the probe and point P on the sample surface is $z' = z_{sample}$. The vertical distance $z_{sample} - z'$ between the tip of the probe and point P, i.e. the tip-sample distance in Figure 13.1, determines the atomic force acting on the tip of the cantilever from the sample surface. The equilibrium state of the probe is changed by the modulation of the tip-sample distance due to the surface profile while scanning $x_0(t)$. The atomic force is not proportional to the tip-sample distance but expressed as a nonlinear function of distance, and this feature enables us to measure the profile as mentioned later. Using the concepts of equilibrium state and the linearization in the neighborhood of the equilibrium state, as mentioned in Chapter 1, we analyze the dynamics of the probe subject to the atomic force and show the measurement principle of FM–AFM.

By considering only the first mode of the motion of the probe, the microcantilever probe can be regarded as a spring-mass system as in Figure 13.2; the spring has an equivalent stiffness k ($k > 0$) and its natural length is zero. When the position of the tip of the cantilever

Linear and Nonlinear Instabilities in Mechanical Systems: Analysis, Control and Application, First Edition. Hiroshi Yabuno.

Companion website: www.wiley.com/go/yabuno/instabilitiesinmechanicalsystems

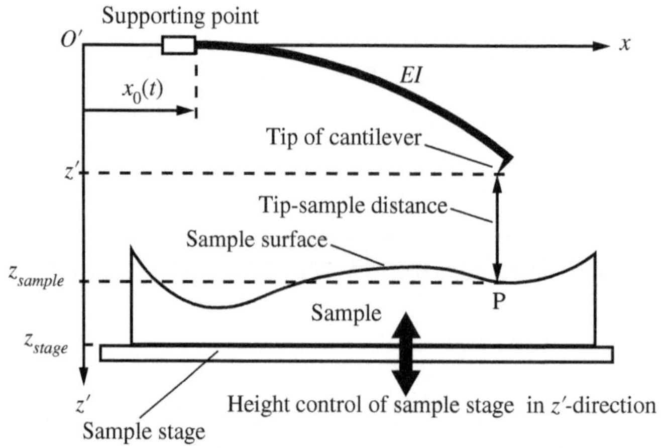

Figure 13.1 Configuration of microcantilever probe and sample surface in AFM.

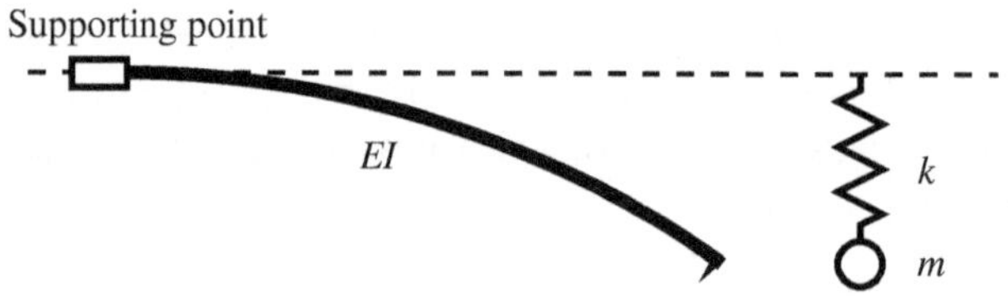

Figure 13.2 Microcantilever probe in AFM and its representation as a sprig-mass model.

or the mass of the spring-mass system is z', the spring elongation in the spring mass model is z'. Then, the restoring force of the spring is expressed as

$$F_s = -kz'. \tag{13.1}$$

Viscous damping effect on the dynamics of probe due to measurement environments is assumed as

$$F_d = -c\frac{dz'}{dt}, \tag{13.2}$$

where c is a positive constant.

We introduce an additional coordinate ζ to directly express the tip-sample distance as shown in Figure 13.3a. The tip-sample force F_a depends on the range of distance $\zeta(= z_{sample} - z')$ between the tip and sample; it is classified into the long and short ranges that are up to 100 nm and fractions of nm, respectively. We consider here chemical forces with short range contribution. Then, the atomic force acting on the tip of the probe or on the mass in the model from the sample surface $F_a(\zeta)$ in Figure 13.3b is related to an interatomic potential first proposed in 1924 by John Lennard-Jones, which is expressed as follows (for example, Israelachvili 1991; Morita et al. 2012):

$$U(\zeta) = 4\epsilon\left[\left(\frac{\sigma}{\zeta}\right)^{12} - \left(\frac{\sigma}{\zeta}\right)^{6}\right], \tag{13.3}$$

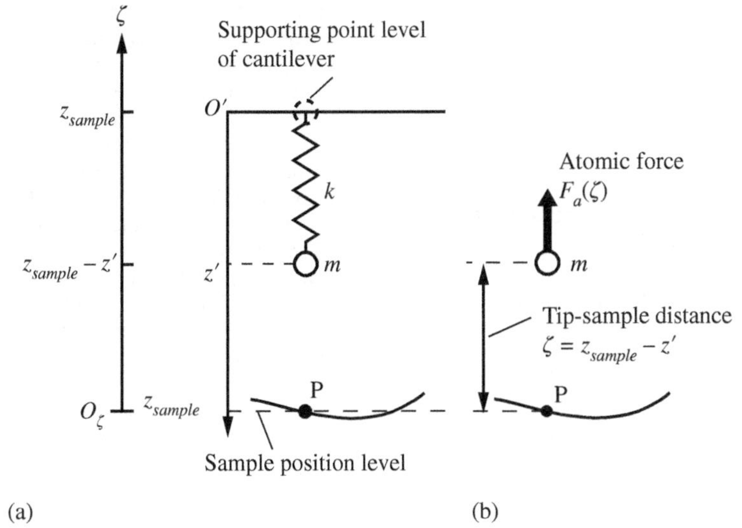

Figure 13.3 Analytical model and tip-sample interaction force: (a) mass-spring system subject to atomic force; (b) tip-sample distance and atomic force.

where ϵ and σ are fitting parameters depending on the materials, which express the depth of the potential well and the distance at which the inter-particle potential energy is zero, respectively. The first and second terms correspond to repulsive and absorptive forces, respectively. The schematic diagram is shown in Figure 13.4a. Then, the corresponding schematic diagram of the atomic force which is obtained by

$$F_a(\zeta) = -\frac{dU}{d\zeta}, \tag{13.4}$$

is described as the solid line in Figure 13.4b. The positive and negative atomic forces show the repulsive and absorptive ones between the tip and the sample, respectively. Figure 13.5 expresses the atomic force and the spring forces acting on the mass by using z'. These forces are balanced at $z' = z_{st}$. The dashed line denotes the restoring force of the spring by Eq. (13.1).

Now, we can write the equation of motion of the mass in Figure 13.3a as

$$\begin{aligned} m\frac{d^2z'}{dt^2} &= F_s(z') + F_d\left(\frac{dz'}{dt}\right) - F_a(\zeta) \\ &= -kz' - c\frac{dz'}{dt} - F_a(\zeta), \end{aligned} \tag{13.5}$$

where it is considered that $\zeta = z_{sample} - z'$ and the direction of the atomic force $F_a(\zeta)$ is opposite to that of z'-axis (see Figure 13.3b) . From the definition in Chapter 1, the equilibrium state $z' = z_{st}$ satisfies the following equilibrium equation:

$$0 = -kz_{st} - F_a(\zeta_{st}), \tag{13.6}$$

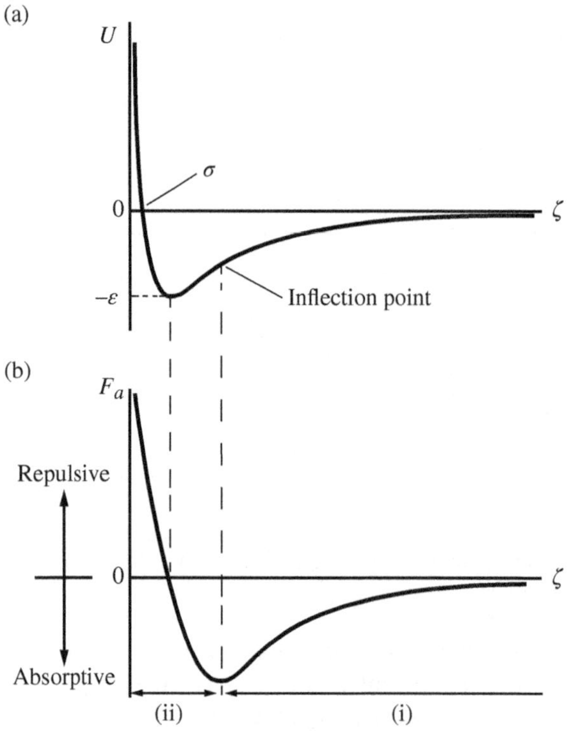

Figure 13.4 (a) Lennard-Jones potential. (b) Corresponding nonlinear atomic force with respect to the tip-sample distance ζ: positive and negative forces show the repulsive and absorptive ones between the tip and the sample, respectively. In the neighborhood of a distance ζ in the ranges of (i) and (ii), which correspond to those in Figure 13.5, the atomic force behaves as a linear spring with positive and negative stiffness, respectively.

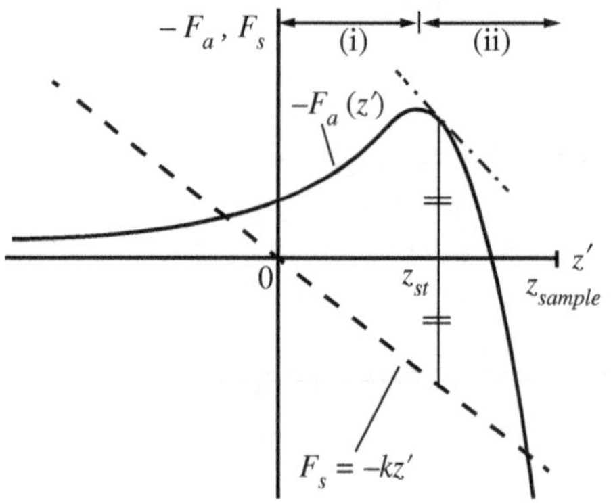

Figure 13.5 Atomic and spring forces. These forces are balanced at the equilibrium state $z' = z_{st}$. The atomic force gives the equivalent positive and negative stiffness in regions (i) and (ii) to the probe, respectively.

where $\zeta_{st} \stackrel{\text{def}}{=} z_{sample} - z_{st}$. We can linearize the atomic force in the neighborhood of the equilibrium state as follows:

$$-F_a(\zeta) = -F_a(\zeta_{st}) - \left.\frac{dF_a}{d\zeta}\right|_{\zeta=\zeta_{st}} (\zeta - \zeta_{st}) + \cdots \tag{13.7}$$

or

$$-F_a(z_{sample} - z') = -F_a(z_{sample} - z_{st}) - \left.\frac{dF_a}{dz'}\right|_{z'=z_{st}} (z' - z_{st}) + \cdots , \tag{13.8}$$

where $-\left.\frac{dF_a}{dz'}\right|_{z'=z_{st}}$ is the gradient of the dashed and dotted lines at $z = z_{st}$ in Figure 13.5. Substituting Eq. (13.8) into Eq. (13.5), considering Eq. (13.6), and letting

$$z' = z_{st} + z \tag{13.9}$$

yields

$$m\frac{d^2z}{dt^2} + c\frac{dz}{dt} + \left(k + \left.\frac{dF_a}{dz'}\right|_{z'=z_{st}}\right) z + \cdots = 0. \tag{13.10}$$

Then, the equivalent natural frequency is as follows:

$$\omega(z_{st}) = \sqrt{\frac{k + \left.\frac{dF_a}{dz'}\right|_{z'=z_{st}}}{m}}. \tag{13.11}$$

Because F_a is not a linear function of z' as shown in Figure 13.5, the force gradient $\frac{dF_a}{dz'}|_{z'=z_{st}}$ depends on z_{st}. Then, because z_{st} is determined by Eq. (13.6), the equivalent natural frequency is instantaneously modulated by the change of z_{sample} (see Figures 13.1 and 13.3), while scanning x-direction. Furthermore, Eq. (13.10) is rewritten as

$$\frac{d^2z}{dt^2} + 2\gamma\omega(z_{st})\frac{dz}{dt} + \omega(z_{st})^2 z = 0, \tag{13.12}$$

where $\gamma = \frac{c}{2m\omega(z_{st})}$ is damping ratio.

To measure the concavity and convexity of a sample surface, while scanning in x-direction, we move the sample stage upward or downward in z'-direction in such a way that the equivalent natural frequency $\omega(z_{st})$ of the probe is kept constant, i.e. z_{sample} is kept constant as a result. Then, the motion of the sample stage while scanning is equivalent to the profile of the sample surface. This is the essence of measurement by AFM and the high accurate measurement of the profile is carried out by the measurement of the equivalent natural frequency $\omega(z_{st})$ with high accuracy. It should be noted that the graphs in Figures 13.4 and 13.5 are not described quantitatively. The measurement of the profile of sample surface does not need them.

13.2 Detection of Frequency Shift Based on External Excitation

We consider two methods to detect the natural frequency of the cantilever probe based on different kinds of vertical excitation. The vertical excitation is applied at the supporting point of the probe. The effect is equivalent to the displacement $z_e(t)$ in the spring-mass model of Figure 13.6. Equation (13.5) is rewritten as

$$m\frac{d^2z'}{dt^2} = -k\Delta z - c\frac{dz'}{dt} - F_a(z_{sample} - z_{st}) - \left.\frac{dF_a}{dz'}\right|_{z'=z_{st}}(z' - z_{st}), \tag{13.13}$$

where Δz is the elongation of the spring expressed as

$$\Delta z = z' - z_e. \tag{13.14}$$

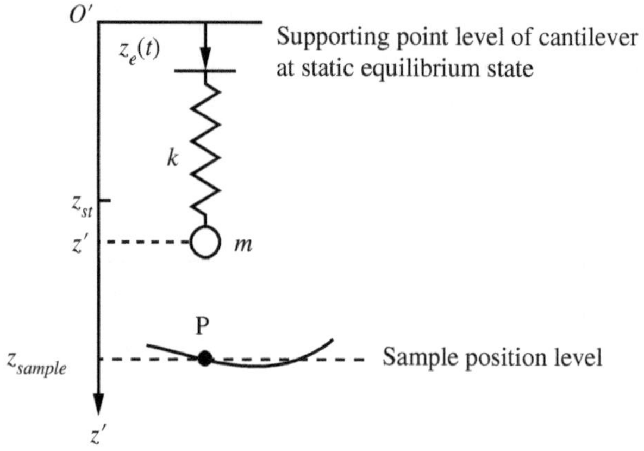

Figure 13.6 Application of displacement at the supporting point of the cantilever.

By using Eqs. (13.6) and (13.9), this is rewritten as

$$m\frac{d^2z}{dt^2} = -k(z - z_e) - c\frac{dz}{dt} - \left.\frac{dF_a}{dz'}\right|_{z'=z_{st}} z, \tag{13.15}$$

or equivalently,

$$\frac{d^2z}{dt^2} + 2\gamma\omega(z_{st})\frac{dz}{dt} + \omega(z_{st})^2 z = \omega_c^2 z_e, \tag{13.16}$$

where the nonlinear terms with respect to $z = z' - z_{st}$ are neglected and $\omega_c = \sqrt{\frac{k}{m}}$ is the natural frequency of the probe or equivalent mass–spring model in the case without atomic force.

In this section, we consider the use of the frequency-response of a probe under a harmonically external excitation. Applying the following displacement on the supporting point:

$$z_e = a_e \cos \nu t, \tag{13.17}$$

where ν and a_e are the excitation frequency and amplitude, respectively. Then, Eq. (13.16) leads to

$$\frac{d^2z}{dt^2} + 2\gamma\omega(z_{st})\frac{dz}{dt} + \omega(z_{st})^2 \zeta = a_e\omega_c^2 \cos \nu t. \tag{13.18}$$

Since the associated homogeneous problem corresponds to the case with $c' > 0$ and $k' > 0$ in Eq. (2.113), the homogeneous solution decays with time. Therefore, the steady-state oscillation of the probe can be expressed by a particular solution of Eq. (13.18). Using the analytical result in Section C.1, we obtain a particular solution as

$$z_p = a_p \cos(\nu t + \phi), \tag{13.19}$$

where

$$a_p = \frac{a_e\omega_c^2}{\sqrt{\{\omega(z_{st})^2 - \nu^2\}^2 + \{2\gamma\omega(z_{st})\nu\}^2}} \tag{13.20}$$

and

$$\phi = -\arctan \frac{2\gamma\omega(z_{st})\nu}{\omega(z_{st})^2 - \nu^2} \quad (-\pi < \phi \leq 0). \tag{13.21}$$

Furthermore, when the excitation frequency ν is

$$\nu = \sqrt{1 - 2\gamma^2}\omega(z_{st}),$$

the response amplitude becomes maximum as

$$a_{p-max} = \frac{a_e\omega_c^2}{2\gamma\omega(z_{st})^2\sqrt{1-\gamma^2}}. \tag{13.22}$$

In the case of $\gamma > \frac{1}{\sqrt{2}}$, there is no peak.

Now, let us consider the natural frequency detection in an ultrahigh vacuum (UHV) measurement environment (for example, Kitamura and Iwatsuki 1998) that can be represented by a very small damping $0 < \gamma \ll 1$. Then, the steady-state amplitude has the peak value when $\nu \approx \omega(z_{st})$. Therefore, by experimentally measuring the excitation frequency ν_{peak}, where the frequency response curve has the peak value, we can detect the equivalent natural frequency of the probe $\omega(z_{st})$. However, the more the damping effect increases, the more the excitation frequency ν_{peak} deviates from the natural frequency $\omega(z_{st})$ to be desired, and the detection accuracy of $\omega(z_{st})$ becomes worse. Furthermore, in the case of $\gamma > \frac{1}{\sqrt{2}}$, the detection of $\omega(z_{st})$ is completely impossible according to this method based on the externally harmonic excitation.

13.3 Detection of Frequency Shift Based on Self-Excitation

In the nanobiological science, high-resolution imaging of native biological samples has been required in liquid environments, i.e. high-viscosity environments. The method based on the externally harmonic excitation mentioned above is not applicable. To avoid the decrease of the quality factor Q, i.e. the increase of γ, owing to the viscosity of the measurement environments, the velocity feedback control is employed to keep positive small γ (for example, Varol et al. 2008). Furthermore, the method using the self-excited oscillation produced in a resonator is well known and applied by setting velocity feedback to be larger and getting a negative γ (for example, Albrecht et al. 1991).

Before proceeding further, let us briefly review the utilization of self-excited oscillation compared to that of external excitation. The response frequency of the self-excited oscillation always equals the natural frequency of a resonator in case of small response amplitude as shown in Eq. (10.32). This characteristic which can detect the natural frequency shift of the resonator has been widely used to enhance accuracy of sensors as not only AFMs but also mass sensors (for example, Davis and Boisen 2005; Ekinci et al. 2004; Gupta et al. 2004; Ono et al. 2003; ZHANG et al. 2012) and stiffness sensors (for example, Greenleaf et al. 2003; Sacks 2000). Another characteristic of the self-excited oscillation is utilized for viscometers. While increasing the velocity feedback gain, the damping of the resonator is changed from positive to negative as arrow (2) in Figure 2.7 and the self-excited oscillation occurs. The effect of the velocity feedback at the critical gain equals the viscous damping acting on the resonator. Through the critical feedback gain, we can measure the viscosity of a sample liquid into which the resonator is immersed (Higashino et al. 2015; Mouro et al. 2017; Woodward 1953). This measurement method is applicable even to highly viscous fluids in contrast to the method based on the half-bandwidth of the frequency response curve under the external harmonic excitation (Lee et al. 2012).

By the way, when the natural frequency is shifted, to maintain the resonance in the resonator by external excitation, we have to tune the excitation frequency. On the other hand, the self-excited oscillation maintains the resonance without tuning; this characteristic is called *autoresonant* (Babitsky 1995). For example, the autoresonance is used in vibrating machines which perform ultrasonically assisted cutting because the resonance is kept regardless of the variation of cutting load (Babitsky et al. 2004). In addition to that, many other applications of the self-excited oscillation have been proposed (Malas and Chatterjee 2014, 2016).

Let us return to the topic of AFM. Instead of the input displacement of Eq. (13.17), we apply the displacement by the feedback with respect to the probe velocity $\frac{dz}{dt}$ as

$$z_e = k_{v-lin}\frac{dz}{dt}, \tag{13.23}$$

where k_{v-lin} is a linear velocity feedback gain. Then, Eq. (13.16) leads to

$$\frac{d^2z}{dt^2} + \{2\gamma\omega(z_{st}) - k_{v-lin}\omega_c^2\}\frac{dz}{dt} + \omega(\zeta_{st})^2 z = 0. \tag{13.24}$$

The homogeneous problem corresponds to the case with $c' = 2\gamma\omega(z_{st}) - k_{v-lin}\omega_c^2$ and $k' = \omega(z_{st})^2 > 0$ in Eq. (2.115). Increasing feedback gain k_{v-lin}, the damping of the system is changed from positive ($c' > 0$) to negative ($c' < 0$), so that the trivial equilibrium state is destabilized at $c' = 0$ along the arrow (2) in Figure 2.7 and self-excited oscillation is produced. The phase space is changed from Figure 2.5a ($c' > 0$: positive damping) to (c) ($c' < 0$: negative damping) through (b) ($c' = 0$: no damping). When k_{v-lin} is set to be near the critical gain $2\gamma\omega(z_{st})/\omega_c^2$, the eigenvalues λ_i described as

$$\lambda_i = \frac{k_{v-lin}\omega_c^2 - 2\gamma\omega(z_{st})}{2} \pm \omega(z_{st})\sqrt{1 - \left(\frac{2\gamma\omega(z_{st}) - k_{v-lin}\omega_c^2}{2\omega(z_{st})}\right)^2}\, i, \tag{13.25}$$

can be approximated to

$$\lambda = \frac{k_{v-lin} - 2\gamma\omega(z_{st})}{2} \pm \omega(z_{st}) i, \tag{13.26}$$

in the case of

$$\left|\frac{2\gamma\omega(z_{st}) - k_{v-lin}\omega_c^2}{2\omega(z_{st})}\right| \ll 1. \tag{13.27}$$

Therefore, if we set the feedback gain k_{v-lin} to be near to and greater than the critical gain $k_{v-lin-cr} \overset{\text{def}}{=} \frac{2\gamma\omega(z_{st})}{\omega_c^2}$, we can detect the frequency $\omega(z_{st})$ from the experimentally observed self-excited oscillation produced. This is the essence of the frequency detection method based on self-excitation.

13.4 Amplitude Control for Self-Excited Microcantilever Probe

By using a self-excited probe, the frequency detection becomes possible even in high-viscosity environment. For the measurement of usually delicate native biological samples, it is necessary to prevent damage to the soft materials by nondestructive measurement without contact, i.e. to realize a constant low response amplitude of the microcantilever. To this end, in general, the gain of the velocity positive feedback can be automatically changed in real time; this is called automatic gain control" (AGC) (Albrecht et al. 1991). The feedback gain in this method is set near the critical value $k_{v-lin-cr}$ to produce self-excited oscillation. Here, we consider another method, that is, by using nonlinear dynamics of van der Pol oscillator. This method can set the feedback gain far from the critical value. Therefore, the self-excited oscillation is not easily stopped even if the viscous effect of the environment on the probe is increased (Yabuno et al. 2008).

In addition to the linear feedback in Eq. (13.23), we apply the nonlinear feedback as

$$z_e = k_{v-lin}\frac{dz}{dt} + k_{v-non}z^2\frac{dz}{dt}, \tag{13.28}$$

where k_{v-non} is a nonlinear velocity feedback gain. Then, Eq. (13.16) leads to

$$\frac{d^2z}{dt^2} + \{2\gamma\omega(z_{st}) - k_{v-lin}\omega_c^2 - k_{v-non}\omega_c^2 z^2\}\frac{dz}{dt} + \omega(z_{st})^2 z = 0. \tag{13.29}$$

This equation is nondimensionalized as

$$\frac{d^2z^*}{dt^{*2}} + \left(2\gamma - \frac{k_{v-lin}\omega_c^2}{\omega(z_{st})} - \frac{k_{v-non}\omega_c^2 L^2}{\omega(z_{st})}z^{*2}\right)\frac{dz^*}{dt^*} + z^* = 0, \tag{13.30}$$

where z and t are nondimensionalized as $z^* = \frac{z}{L}$ and $t^* = \frac{t}{T}$ by introducing the representative length and time, and in particular, T is set as $T = \frac{1}{\omega(z_{st})}$. By regarding the dimensionless parameters $2\gamma - \frac{k_{v-lin}\omega_c^2}{\omega(z_{st})}$ and $-\frac{k_{v-non}\omega_c^2 L^2}{\omega(z_{st})}$ in Eq. (13.30) as γ and β_3 in Eq. (10.7), Eq. (13.30) is equivalent to Eq. (10.7) in case of $\gamma_3 = 0$. Therefore, by setting the linear and nonlinear feedback gains, k_{v-lin} and k_{v-non} so that $\gamma < 0$ and $\beta_3 > 0$ in Eq. (10.7), we produce the self-excited oscillation through supercritical Hopf bifurcation with nontrivial stable steady-state amplitude. Furthermore, by letting $\Gamma(= \beta_3)$ in Eq. (10.26) be $-\frac{k_{v-non}\omega_c^2 L^2}{\omega(z_{st})}$, the steady-state amplitude can be decreased by setting the nonlinear feedback gain k_{v-non} to be negative and its absolute value to be large.

References

Albrecht, T., P. Grütter, D. Horne, and D. Rugar (1991). Frequency modulation detection using high-q cantilevers for enhanced force microscope sensitivity. *Journal of Applied Physics* 69(2), 668–673.

Babitsky, V. (1995). Autoresonant mechatronic systems. *Mechatronics* 5(5), 483–495.

Babitsky, V., A. Kalashnikov, and F. Molodtsov (2004). Autoresonant control of ultrasonically assisted cutting. *Mechatronics* 14(1), 91–114.

Davis, Z. J. and A. Boisen (2005). Aluminum nanocantilevers for high sensitivity mass sensors. *Applied Physics Letters* 87(1), 013102.

Ekinci, K., X. Huang, and M. Roukes (2004). Ultrasensitive nanoelectromechanical mass detection. *Applied Physics Letters* 84(22), 4469–4471.

Greenleaf, J. F., M. Fatemi, and M. Insana (2003). Selected methods for imaging elastic properties of biological tissues. *Annual review of biomedical engineering* 5(1), 57–78.

Gupta, A., D. Akin, and R. Bashir (2004). Single virus particle mass detection using microresonators with nanoscale thickness. *Applied Physics Letters* 84(11), 1976–1978.

Higashino, K., H. Yabuno, K. Aono, Y. Yamamoto, and M. Kuroda (2015). Self-excited vibrational cantilever-type viscometer driven by piezo-actuator. *Journal of Vibration and Acoustics* 137(6), 061009.

Israelachvili, J. (1991). *Intermolecular and Surface Forces* (2nd Edn. ed.). Academic Press.

Kitamura, S. and M. Iwatsuki (1998). High-resolution imaging of contact potential difference with ultrahigh vacuum noncontact atomic force microscope. *Applied Physics Letters* 72(24), 3154–3156.

Lee, I., K. Park, and J. Lee (2012). Note: Precision viscosity measurement using suspended microchannel resonators. *Review of Scientific Instruments* 83(11), 116106.

Malas, A. and S. Chatterjee (2014). Generating self-excited oscillation in a class of mechanical systems by relay-feedback. *Nonlinear Dynamics* 76(2), 1253–1269.

Malas, A. and S. Chatterjee (2016). Modal self-excitation in a class of mechanical systems by nonlinear displacement feedback. *Journal of Vibration and Control*, 1077546316651786.

Morita, S., R. Wiesendanger, and E. Meyer (2012). *Noncontact Atomic Force Microscopy. NanoScience and Technology*. Springer Berlin Heidelberg.

Mouro, J., B. Tiribilli, and P. Paoletti (2017). Nonlinear behaviour of self-excited microcantilevers in viscous fluids. *Journal of Micromechanics and Microengineering* 27(9), 095008.

Ono, T., X. Li, H. Miyashita, and M. Esashi (2003). Mass sensing of adsorbed molecules in sub-picogram sample with ultrathin silicon resonator. *Review of scientific instruments* 74(3), 1240–1243.

Sacks, M. S. (2000). Biaxial mechanical evaluation of planar biological materials. *Journal of elasticity and the physical science of solids* 61(1-3), 199.

Varol, A., I. Gunev, B. Orun, and C. Basdogan (2008). Numerical simulation of nano scanning in intermittent-contact mode afm under q control. *Nanotechnology* 19(7), 075503.

Woodward, J. (1953). A vibrating-plate viscometer. *The Journal of the Acoustical Society of America* 25(1), 147–151.

Yabuno, H., H. Kaneko, M. Kuroda, and T. Kobayashi (2008). Van der pol type self-excited micro-cantilever probe of atomic force microscopy. *Nonlinear Dynamics* 54(1), 137–149.

ZHANG, H.-Y., P. Hong-Qing, B.-L. ZHANG, and T. Ji-Lin (2012). Microcantilever sensors for chemical and biological applications in liquid. *Chinese Journal of Analytical Chemistry* 40(5), 801–808.

14

High-Sensitive Mass Sensing by Eigenmode Shift

Due to the need of fast response probes providing instantaneous and continuous measurements, vibrational sensors are suited for online measurement. As mentioned in Chapter 13, the atomic force microscopy (AFM) measures the surface profile of samples by detecting the modulation of the natural frequency of a resonator. As will be mentioned in the first section of this chapter, the conventional mass sensor also utilizes the natural frequency shift occurring when an additional mass is settled on the resonator (Davis and Boisen (2005); Gupta et al. (2004); Ono et al. (2003)). The sensing mechanism is very easy to understand since it is based on the elementally linear vibration theory, which is mentioned in Chapter 1 and Section 2.4.4. After Section 14.2, toward the development of sensors featuring much higher sensitivity, another measurement method based on eigenmode shift produced in coupled resonators by the measured mass is introduced (Spletzer et al. (2008 ,2006)). This method is also interesting as it is related to the mode localization phenomena in coupled cantilevers (Anderson (1958); Dick et al. (2008); Pierre (1988); Sato et al. (2003)).

14.1 Conventional Mass Sensing by Frequency Shift of Resonator

We consider the mass-spring system as an analytical model of the resonator for mass sensing as shown in Figure 14.1, where Δm is the measured mass, m and k are the equivalent mass and stiffness of the resonator, respectively, and c is the equivalent damping effect of the resonator due to the environmental conditions in which the measurement takes place. The equation of motion of the resonator is

$$(m + \Delta m)\frac{d^2x}{dt^2} + c\frac{dx}{dt} + kx = 0. \tag{14.1}$$

In the situation where the damping term is neglected, for example where the damping effect is compensated by the velocity feedback control as mentioned in Section 13.3, the natural frequency ω is

$$\omega' = \sqrt{\frac{k}{m + \Delta m}} \approx \omega\left(1 - \frac{1}{2}\delta\right) \tag{14.2}$$

Linear and Nonlinear Instabilities in Mechanical Systems: Analysis, Control and Application, First Edition. Hiroshi Yabuno.

Companion website: www.wiley.com/go/yabuno/instabilitiesinmechanicalsystems

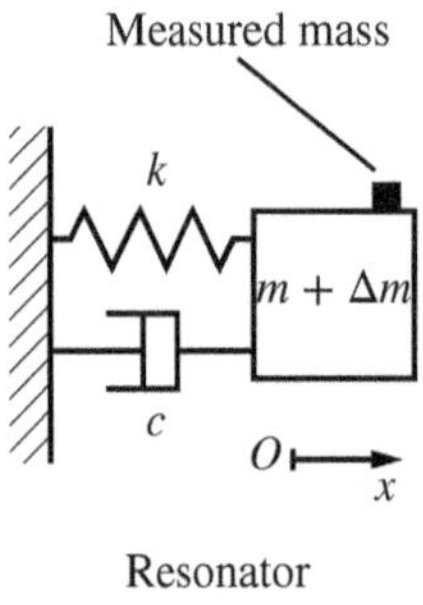

Figure 14.1 Single resonator for mass sensing: m and k are the equivalent mass and stiffness of a resonator, respectively; c expresses the damping effect of the resonator due to the kind of measurement environment; Δm is the measured mass. Source: Reprinted with permission (Yabuno (2019)).

or

$$\frac{\Delta\omega}{\omega} \approx -\frac{1}{2}\delta, \tag{14.3}$$

where $\omega = \sqrt{\frac{k}{m}}$ is the natural frequency of the resonator in the case without the measured mass, and $\delta = \frac{\Delta m}{m}$ is the ratio of the measured mass to the equivalent mass of the resonator, which is generally less than 1. $\Delta\omega = \omega' - \omega$ is the frequency shift by the measured mass. The sensitivity is expressed as

$$\left|\frac{d\omega}{dm}\right| = \frac{\omega}{2m}. \tag{14.4}$$

By increasing the natural frequency of the resonator ω or decreasing the mass of the resonator m, the sensitivity can be increased. Special fabrication processes as nanolithography techniques are now available to manufacture micro-resonators. Although such a fabrication technique has been advanced, there is the limitation of miniaturizing the resonator beyond the micro-scale. To break through the problem, a new measurement principle is proposed which uses an additional resonator, i.e. coupled twin resonators (Spletzer et al. (2008, 2006)). The method does not rely on the natural frequency shift, but the eigenmode shift, which is related to mode localization in coupled cantilevers (Anderson (1958); Dick et al. (2008); Pierre (1988); Sato et al. (2003)).

14.2 High-Sensitive Mass Sensing by Coupled Resonators

We consider the system with two coupled resonators as shown in Figure 14.2, where m and k are the equivalent mass and stiffness of the resonators, k_c is the coupled stiffness, and Δm is a measured mass which is deposited on the resonator II. This system is similar to Figure 3.4 and the equations governing the displacements x_1 and x_2 of the resonators I and

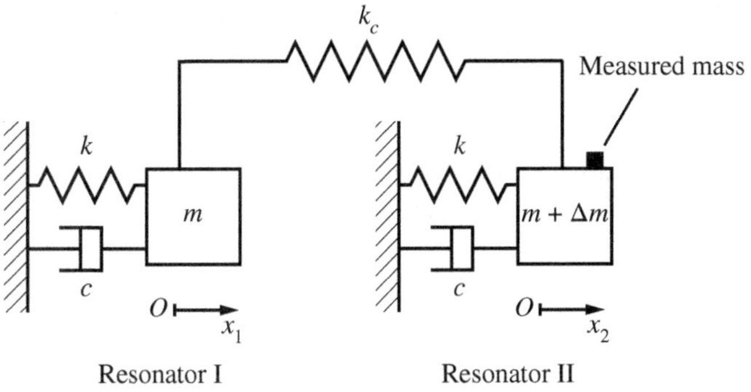

Figure 14.2 Coupled resonators with damping for mass sensing: two resonators I and II have the same physical parameters, *m*, *c*, and *k*; the coupling stiffness is k_c; Δm is the measured mass settled on resonator II. Source: Reprinted with permission (Yabuno (2019)).

II, respectively, are expressed as

$$\begin{cases} m\dfrac{d^2x_1}{dt^2} + c\dfrac{dx_1}{dt} + (k+k_c)x_1 - k_cx_2 = 0, & (14.5)\\ (m+\Delta m)\dfrac{d^2x_2}{dt^2} + c\dfrac{dx_2}{dt} - k_cx_1 + (k+k_c)x_2 = 0, & (14.6)\end{cases}$$

where c expresses the damping effects of the resonators due to a viscous environment. Introducing the dimensionless time $t^* = \sqrt{\frac{k}{m}}t$, we obtain the corresponding dimensionless equations of motion

$$\begin{cases} \dfrac{d^2x_1}{dt^{*2}} + 2\gamma\dfrac{dx_1}{dt^*} + (1+\kappa)x_1 - \kappa x_2 = 0, & (14.7)\\ \dfrac{d^2x_2}{dt^{*2}} + \dfrac{2\gamma}{1+\delta}\dfrac{dx_2}{dt^*} - \dfrac{\kappa}{1+\delta}x_1 + \dfrac{1+\kappa}{1+\delta}x_2 = 0, & (14.8)\end{cases}$$

where $\gamma = \frac{c}{2\sqrt{mk}}$, $\kappa = \frac{k_c}{k}$, and $\delta = \frac{\Delta m}{m}$ are the damping ratio, the dimensionless coupling stiffness, and the mass ratio, respectively.

14.3 Solution of Equations of Motion

Neglecting the damping effect ($\gamma = 0$), we obtain the mode. Substituting $x_1 = Ae^{\lambda t}$ and $x_2 = Be^{\lambda t}$ into Eqs. (14.7) and (14.8) yields

$$\begin{bmatrix} \Lambda + 1 + \kappa & -\kappa \\ -\dfrac{\kappa}{1+\delta} & \Lambda + \dfrac{1+\kappa}{1+\delta} \end{bmatrix} \begin{bmatrix} A \\ B \end{bmatrix} = 0, \qquad (14.9)$$

where $\Lambda = \lambda^2$. From the condition that the complex values, A and B, have nontrivial solutions, we have the following characteristic equation:

$$\Lambda^2 + \frac{(1+\kappa)(2+\delta)}{1+\delta}\Lambda + \frac{1+2\kappa}{1+\delta} = 0. \tag{14.10}$$

Because the two solutions Λ_1 and Λ_2 are real and negative, we obtain $\lambda = \pm i\omega_1$ and $\lambda = \pm i\omega_2$, where $\omega_1 = \sqrt{-\Lambda_1}$ and $\omega_2 = \sqrt{-\Lambda_2}$ $(|\Lambda_1| < |\Lambda_2|)$, as

$$\begin{cases} \Lambda_1 = \frac{1}{2(1+\delta)}\{-(2+\delta)(1+\kappa) + \sqrt{\delta^2(1+\kappa)^2 + 4\kappa^2(1+\delta)}\}, & (14.11) \\ \Lambda_2 = \frac{1}{2(1+\delta)}\{-(2+\delta)(1+\kappa) - \sqrt{\delta^2(1+\kappa)^2 + 4\kappa^2(1+\delta)}\}. & (14.12) \end{cases}$$

The eigenmodes corresponding to Λ_i $(i = 1, 2)$ are

$$\begin{bmatrix} A_i \\ B_i \end{bmatrix} = \begin{bmatrix} 1 \\ \left(\frac{\kappa}{1+\delta}\right) \Big/ \left(\Lambda_i + \frac{1+\kappa}{1+\delta}\right) \end{bmatrix}. \tag{14.13}$$

Since the ratio between A_i and B_i is real, we express A_i and B_i as

$$\begin{aligned} \begin{bmatrix} x_1 \\ x_2 \end{bmatrix} &= a_1 \begin{bmatrix} 1 \\ \left(\frac{\kappa}{1+\delta}\right) \Big/ \left(\Lambda_1 + \frac{1+\kappa}{1+\delta}\right) \end{bmatrix} \cos(\omega_1 t + \phi_1) \\ &+ a_2 \begin{bmatrix} 1 \\ \left(\frac{\kappa}{1+\delta}\right) \Big/ \left(\Lambda_2 + \frac{1+\kappa}{1+\delta}\right) \end{bmatrix} \cos(\omega_2 t + \phi_2), \end{aligned} \tag{14.14}$$

where a_1, a_2, ϕ_1, and ϕ_2 are the real numbers determined from the initial condition. The first and second terms are the first and second modes, respectively.

14.4 Mode Shift Due to Measured Mass

We calculate the mode shift when the measured mass is attached to resonator II. Substituting Eqs. (14.11) and (14.12) into Eq. (14.13), we obtain the first and second modes as

$$\begin{bmatrix} A_1 \\ B_1 \end{bmatrix} = \begin{bmatrix} 1 \\ 1 \Big/ \left(-\frac{\delta(1+\kappa)}{2\kappa} + \sqrt{\frac{\delta^2}{4\kappa^2}(1+\kappa)^2 + 1 + \delta}\right) \end{bmatrix}, \tag{14.15}$$

and

$$\begin{bmatrix} A_2 \\ B_2 \end{bmatrix} = \begin{bmatrix} 1 \\ 1 \Big/ \left(-\frac{\delta(1+\kappa)}{2\kappa} - \sqrt{\frac{\delta^2}{4\kappa^2}(1+\kappa)^2 + 1 + \delta}\right) \end{bmatrix}, \tag{14.16}$$

respectively. When the measured mass is not put $(\delta = 0)$, the amplitude ratios in the first and second modes are 1 to 1 and 1 to -1, respectively. When the measured mass is very small $(\delta \ll 1)$, Eqs. (14.15) and (14.16) are approximated as

$$\begin{bmatrix} A_1 \\ B_1 \end{bmatrix} = \begin{bmatrix} 1 \\ 1 + \frac{\delta}{2\kappa} \end{bmatrix} + O(\delta^2) \tag{14.17}$$

and

$$\begin{bmatrix} A_2 \\ B_2 \end{bmatrix} = \begin{bmatrix} 1 \\ -1 + \dfrac{\delta}{2\kappa} + \delta \end{bmatrix} + O(\delta^2), \tag{14.18}$$

respectively. The mode shifts due to the measured mass are $\frac{\delta}{2\kappa}$ and $\frac{\delta}{2\kappa} + \delta$ for the first and second modes, respectively. If the coupled stiffness is set to be the same order as the stiffness of resonators I and II, i.e. $\kappa = O(1)$, the mode shift is $O(\delta)$ and the sensitivity of measurement is very low. However, if we set the coupled stiffness κ to be very small, i.e. $\delta \ll \kappa \ll 1$, it is noticed that the mode shift of the first or second mode is very large even in the case of $\delta \ll 1$. Under the condition $\delta \ll \kappa \ll 1$, Eqs. (14.17) and (14.18) are rewritten as

$$\begin{bmatrix} A_1 \\ B_1 \end{bmatrix} = \begin{bmatrix} 1 \\ 1 + \dfrac{\delta}{2\kappa} \end{bmatrix} + O(\delta), \tag{14.19}$$

and

$$\begin{bmatrix} A_2 \\ B_2 \end{bmatrix} = \begin{bmatrix} 1 \\ -1 + \dfrac{\delta}{2\kappa} \end{bmatrix} + O(\delta), \tag{14.20}$$

respectively. The sensitivity is remarkably increased and much greater than that in the measurement based on the natural frequency shift mentioned in Section 14.1 (Spletzer et al. (2006, 2008)).

14.5 Experimental Detection Methods for Mode Shift

Using weakly coupled resonators ($\kappa \ll 1$), we carry out the very high-sensitive mass sensing from the mode shift due to measured mass. In this section, we consider methods to experimentally detect the mode shift in practical measurement systems by applying a displacement excitation Δx to resonators, I and II, as shown in Figure 14.3.

Then, the equations of motion are expressed as

$$\begin{cases} m\dfrac{d^2x_1}{dt^2} + c\dfrac{dx_1}{dt} + (k + k_c)x_1 - k_c x_2 = k\Delta x, & (14.21) \\[2ex] (m + \Delta m)\dfrac{d^2x_2}{dt^2} + c\dfrac{dx_2}{dt} - k_c x_1 + (k + k_c)x_2 = k\Delta x, & (14.22) \end{cases}$$

where x_1 and x_2 are the absolute displacements of resonators, I and II. The dimensionless equations of motion can be written as

$$\begin{cases} \dfrac{d^2x_1}{dt^{*2}} + 2\gamma\dfrac{dx_1}{dt^*} + (1 + \kappa)x_1 - \kappa x_2 = \Delta x, & (14.23) \\[2ex] \dfrac{d^2x_2}{dt^{*2}} + \dfrac{2\gamma}{1 + \delta}\dfrac{dx_2}{dt^*} - \dfrac{\kappa}{1 + \delta}x_1 + \dfrac{1 + \kappa}{1 + \delta}x_2 = \dfrac{\Delta x}{1 + \delta}. & (14.24) \end{cases}$$

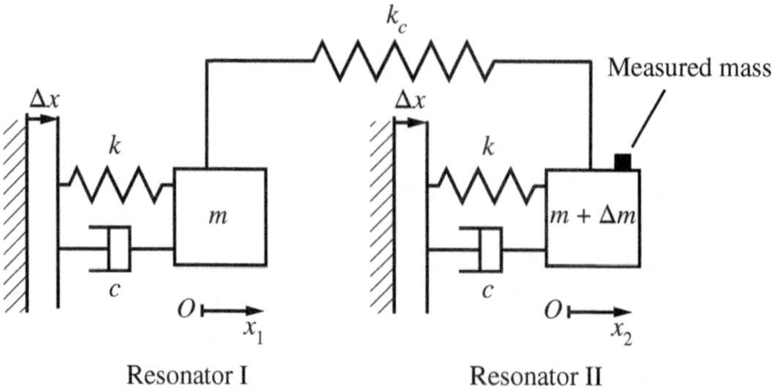

Figure 14.3 Coupled resonators with damping for mass sensing under displacement excitation.

14.5.1 Use of External Excitation

To apply the external excitation, Δx is set as

$$\Delta x = a_e \cos \nu t, \tag{14.25}$$

where a_e and ν are the excitation amplitude and frequency, respectively. If the damping effect in the measurement environment is very small, the ratios of the response amplitudes of resonators I and II, at the first and second resonance points approximately correspond to the first and second modes, respectively. However, the amplitude ratios of the cantilevers at the resonance peaks do not exactly reflect the eigenmodes due to the viscosity in the measurement environment. Namely, the two oscillators don't have in- or anti- phase unless the damping is the so-called proportional viscous damping (Caughey (1960)) and the amplitude ratio at the peaks of the frequency response curves under the external excitation deviates from the amplitude ratio of the eigenmodes. Moreover, in much-higher-viscosity environments where the resonance peak itself does not exist in the frequency response curves (for example, highly viscous liquid environments), the external excitation method is not applicable to the detection of the eigenmode. In order to compensate for the viscosity in measurement environments, we can propose the utilization of self-excited oscillation as in AFM (Endo et al. (2015, 2018); Yabuno et al. (2013)).

14.6 Use of Self-Excitation

We set Δx by the velocity feedback of resonator I as

$$\Delta x = \alpha \frac{dx_1}{dt}, \tag{14.26}$$

where α is a velocity feedback gain. Then, the dimensionless equations of motion corresponding to Eqs. (14.23) and (14.24) are

$$\begin{cases} \dfrac{d^2x_1}{dt^{*2}} + 2\gamma \dfrac{dx_1}{dt^*} + (1+\kappa)x_1 - \kappa x_2 = \beta \dfrac{dx_1}{dt^*}, & (14.27) \\[2ex] \dfrac{d^2x_2}{dt^{*2}} + \dfrac{2\gamma}{1+\delta}\dfrac{dx_2}{dt^*} - \dfrac{\kappa}{1+\delta}x_1 + \dfrac{1+\kappa}{1+\delta}x_2 = \dfrac{\beta}{1+\delta}\dfrac{dx_1}{dt^*}, & (14.28) \end{cases}$$

where $\beta = \alpha\sqrt{\frac{k}{m}}$. These equations are expressed in the matrix form as

$$\begin{bmatrix}\ddot{x}_1\\ \ddot{x}_2\end{bmatrix} + \begin{bmatrix}2\gamma-\beta & 0\\ -\dfrac{\beta}{1+\delta} & \dfrac{2\gamma}{1+\delta}\end{bmatrix}\begin{bmatrix}\dot{x}_1\\ \dot{x}_2\end{bmatrix} + \begin{bmatrix}1+\kappa & -\kappa\\ -\dfrac{\kappa}{1+\delta} & \dfrac{1+\kappa}{1+\delta}\end{bmatrix}\begin{bmatrix}x_1\\ x_2\end{bmatrix} = 0, \tag{14.29}$$

where dot denotes the differentiation with respect to the dimensionless time t^*. We introduce the coordinate transformation based on the eigenmodes

$$\begin{bmatrix}x_1\\ x_2\end{bmatrix} = P\begin{bmatrix}y_1\\ y_2\end{bmatrix}, \tag{14.30}$$

where

$$P = \begin{bmatrix}A_1 & A_2\\ B_1 & B_2\end{bmatrix} = \begin{bmatrix}1 & 1\\ 1+\dfrac{\delta}{2\kappa} & -1+\dfrac{\delta}{2\kappa}\end{bmatrix}. \tag{14.31}$$

Equation (14.30) is substituted into Eq. (14.29) and the result premultiplied by P^{-1}. Furthermore, by neglecting $O(\delta)$, we obtain

$$\begin{bmatrix}\ddot{y}_1\\ \ddot{y}_2\end{bmatrix} + \begin{bmatrix}c_{11} & c_{12}\\ c_{21} & c_{22}\end{bmatrix}\begin{bmatrix}\dot{y}_1\\ \dot{y}_2\end{bmatrix} + \begin{bmatrix}\omega_1'^2 & 0\\ 0 & \omega_2'^2\end{bmatrix}\begin{bmatrix}y_1\\ y_2\end{bmatrix} = 0, \tag{14.32}$$

where

$$c_{11} = 2\gamma - \beta + \frac{\beta\delta}{4\kappa}, \tag{14.33}$$

$$c_{12} = -\beta + \frac{\beta\delta}{4\kappa}, \tag{14.34}$$

$$c_{21} = -\frac{\beta\delta}{4\kappa}, \tag{14.35}$$

$$c_{22} = 2\gamma - \frac{\beta\delta}{4\kappa}, \tag{14.36}$$

and $\omega_1'^2$ and $\omega_2'^2$ are the approximate first and second natural frequencies, which are obtained by neglecting the small term of $O(\delta)$ in Eqs. (14.11) and (14.12), respectively.

In order to produce the self-excited oscillation in the coupled resonators, we set the feedback gain β as

$$\beta = 2\gamma + \Delta\beta, \tag{14.37}$$

where $\Delta\beta$ is very small and is represented as

$$\Delta\beta = \epsilon\Delta\hat{\beta}, \tag{14.38}$$

where $0 < \epsilon \ll 1$. Also, the smallness of $\frac{\delta}{\kappa}$ is expressed as

$$\frac{\delta}{\kappa} = \epsilon\hat{\delta}. \tag{14.39}$$

Then, Eqs. (14.33)–(14.36) are expressed respectively as

$$c_{11} = \left(\frac{\gamma\hat{\delta}}{2} - \Delta\hat{\beta}\right)\epsilon, \tag{14.40}$$

$$c_{12} = -2\gamma + \left(\frac{\gamma\hat{\delta}}{2} - \Delta\hat{\beta}\right)\epsilon, \tag{14.41}$$

$$c_{21} = -\frac{\gamma\hat{\delta}}{2}\epsilon, \tag{14.42}$$

$$c_{22} = 2\gamma - \frac{\gamma\hat{\delta}}{2}\epsilon. \tag{14.43}$$

Substituting $y_1 = Ae^{\lambda t}$ and $y_2 = Be^{\lambda t}$ into Eq. (14.32) yields the characteristic equation

$$\lambda^4 + C_3\lambda^3 + C_2\lambda^2 + C_1\lambda + C_0 = 0, \tag{14.44}$$

where

$$C_3 = 2\gamma - \Delta\hat{\beta}\epsilon \stackrel{\text{def}}{=} C_{30} + \hat{C}_{31}\epsilon, \tag{14.45}$$

$$C_2 = \omega_1'^2 + \omega_2'^2 - 2\gamma\Delta\hat{\beta}\epsilon \stackrel{\text{def}}{=} C_{20} + \hat{C}_{21}\epsilon, \tag{14.46}$$

$$C_1 = 2\gamma\omega_1'^2 + \left\{\frac{\gamma}{2}(\omega_2'^2 - \omega_1'^2)\hat{\delta} - \omega_2'^2\Delta\hat{\beta}\right\}\epsilon \stackrel{\text{def}}{=} C_{10} + \hat{C}_{11}\epsilon, \tag{14.47}$$

$$C_0 = \omega_1'^2\omega_2'^2 \stackrel{\text{def}}{=} C_{00}. \tag{14.48}$$

We express the solution of Eq. (14.44) as

$$\lambda = \lambda_0 + \epsilon\lambda_1. \tag{14.49}$$

Substituting this equation into the left-hand side of Eq. (14.44) and equating like powers of ϵ to zero yields

$$O(\epsilon^0):$$

$$(\lambda_0^2 + \omega_{1'}^2)(\lambda_0^2 + 2\gamma\lambda_0 + \omega_2'^2) = 0, \tag{14.50}$$

$$O(\epsilon):$$

$$(4\lambda_0^3 + 3C_{30}\lambda_0^2 + 2C_{20}\lambda_0 + C_{10})\lambda_1 + (\hat{C}_{31}\lambda_0^3 + \hat{C}_{21}\lambda_0^2 + \hat{C}_{11}\lambda_0) = 0. \tag{14.51}$$

By solving Eq. (14.50), the eigenvalues corresponding to the first mode are $\lambda_{1st-0} = \pm i\omega_1'$. Such first mode is neutral at this level of approximation. On the other hand, considering $\gamma > 0$, because the eigenvalues corresponding to the second mode are

$$\lambda_{2nd-0} = -\gamma \pm \sqrt{\gamma^2 - \omega_2'^2}, \tag{14.52}$$

the second mode is stable.

The stability of the first mode is determined by the analysis at $O(\epsilon)$. Substituting $\lambda_{1st-0} = \pm\omega_1'$ into λ_0 of Eq. (14.51) yields

$$\begin{bmatrix} -3C_{30}\omega_1'^2 + C_{10} & \pm 4\omega_1'^3 \mp 2C_{20}\omega_1' \\ \mp 4\omega_1'^3 \pm 2C_{20}\omega_1' & -3C_{30}\omega_1'^2 + C_{10} \end{bmatrix} \begin{bmatrix} \lambda_{1r} \\ \lambda_{1i} \end{bmatrix} = \begin{bmatrix} \hat{C}_{21}\omega_1'^2 \\ \pm\hat{C}_{31}\omega_1'^3 \mp \hat{C}_{11}\omega_1' \end{bmatrix}, \tag{14.53}$$

where $\lambda_{1r} + i\lambda_{1i} \stackrel{\text{def}}{=} \lambda_{1st-1}$. The solution of λ_{1r} is

$$\lambda_{1r} = \frac{2\omega_1'^2\{4\gamma^2\omega_1'^2 + (\omega_1'^2 - \omega_2'^2)^2\}\Delta\hat{\beta} - \omega_1'^2\gamma\hat{\delta}(\omega_1'^2 - \omega_2'^2)^2}{(-3C_{30}\omega_1'^2 + C_{10})^2 + (4\omega_1'^3 - 2C_{20}\omega_1')^2}. \tag{14.54}$$

Because the eigenvalue of the first mode is expressed from Eq. (14.49) as

$$\lambda_{1st} = \lambda_{1st-0} + \lambda_{1st-1}\epsilon, \tag{14.55}$$

we obtain

$$\lambda_{1st} = \frac{2\omega_1'^2\{4\gamma^2\omega_1'^2 + (\omega_1'^2 - \omega_2'^2)^2\}\kappa\Delta\beta - \omega_1'^2\gamma\delta(\omega_1'^2 - \omega_2'^2)^2}{\kappa\{(-3C_{30}\omega_1'^2 + C_{10})^2 + (4\omega_1'^3 - 2C_{20}\omega_1')^2\}} + i(\omega_1' + \epsilon\lambda_{1i}). \tag{14.56}$$

Therefore, if we set the feedback gain as

$$\Delta\beta > \frac{\gamma(\omega_1'^2 - \omega_2'^2)^2\delta}{2\kappa\{4\gamma^2\omega_1'^2 + (\omega_1'^2 - \omega_2'^2)^2\}}, \tag{14.57}$$

or

$$\beta > 2\gamma + \frac{\gamma\delta(\omega_1'^2 - \omega_2'^2)^2}{2\kappa\{4\gamma^2\omega_1'^2 + (\omega_1'^2 - \omega_2'^2)^2\}}, \tag{14.58}$$

the resonator undergoes self-excited oscillation due to the fact that the real part of the first mode eigenvalue is positive.

References

Anderson, P. W. (1958). Absence of diffusion in certain random lattices. *Physical review* 109(5), 1492.

Caughey, T. (1960). Classical normal modes in damped linear dynamic systems. *Journal of Applied Mechanics* 27(2), 269–271.

Davis, Z. J. and A. Boisen (2005). Aluminum nanocantilevers for high sensitivity mass sensors. *Applied Physics Letters* 87(1), 013102.

Dick, A., B. Balachandran, and C. Mote (2008). Intrinsic localized modes in microresonator arrays and their relationship to nonlinear vibration modes. *Nonlinear Dynamics* 54(1-2), 13–29.

Endo, D., H. Yabuno, K. Higashino, Y. Yamamoto, and S. Matsumoto (2015). Self-excited coupled-microcantilevers for mass sensing. *Applied Physics Letters* 106(22), 223105.

Endo, D., H. Yabuno, Y. Yamamoto, and S. Matsumoto (2018). Mass sensing in a liquid environment using nonlinear self-excited coupled-microcantilevers. *Journal of Microelectromechanical Systems* (99), 1–6.

Gupta, A., D. Akin, and R. Bashir (2004). Single virus particle mass detection using microresonators with nanoscale thickness. *Applied Physics Letters* 84(11), 1976–1978.

Ono, T., X. Li, H. Miyashita, and M. Esashi (2003). Mass sensing of adsorbed molecules in sub-picogram sample with ultrathin silicon resonator. *Review of scientific instruments* 74(3), 1240–1243.

Pierre, C. (1988). Mode localization and eigenvalue loci veering phenomena in disordered structures.

Sato, M., B. Hubbard, A. Sievers, B. Ilic, D. Czaplewski, and H. Craighead (2003). Observation of locked intrinsic localized vibrational modes in a micromechanical oscillator array. *Physical review letters* 90(4), 044102.

Spletzer, M., A. Raman, H. Sumali, and J. P. Sullivan (2008). Highly sensitive mass detection and identification using vibration localization in coupled microcantilever arrays. *Applied Physics Letters* 92(11), 114102.

Spletzer, M., A. Raman, A. Q. Wu, X. Xu, and R. Reifenberger (2006). Ultrasensitive mass sensing using mode localization in coupled microcantilevers. *Applied Physics Letters* 88(25), 254102.

Yabuno, H. (2019). Review of applications of self-excited oscillations to highly sensitive vibrational sensors. *ZAMM-Journal of Applied Mathematics and Mechanics/Zeitschrift für Angewandte Mathematik und Mechanik*, e201900009.

Yabuno, H., Y. Seo, and M. Kuroda (2013). Self-excited coupled cantilevers for mass sensing in viscous measurement environments. *Applied Physics Letters* 103(6), 063104.

15

Motion Control of Underactuated Manipulator Without State Feedback Control

In Chapter 12, it is clarified that the static equilibrium state in a pendulum can be stabilized by high-frequency excitation and the stable equilibrium states can be produced by perturbing the bifurcation produced in the pendulum. In this chapter, this nonlinear phenomena under high-frequency excitation are applied to the motion control of an underactuated manipulator. This is an example of how the positive utilization of nonlinear phenomena enables a motion control which cannot be achieved from the viewpoint of linear theory. The effectiveness of the control method is verified by experiments compared with theoretical results. A dedicated video will be introduced in Section 16.10.

15.1 What is an Underactuated Manipulator

Manipulators with free or passive joints (links) are called underactuated manipulators. In these systems, the number of generalized coordinates is larger than that of control inputs (the number of actuators). Figure15.1 is a two-link underactuated manipulator, where an actuator is attached only to the first joint and the second link has no actuator. Therefore, the configuration of the first active link is freely controlled by the actuator, while the motion control of the second free link is more complex. From the practical point of view, the study of the motion control of an underactuated manipulator can be very interesting not only for the benefit of a reduction in weight, energy consumption, and cost of the manipulator, but also for the effective possibility to overcome the actuator failure due to unexpected accidents. Comprehensive references on underactuated manipulators can be found in several studies (Arai et al. (1998); Hong (2002); Liu and Yu (2013)).

In most studies on underactuated manipulators, the control of the free link is carried out by actuating the active joint based on the motion of the free link. Hence, the information related to the motion of the free link is essential, this involves the assumption that the measurement of the angle of the free joint has to be carried out by a sensor. The method introduced in this chapter is for a two-link underactuated manipulator under gravity effect, in which the second joint (free joint) lacks not only an actuator but also a sensor; this method can also be regarded as a control strategy for the case when not only the actuator but also the sensor breaks down.

Linear and Nonlinear Instabilities in Mechanical Systems: Analysis, Control and Application,
First Edition. Hiroshi Yabuno.

Companion website: www.wiley.com/go/yabuno/instabilitiesinmechanicalsystems

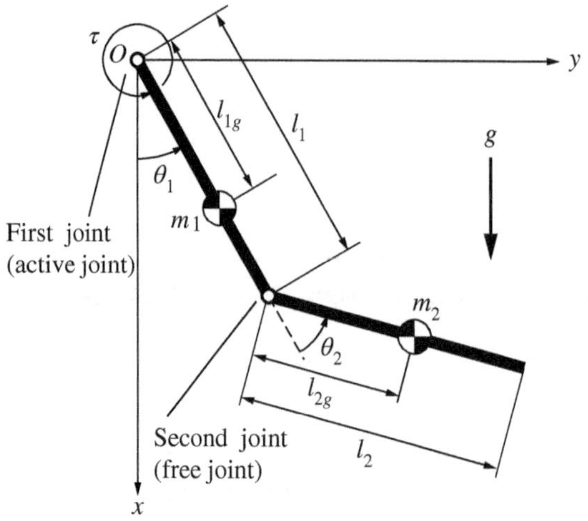

Figure 15.1 Analytical model of underactuated manipulator. An actuator is set at the first joint, but the section joint is free without actuator. Source: Reprinted with permission Yabuno et al. (2005).

15.2 Equation of Motion

The equations of motion of the two-link underactuated manipulator in Figure 15.1 governing the angle θ_1 of the first link and the relative angle θ_2 of the second link with respect to the first link are

$$(A_1 + A_2 + 2\beta_1 \cos\theta_2)\frac{d^2\theta_1}{dt^2} + (A_2 + \beta_1 \cos\theta_2)\frac{d^2\theta_2}{dt^2}$$
$$-\left(2\frac{d\theta_1}{dt} + \frac{d\theta_2}{dt}\right)\frac{d\theta_2}{dt}\beta_1 \sin\theta_2 + \beta_2 \sin\theta_1 + \beta_3 \sin(\theta_1 + \theta_2) = \tau, \tag{15.1}$$

$$A_2\frac{d^2\theta_2}{dt^2} + (A_2 + \beta_1 \cos\theta_2)\frac{d^2\theta_1}{dt^2} + \beta_1\left(\frac{d\theta_1}{dt}\right)^2 \sin\theta_2$$
$$+\beta_3 \sin(\theta_1 + \theta_2) = 0. \tag{15.2}$$

The constant coefficients $A_1, A_2, \beta_1, \beta_2$, and β_3 are expressed as

$$A_1 = I_1 + m_1 l_{1g}^2 + m_2 l_1^2, \quad A_2 = I_2 + m_2 l_{2g}^2,$$
$$\beta_1 = m_2 l_1 l_{2g}, \quad \beta_2 = (m_1 l_{1g} + m_2 l_1)g, \quad \beta_3 = m_2 l_{2g} g, \tag{15.3}$$

where parameters, m_i, l_i, and l_{ig} in Figure 15.1 show the ith mass, the length of the ith link, and the distance between the ith joint and the center of gravity of the ith link, respectively. Parameters I_i are as a mass moment of inertia about the center of the ith link (Yabuno et al. (2005)).

We propose a unified strategy for the swing-up and stabilization at the upright position without state feedback control of the second link state, by using the bifurcations produced in the second link under the high-frequency excitation, which are equivalent to those produced in the pendulum in Chapter 12. For the motion control, we excite the angle of the first link θ_1 sinusoidally. The motion of the first link is influenced by the motion of the second link, as shown in Eq. (15.1). However, it is assumed here that the position control of the first link can be approximately realized by the application of a torque τ to the first joint according to a feedback control only with respect to angle θ_1 and an angular velocity $\frac{d\theta_1}{dt}$ of the first link. Hence, we assume that the motion of the first link is not affected by the motion of the second link, and then we can ignore Eq. (15.1). Now we can set the position of the first link as follows:

$$\theta_1 = a_e \cos \nu t + \theta_{1off}, \tag{15.4}$$

where the first term is set to provide high-frequency excitation with excitation frequency ν to the second link at the free joint and the second term expresses the configuration of the first link, with respect to the gravity direction, i.e. the offset of the excitation for producing the perturbations in the bifurcations mentioned in Chapter 12.

Substituting Eq. (15.4) into Eq. (15.2), and nondimensionalizing by the representative time $\frac{1}{\nu}$, yield the nondimensional equation of motion of the second link

$$\begin{aligned}\frac{d^2\theta_2}{dt^{*2}} &+ \mu\frac{d\theta_2}{dt^*} + \{a_e^2 c\sin^2 t^* + \sigma\cos(a_e\cos t^* + \theta_{1off})\}\sin\theta_2 \\ &+\{\sigma\sin(a_e\cos t^* + \theta_{1off}) - a_e c\cos t^*\}\cos\theta_2 = a_e\cos t^*,\end{aligned} \tag{15.5}$$

where the viscous damping of the second free joint is assumed as $\mu\dot{\theta}_2$ with damping coefficient μ, and the dimensionless parameters, c and σ, are expressed as follows:

$$c = \frac{\beta_1}{A_2}, \quad \sigma = \frac{\beta_3/A_2}{\nu^2}. \tag{15.6}$$

15.3 Averaging by the Method of Multiple Scales and Bifurcation Analysis

Based on the same ordering as in Section 12.2, i.e. $\sigma = \epsilon^2\hat{\sigma}$, $\mu = \epsilon\hat{\mu}$, $a_e = \epsilon\hat{a}_e$, we introduce the method of multiple scales with three time scales, $t_0 = t$, $t_1 = \epsilon t$, and $t_2 = \epsilon^2 t$, and derive the averaging equation (Yabuno et al. (2005)). We assume the uniform expansion to be

$$\theta_2 = \theta_{20} + \epsilon\theta_{21} + \cdots. \tag{15.7}$$

By the same process as in Section 12.2, we obtain the autonomous equation governing the motion of the second link as

$$\frac{d^2\theta_2}{dt^2} + \mu\frac{d\theta_2}{dt} + \sigma\sin(\theta_{1off} + \theta_2) - \frac{a_e^2 c^2}{2}\sin\theta_2\cos\theta_2 = 0. \tag{15.8}$$

Therefore, the equilibrium states θ_{2st} of the second link satisfy the following equilibrium equation:

$$f(\sigma, \theta_{1off}, \theta_{2st}) \stackrel{\text{def}}{=} \sigma\sin(\theta_{1off} + \theta_{2st}) - \frac{a_e^2 c^2}{2}\sin\theta_{2st}\cos\theta_{2st} = 0. \tag{15.9}$$

The equilibrium states are changed by setting θ_{1off} to be different values. To determine the stability of the equilibrium states, substituting

$$\theta_2 = \theta_{2st} + \Delta\theta_2 \tag{15.10}$$

into Eq. (15.8) yields

$$\frac{d^2\Delta\theta_2}{dt^2} + \mu\frac{d\Delta\theta_2}{dt} + \left\{\sigma - \frac{c^2 a_e^2}{2}(\cos\theta_{2st} - \sin\theta_{2st})\right\}\Delta\theta_2 = 0. \tag{15.11}$$

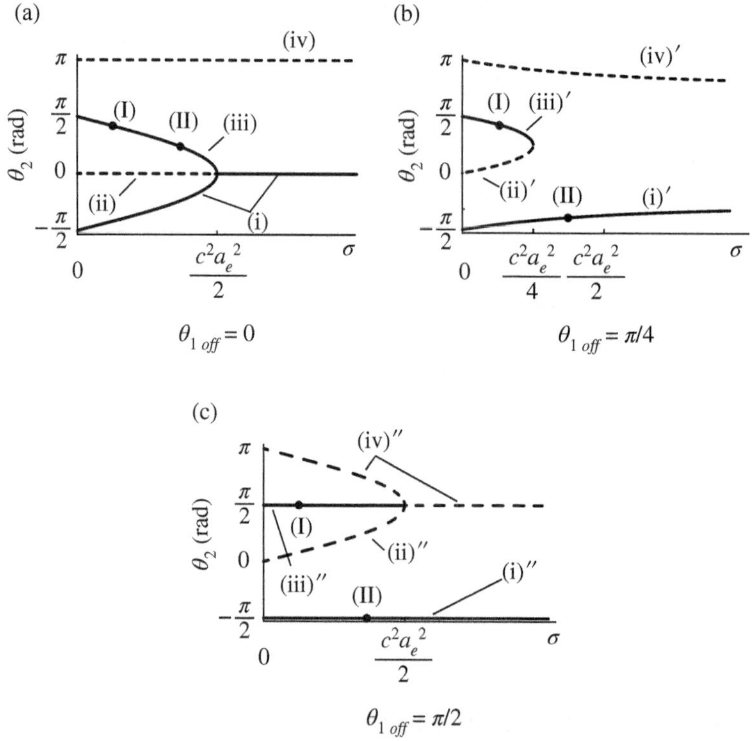

Figure 15.2 Bifurcation diagrams depending on θ_{1off}. (a) In case of $\theta_{1off} = 0$: the branches consist of the equilibrium states of supercritical pitchfork bifurcation. (b) In case of $\theta_{1off} = \pi/4$: the branches consist of the equilibrium states of perturbed supercritical and subcritical pitchfork bifurcations. (c) In case of $\theta_{1off} = \pi/2$: the branches consist of the equilibrium states of subcritical pitchfork bifurcation. The symbols in the figures, (i), (ii), and so on, correspond to those in Figure 15.4. Source: Reprinted with permission Yabuno et al. (2003).

For the special cases of $\theta_{1off} = 0$, $\theta_{1off} = \pi/4$, and $\theta_{1off} = \pi/2$, we show the variation of the equilibrium states, depending on the dimensionless parameter σ expressing the magnitude of the excitation frequency of the first link in Figure 15.2.

Figure 15.2a is the bifurcation diagram which shows the relationship between the excitation frequency of the first link and the equilibrium states of the second link in the case of $\theta_{1off} = 0$, where the solid and dashed lines denote the stable and unstable equilibrium states, respectively. Increasing the excitation frequency ν, i.e. decreasing σ, the second link undergoes a supercritical pitchfork bifurcation at $\sigma = c^2a_e^2/2$ and the trivial equilibrium state θ_{2st} is changed to an unstable equilibrium state. Also, stable nontrivial equilibrium states bifurcate from this bifurcation point. Therefore, the second link has the stable nontrivial equilibrium states, and is swung up from the downward vertical direction. Because the second joint is excited approximately in the horizontal direction, we can regard the second link as the pendulum whose supporting point is horizontally excited as the lower Figure 15.3a. Therefore, the supercritical pitchfork bifurcation in Figure 15.2a is equivalent to that in Figure 12.5a.

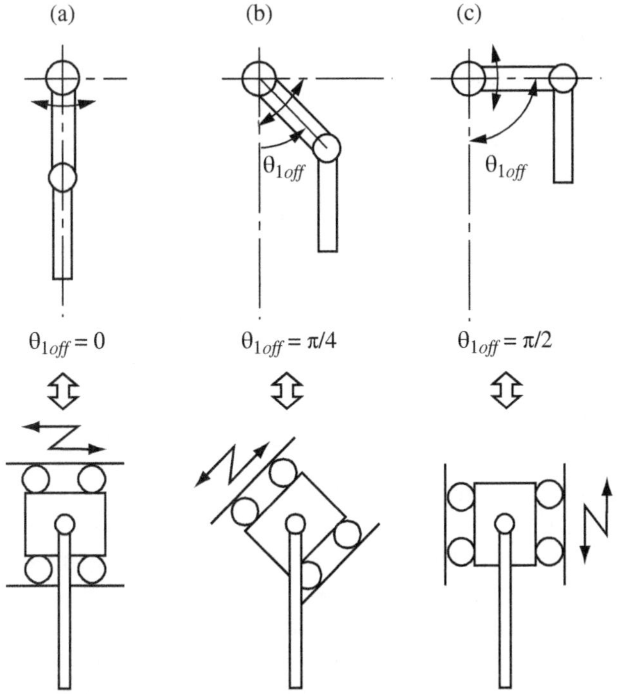

Figure 15.3 Relationship between the offset θ_{1off} and the excitation direction of a pendulum. Source: Reprinted with permission Yabuno (2004).

Next, we consider the behavior of the second link in the case of $\theta_{1off} = \pi/2$. Figure 15.2c is the bifurcation diagram which shows the relationship between the excitation frequency of the first link and the equilibrium states of the second link. In the case of $\sigma > c^2a_e^2/2$, i.e. when the excitation frequency ν is not sufficiently high, the second link has only one stable equilibrium state $\theta_{2st} = -\pi/2$, where the second link hangs down in the direction of the gravity effect (the absolute angle of the second link is $\theta_{2abs} = \theta_1 + \theta_{2st} = \pi/2 - \pi/2 = 0$). However, the second link undergoes a subcritical pitchfork bifurcation at $\sigma = c^2a_e^2/2$ and then another equilibrium state, $\theta_{2st} = \pi/2$, where the absolute angle of the second link is $\theta_{2abs} = \theta_1 + \theta_{2abs} = \pi/2 + \pi/2 = \pi$ (upright position), is changed to be stable. Because the second joint is excited approximately in the vertical direction we can regard the second link as the pendulum whose supporting point is vertically excited as the lower Figure 15.3c. Therefore, the subcritical pitchfork bifurcation in Figure 15.2c is equivalent to that in Figure 12.2a.

Furthermore, we consider the behavior of the second link in the case of $\theta_{1off} = \pi/4$ which is not horizontal or vertical to the direction of the gravity effect. The second joint is excited approximately in the vertical direction as Figure 15.3b. Then, the bifurcation diagram is Figure 15.2b which includes simultaneously the perturbed supercritical and perturbed pitchfork bifurcations. The combination of the equilibrium states, (i)′, (ii)′, and (iii)′, can be regarded as the perturbed supercritical pitchfork bifurcation diagram that is the reflection of Figure 7.10a with respect to the vertical line $\epsilon = 0$. The combination of the equilibrium states, (ii)′, (iii)′, and (iv)′, can be regarded as the perturbed subcritical pitchfork bifurcation diagram in Figure 7.10b.

It is the control objective to swing up the second link to the downward vertical position from the upright one. When $\theta_{1off} = \pi/2$ and $\sigma < c^2a_e^2/2$, the upright position is stable from Figure 15.2c. However, since the state in which the second link hangs down ($\theta_{2abs} = \theta_{1off} + \theta_{2st} = \pi/2 - \pi/2 = 0$) is also a stable equilibrium state independent of excitation frequency in Figure 15.2c, the swing-up from $\theta_{2abs} = 0$ cannot be realized by the excitation of the first link with $\theta_{1off} = \pi/2$, even if we apply the excitation frequency satisfying $\sigma < c^2a_e^2/2$. Therefore, we swing up the second link by using the variation of the stable equilibrium state, depending on the value of θ_{1off} as shown in Figure 15.2a–c.

15.4 Motion Control of Free Link

We show the continuous variation of stable equilibrium states by continuously changing the configuration of the first link θ_{1off} from 0 to $\frac{\pi}{2}$ in Figure 15.4. This figure is the so-called equilibrium surface (Poston and Stewart (2014)), which includes all the bifurcation diagrams in Figure 15.2a–c. The cross sections of the surface with $\theta_{1off} = 0$, $\theta_{1off} = \frac{\pi}{4}$, and $\theta_{1off} = \frac{\pi}{2}$ are the bifurcation diagrams in Figure 15.2a,b,c, which include supercritical pitchfork bifurcation, perturbed pitchfork bifurcations, and subcritical pitchfork bifurcation, respectively. The branches indicated by alphabets with no prime, single prime, and double prime correspond to the branches of the supercritical, subcritical, and perturbed pitchfork bifurcations shown in Figure 15.2a–c, respectively. From Figure 15.4, we notice that the upper branch (iii) of the bifurcation diagram in Figure 15.2a, which expresses the stable equilibrium state, continuously leads to the trivial stable equilibrium state (iii)″ in Figure 15.2c through stable equilibrium state (iii)′ in Figure 15.2b, with increasing θ_{1off}. Therefore, the swing-up of the second link can be realized by changing the θ_{1off} of the first link under the constant excitation frequency ν, i.e. under the constant σ.

Under a constant excitation frequency satisfying $\sigma = \sigma_a < \frac{c^2a_e^2}{4}$ (the value of $\frac{c^2a_e^2}{4}$ is mentioned later), the stable equilibrium state of the second link is changed from state (I) in Figure 15.2a to state (I) in Figure 15.2c through state (I) in Figure 15.2b. The relative angle of the second link is changed along arrow (I) in Figure 15.4 and the second link is swung up to the upright position: $\theta_{2abs} = \theta_1 + \theta_{2st} = \pi/2 + \pi/2 = \pi$. On the other hand, under a lower constant excitation frequency satisfying $\sigma > \frac{c^2a_e^2}{4}$, the stable equilibrium state of the second link is changed from state (II) in Figure 15.2a to state (II) in Figure 15.2c through state (II) in Figure 15.2b. The relative angle of the second link is changed along arrow (II) in Figure 15.4 and the second link cannot be swung up to the upright position.

Furthermore, let us examine these situation using Figure 15.5: (a) $\sigma_a < \frac{c^2a_e^2}{4}$, (b) $\sigma_b > \frac{c^2a_e^2}{4}$. It is noticed that the swing-up is accomplished if there continuously exist four equilibrium states while changing θ_{1off} from 0 to $\frac{\pi}{2}$. The disconnection of equilibrium states as Figure 15.5 occurs at the saddle-node bifurcation points in the perturbed pitchfork bifurcations. The bifurcation set obtained from Eq. (15.9) and $\frac{\partial f(\sigma,\theta_{1off},\theta_{2st})}{\partial \theta_{2st}} = 0$ is shown in Figure 15.6 and describes the combination of parameters σ and θ_{1off}: the saddle-node bifurcations occur on the bifurcation set; there are four and two equilibrium states, in

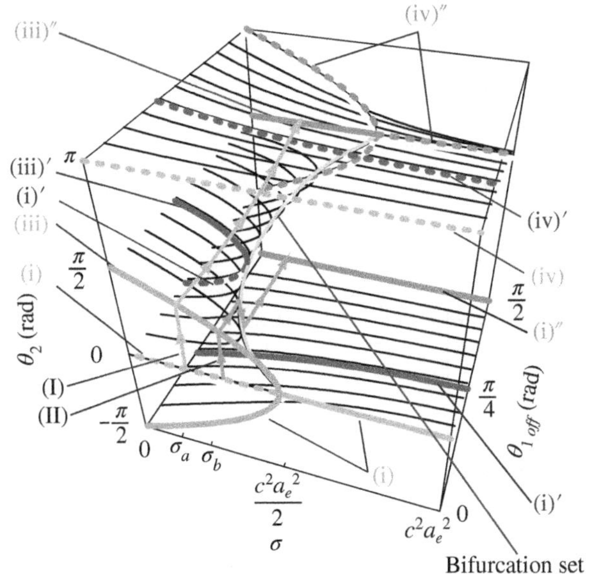

Figure 15.4 Equilibrium surface with parameters σ and θ_{1off}. The lowercase numbers without prime as (i), (ii), and so on denote the branches of the bifurcations in case $\theta_{1off} = 0$. The numbers with prime and double prime denote the branches in cases of $\theta_{1off} = \pi/4$ and $\theta_{1off} = \pi/2$, respectively. The arrows with (I) and (II) denote the change of the relative angle of the second link θ_2 while increasing θ_{1off}. In the case of sufficient excitation frequency as $\sigma = \sigma_a$, the swing-up is achieved. In the case of insufficient excitation frequency as $\sigma = \sigma_b$, the swing-up is impossible. These motions of the second link are described in the bifurcation diagrams of Figure 15.5a,b, respectively. The curve denoted with "bifurcation set" corresponds to the bifurcation set in Figure 15.6. Source: Reprinted with permission Yabuno et al. (2003).

the hatched and nonhatched regions, respectively. The bifurcation set is expressed also in Figure 15.4. As seen from this bifurcation set, if we set $\sigma < \frac{c^2 a_e^2}{4}$ as σ_a, the system does not undergo any saddle-node bifurcations and the stable equilibrium states continuously connect from $\theta_{1off} = 0$ to $\theta_{1off} = \frac{\pi}{2}$, we accomplish to swing up the free link to the upright position.

15.5 Experimental Results

We show an experimental result about the motion control of the second link from the downward vertical position to the upright position in Figure 15.7. First, we set $\theta_{1off} = 0$ and produce the supercritical pitchfork bifurcation. As shown in Figure 15.7a, the second link is swung and the relative angle of the second link corresponds to point I in Figure 15.2a. Then, we gradually increase θ_{1off}. Figure 15.7b shows the state of the links at $\theta_{1off} = \pi/4$. The second link is swung up more by the perturbations of pitchfork bifurcation. The relative angle of the second link corresponds to point I in Figure 15.2b. Furthermore, we increase θ_{1off}. Figure 15.7c shows that the second link is swung up to the upright position and stabilized by the subcritical pitchfork bifurcation. The relative angle of the second link corresponds to

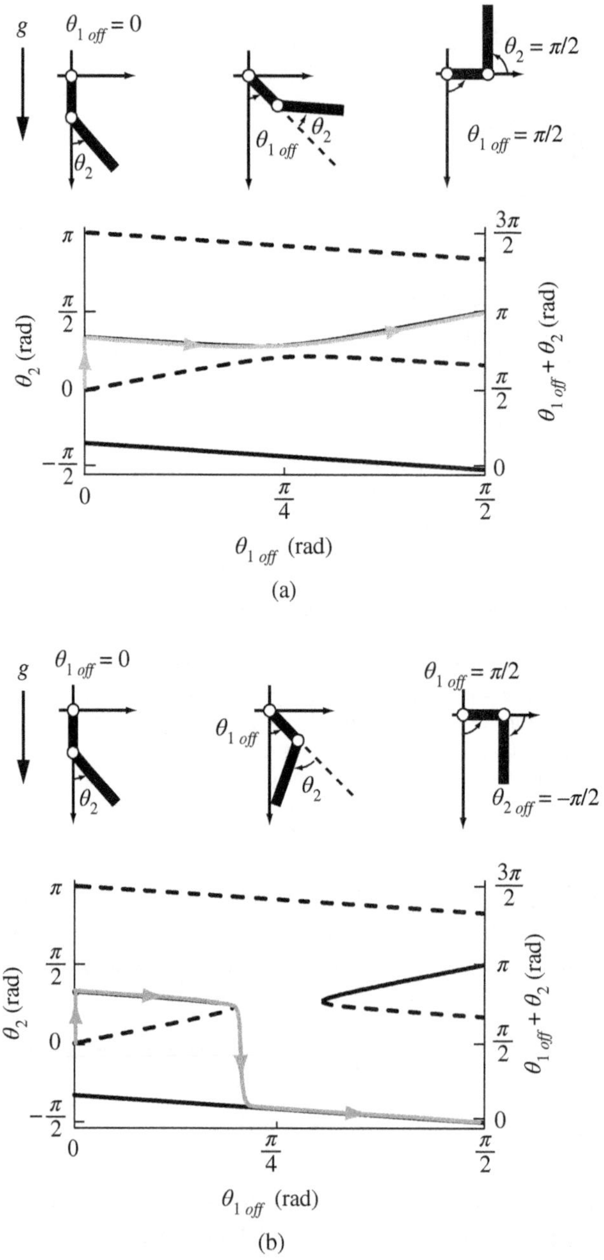

Figure 15.5 Bifurcation diagrams depending on σ. (a) is the case when the excitation frequency is relatively high ($\sigma = \sigma_a$ is relatively small). Since the stable equilibrium state of the second link is connected to the upright position with increasing θ_{1off}, the swing-up to the upright position is achieved. (b) is the case when the excitation frequency is relatively low ($\sigma = \sigma_b$ is relatively large). Since the stable equilibrium state of the second link are separated while increasing θ_{1off}, the swing-up to the upright position is impossible. Source: Reprinted with permission Yabuno et al. (2003).

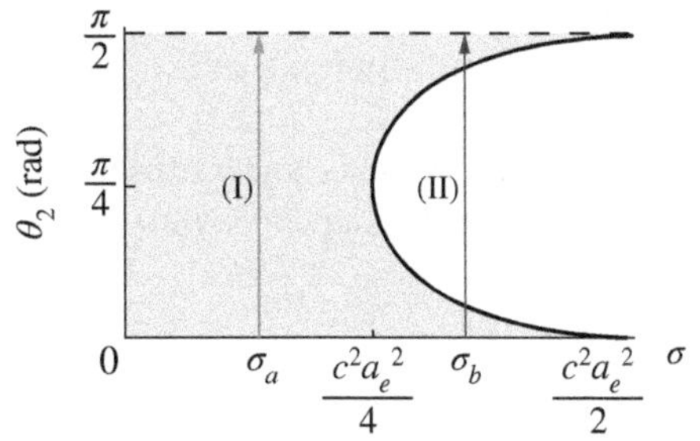

Figure 15.6 Bifurcation set for saddle-node bifurcation in the second link: at the combination of parameters σ and θ_{off} on the curve. In the hatched and nonhatched regions, there are four and two equilibrium states, respectively. Source: Reprinted with permission Yabuno et al. (2003).

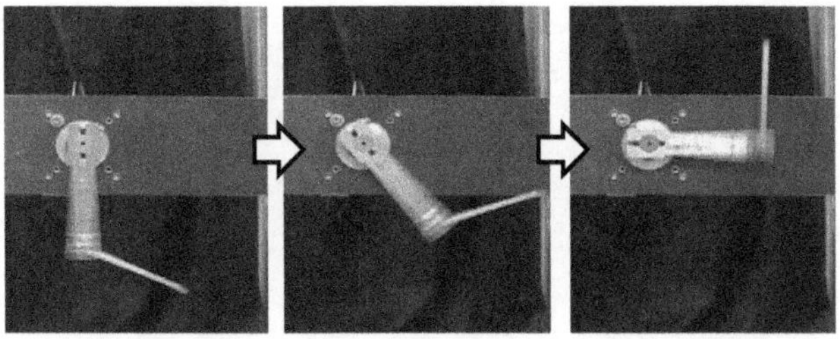

Figure 15.7 Motion control of underactuated manipulator by using pitchfork bifurcations. (a) The configuration of the first arm is $\theta_{1off} = 0$ and corresponds to Figure 15.2a. The second link is swung up by supercritical pitchfork bifurcation. (b) The configuration of the first arm is $\theta_{1off} = \pi/4$ and corresponds to Figure 15.2b. The second link is swung up more by the perturbation of pitchfork bifurcations. (b) The configuration of the first arm is $\theta_{1off} = \pi/2$ and corresponds to Figure 15.2c. The second link is stabilized at the upright position by subcritical pitchfork bifurcation. Source: Reprinted with permission Yabuno et al. (2003).

point I in Figure 15.2c. For the detail of the experiment, refer to Yabuno et al. (2004, 2005) and Section 16.10.

References

Arai, H., K. Tanie, and N. Shiroma (1998). Nonholonomic control of a three-DOF planar underactuated manipulator. *IEEE Transactions on Robotics and Automation* 14(5), 681–695.

Hong, K.-S. (2002). An open-loop control for underactuated manipulators using oscillatory inputs: Steering capability of an unactuated joint. *IEEE Transactions on Control Systems Technology* 10(3), 469–480.

Liu, Y. and H. Yu (2013). A survey of underactuated mechanical systems. *IET Control Theory & Applications* 7(7), 921–935.

Poston, T. and I. Stewart (2014). *Catastrophe theory and its applications*. Courier Corporation.

Yabuno, H. (2004). *Kougaku no tameno Hisenkei-kaiseki Nyuumon (The Elements of Nonlinear Analysis for Engineering)*. Saiensu-sha.

Yabuno, H., K. Goto, and N. Aoshima (2003). Swing-up and stabilization of an underactuated manipulator without state feedback of free joint. *IEEE Transactions on Robotics and Automation* 20(2), 359–365.

Yabuno, H., T. Matsuda, and N. Aoshima (2003). Swing-up and stabilization of a rotating pendulum without feedback control. In *32nd SICE Symposium on Control Theory*, pp. 331–334. The society of Instrument and Control Engineers.

Yabuno, H., T. Matsuda, and N. Aoshima (2005). Reachable and stabilizable area of an underactuated manipulator without state feedback control. *IEEE/ASME Transactions on Mechatronics* 10(4), 397–403.

16

Experimental Observations

16.1 Experiments of a Single Degree-of-Freedom System (Chapters 2 and 6)

16.1.1 Stability of Spring-Mass-Damper System Depending on the Stiffness *k* and the Damping *c*

Using a practical spring-mass-damper system as Figure 16.1, let us observe the qualitative variation of dynamics depending on stiffness and damping. We use a DC voice coil linear motor. The stator is fixed to the vertical wall on the base. The mover is used as the mass of the mass-spring-damper system, which can be freely moved on a linear bearing fixed to the base in the horizontal direction; x is the displacement of the mass from the static equilibrium state. The thrust force F_{cont} is applied to the mass in the horizontal direction by the linear motor. The mass is subject to the restoring force $-kx$ from the spring connecting to the wall and to the equivalent viscous damping force $-c\frac{dx}{dt}$ due to the bearing motion. Then, the corresponding analytical model can be described as Figure 16.2. The equation of motion is

$$m\frac{d^2x}{dt^2} + c\frac{dx}{dt} + kx = F_{cont}, \tag{16.1}$$

where k and c are positive constants in the real spring and damper as characterized in Figure 1.2. The thrust force of a voice coil motor F_{cont} is proportional to the input current I as $F_{cont} = \kappa_i I$, where κ_i is the thrust constant of the linear motor. Setting the input current by the velocity and position feedback

$$I = \frac{k_{cont}}{\kappa_i}x + \frac{c_{cont}}{\kappa_i}\frac{dx}{dt}, \tag{16.2}$$

where the feedback gains of k_{cont} and c_{cont} can be set arbitrarily. Then, Eq. (16.1) is rewritten as

$$m\frac{d^2x}{dt^2} + c_{equiv}\frac{dx}{dt} + k_{equiv}x = 0, \tag{16.3}$$

where $c_{equiv} = c - c_{cont}$ and $k_{equiv} = k - k_{cont}$. This equation is equivalent to Eq. (2.114). By changing the feedback gains of k_{cont} and c_{cont}, k_{equiv} and c_{equiv} can be also set to negative

Linear and Nonlinear Instabilities in Mechanical Systems: Analysis, Control and Application, First Edition. Hiroshi Yabuno.

Companion website: www.wiley.com/go/yabuno/instabilitiesinmechanicalsystems

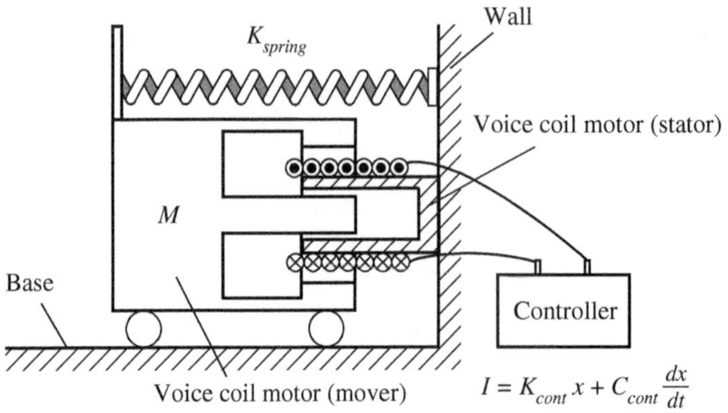

Figure 16.1 Experimental apparatus to observe the qualitative variation of the single degree-of-freedom system depending on parameters.

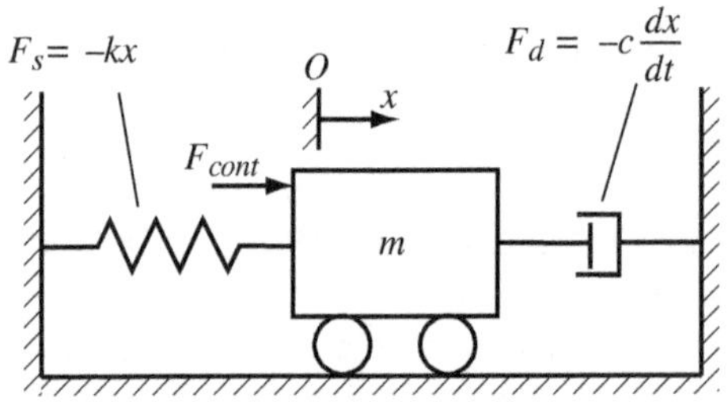

Figure 16.2 Spring-mass-damper system subject to control force. F_s and F_d are the forces by the original spring and damper. F_{cont} is a control force to change the characteristics of the original spring and damper.

stiffness and negative damping, respectively. Also, $\frac{k_{equiv}}{m}$ and $\frac{c_{equiv}}{m}$ correspond to k' and c' in Figure 2.7, respectively. Changing the feedback gains produces the various dynamics shown in Figure 2.7. Let us see the change of the dynamics under the decreasing of c' while k' is kept a positive constant.

1. Videoclip 16.1.1(1) shows the behavior corresponding to (E) in Figure 2.7 under $c' > 0$, $k' > 0$, and $D > 0$. Damping is relatively large and the displacement is over-damped without oscillation.
2. When c' is decreased but still positive, i.e. $c' > 0$, $k' > 0$, and $D < 0$, the displacement oscillatory decays as Videoclip 16.1.1(2); the phase portrait is (G) in Figure 2.7.
3. When c' is decreased more and becomes negative, i.e. $c' < 0$, $k' > 0$, and $D < 0$, the self-excited oscillation is produced; the phase portrait is (D) in Figure 2.7. The amplitude of the oscillation grows with time but in this experiment, does not become large due to the limited range of the movement of the linear bearing as Videoclip 16.1.1(3).

On the other hand, we decrease k' from the condition of $k' > 0$ and $D > 0$ while keeping $c' > 0$. Then, the dynamics change from the above result (item 1) as follows.

4. When k' becomes 0, the buckling occurs; the phase portrait is (J) in Figure 2.7. In practical experiments, it is impossible to set k' to be exactly equal to 0.

5. When k' becomes negative, the system is in the post-buckling state; the phase portrait is (A) in Figure 2.7. The displacement monotonically grows but in this experiment, does not become large due to the limited range of the movement of the linear bearing as Videoclip 16.1.1(4).

16.1.2 Self-Excited Oscillation of a Window Shield Wiper Blade Around the Reversal

Let us observe the vibration of a window shield wiper blade around the reversal in Videoclip 16.1.2. Because of no periodic excitation, it may be regarded as a self-excited oscillation. The friction force $F^{\#}$ between the tip of the blade and the surface of the window depends on the relative velocity v between them, which is schematically shown as Figure 6.2. The friction force shows a negative damping feature in the range of low relative velocity. Therefore, around the reversal, the wiper blade is subject to the negative damping force, which may be related to the dynamical unstable phenomenon of (D) in Figure 2.7. The negative damping is the one of reasons producing the self-excited oscillation in Videoclip 16.1.2.

16.2 Buckling of a Slender Beam Under a Compressive Force

16.2.1 Observation of Pitchfork Bifurcation (Sections 5.1 and 7.1)

We consider the static instability of the clamped–clamped beam as in Videoclips 16.2.1(1)–(3). We apply a compressive force to the left end corresponding to P by using a voice coil linear motor as in the videos; the compressive force is proportional to the input current of the linear motor. In this case, the gravity effect does not affect the lateral motion of the beam and the experimental results correspond to the theoretical results about the system in Figure 5.2.

1. Case when the compressive force is below a critical value, i.e. when $\epsilon < 0$ in the bifurcation diagram of Figure 7.1a ($P < P_{pf}$ in the bifurcation diagram of Figure 16.3a):
   (a) In the case without or with very small compressive force, as Figure 5.3a, the deflection is oscillatory decayed as (G) in Figure 2.7: Videoclip 16.2.1(1).
   (b) Increasing the compressive force, as Figure 5.3c, the deflection is monotonically decayed as (E) in Figure 2.7: Videoclip 16.2.1(2).
2. Case when the compressive force is above a critical value, i.e. when $\epsilon > 0$ in the bifurcation diagram of Figure 7.1a ($P > P_{pf}$ in the bifurcation diagram of Figure 16.3a). Buckling occurs as Videoclip 16.2.1(3). For example, at $\epsilon = \epsilon_2$ ($P = P_{post}$), there are two post-buckling states, (a) and (b) in Fig. 16.3a, and their possibilities are the same.

16.2.2 Observation of Perturbed Pitchfork Bifurcation (Section 7.5)

We consider the fixed–fixed beam under the gravity effect by Videoclips 16.2.2(1) and 16.2.2(2), which corresponds to the system in Figure 7.9b. The behavior expresses the characteristics of the bifurcation diagram of Figure 7.10a. Checking the equivalent bifurcation diagram in Figure 16.3b, we will see the videoclips.

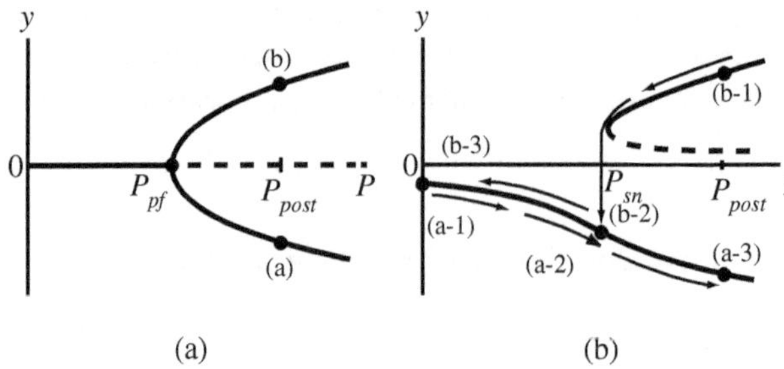

Figure 16.3 Bifurcation diagram of pitchfork bifurcation: (a) complete; (b) perturbed.

1. Case without initial artificial deflection (Videoclip 16.2.2(1)):
   (a) The beam is slightly deflected in the lower direction due to the gravity effect: (a-1).
   (b) Increasing the compressive force P from zero, the deflection continuously increases in the lower direction: (a-1)→(a-2)→(a-3).
   (c) Returning the compressive force to zero, i.e. the input current to zero, the deflection returns to the initial deflection: (a-3)→(a-2)→(a-1).
2. Case with initial artificial upper deflection under a large compressive force (for example, $P = P_{post}$) (Videoclip 16.2.2(2)):
   (a) Decreasing the compressive force P, the upper deflection continuously decreases from (b-1).
   (b) When P reaches P_{sn}, i.e. saddle-node bifurcation point, the deflection jumps to the lower stable deflection: (b-2).
   (c) Further decreasing P, the deflection decreases: (b-2)→(b-3).

16.2.3 Effect of Coulomb Friction on Pitchfork Bifurcation (Section 7.6)

1. The clamped–clamped boundary conditions of the beam in Videoclip 16.2.1 are changed to the hinged-hinged ones as Videoclip 16.2.3(1). The usual radial bearings are used for the hinged–hinged support. Then, the moment due to Coulomb friction against its rotation acts on each hinged supporting point, but it is very small. Considering only the first mode, the dynamics of the deflection equivalent to that of the system we dealt with in Section 7.6 (Hetzler (2014); Yabuno et al. (1999)). The small moment due to the friction means the smallness of $F^*_{cmax} = \frac{F_{max}}{k_1 l}$ in the system expressed by Eq. (7.86).
2. Videoclip 16.2.3(2) shows the behavior in the case without compressive force. In this state, the restoring force existing in the beam is relatively large so that F^*_{cmax} is very small. Then, because the dead zone is very narrow, the free oscillation stops near the straight position, i.e. $x = 0$.
3. Let's observe the behavior in the case with large compressive force. In this state, the restoring force is negative, but the same pitchfork bifurcation as Figure 16.3a (Videoclip 16.2.1(3)) does not occur due to the small Coulomb friction at the bearings. In the neighborhood of the bifurcation point, the equilibrium region becomes large as shown in Figure 7.17. Videoclip 16.2.3(3) shows the dynamics in the state where $P = P_{post} > P_{cr}$. This video highlights that there are many stable equilibrium states in the regions; even

in the case of $P > P_{cr}$, the beam can be in a rest near the unstable trivial steady state in the case without Coulomb friction. In the case of higher compressive force, as shown in Figure 7.17, the equilibrium regions are separated into the region surrounding the trivial unstable equilibrium state (dashed line) in the case without friction and the two regions surrounding the nontrivial stable equilibrium states (solid lines) .

16.3 Hunting Motion of a Railway Vehicle Wheelset (Section 6.3)

Let us observe the hunting motion of a railway vehicle wheelset model using some videos. The wheelset is set on the roller rig. By changing the rotational speed, we provide various running speeds to the wheelset; the meter in the video shows the running speed v (m/s). Let us see Videoclips 16.3(1)–(4) while checking the corresponding schematic bifurcation diagrams Figure 16.4a–d, where the ordinate and abscissa are the representative amplitude obtained from the normal form and the running speed, respectively (see Wei and Yabuno, (2019) in detail). First, we see Videoclip 16.3(1). Under fluctuation noise without large disturbance, when increasing the running speed, the oscillation starts at $v = v_H = 6.55$ m/s. This corresponds to the destabilization of the trivial amplitude at $v = v_H$ in Figure 16.4a which is theoretically obtained without taking into account the nonlinear effect, i.e. from the linear analysis. Increasing the running speed more above $v = v_H = 6.55$ m/s, the amplitude does not continuously increase, but jump up to the nontrivial steady-state amplitude. Therefore, it is predicted that the Hopf bifurcation is not supercritical (Figure 16.4b), but subcritical (Figure 16.4c) under the effect of cubic nonlinearity in the system. To confirm this, we see Videoclips 16.3(2) and 16.3(3). In both the videos, the running speed is below the critical speed $v_H = 6.55$ m/s, $v = 5.00$ m/s, but the artificial disturbance is different; Videoclips 16.3(2) and 16.3(3) are the cases under relatively small and large disturbances, respectively. The dynamics under the small disturbance follow the result of the linear theory in Figure 16.4a and the amplitude decays. On the other hand, under the large disturbance, the amplitude grows and the phenomenon cannot be explained from the bifurcation diagram Figure 16.4a obtained from the linear theory. The dependence of the stability on the magnitude of disturbance implies the unstable steady-state amplitude shown in the bifurcation diagram of the subcritical Hopf bifurcation diagram Figure 16.4c. In the subcritical characteristic, the nontrivial stable steady-state amplitude cannot be theoretically found by only the consideration of cubic nonlinearity. The quintic nonlinearity plays an important role to understand the nonlinear dynamics of hunting motion (for example, see Sedighi and Shirazi (2012); Wei and Yabuno (2019)). The schematic diagram theoretically obtained by taking into account the cubic and quintic nonlinearities is Figure 16.4d, where the saddle-node bifurcation point is found. Let us observe such effect in Videoclip 16.3(4). In the lower running velocity, there is no response. Increasing the running velocity, at the subcritical Hopf bifurcation point ($v = v_H = 6.55$ m/s) as in Videoclip 16.3(1), the resonance occurs and the amplitude jumps up to the finite value. We further increase the running speed until $v_{post} = 6.90$ m/s. Then, the amplitude is changed as (a-1)→(a-2)→(a-3). Next, we decrease the running velocity from v_{post}. The amplitude does not tend to zero at $v = v_H$, but the finite amplitude is kept. For example, at $v = v_H = 4.90$ m/s, different from the case when

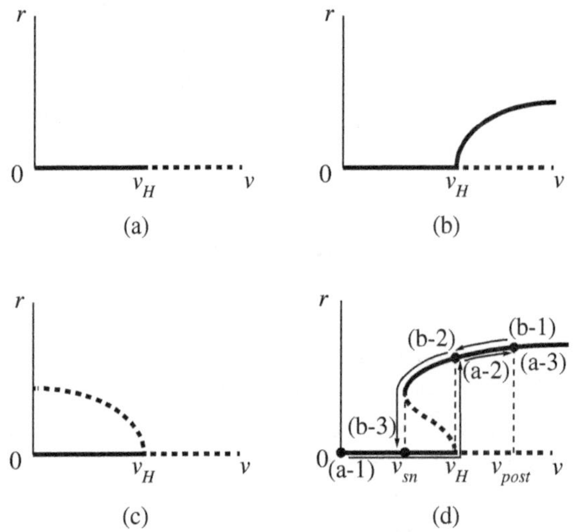

Figure 16.4 Hopf bifurcation diagrams: (a) linear analysis; (b) nonlinear analysis taking into account cubic nonlinearity (supercritical Hopf bifurcation); (c) nonlinear analysis taking into account cubic nonlinearity (subcritical Hopf bifurcation); (d) nonlinear analysis taking into account cubic and quintic nonlinearities (subcritical Hopf and saddle-node bifurcations).

increasing the running speed, the amplitude is not zero. Decreasing the running speed even more, through the saddle node bifurcation at $v = v_{sn} = 3.80$ m/s, the finite amplitude jumps down to zero. The amplitude is changed as (b-1)→(b-2)→(b-3). Hence, the hysteresis in the variation of response amplitude exists depending on the increase and the decrease of running velocity and the phenomenon is caused from the cubic and quintic nonlinearities in the system.

16.4 Stabilization of Hunting Motion by Gyroscopic Damper (Section 6.3)

Using Videoclip 16.4, let us observe stabilization of the hunting motion by using a gyroscopic damper. First, we do not rotate the gyro and increase the running speed. At $v = 5.03$ m/s, the hunting motion is produced. We start the rotation of the gyro. Increasing the rotational speed, the gyro effect increases. Above a rotational speed, the hunting motion is stabilized. Even if we increase the running speed, the stability is kept until a much higher critical speed than that in case without rotation of gyro (for more detail, see Yabuno et al. (2008); Yoshino et al. (2015)).

16.5 Self-Excited Oscillation of Fluid-Conveying Pipe (Section 6.5)

Let us observe a self-excited oscillation of the fluid conveying pipe shown in Figure 16.5. The dynamics essentially corresponds to the two-degree-of-freedom articulated pipes conveying fluid analyzed in Section 6.5. The system is subject to a circulatory term due to the fluid and the stiffness matrix is nonsymmetric. Therefore, the system is nonconservative

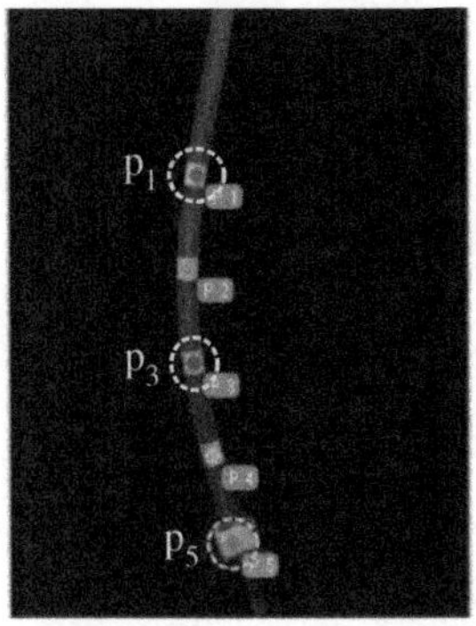

Figure 16.5 Fluid conveying pipe.

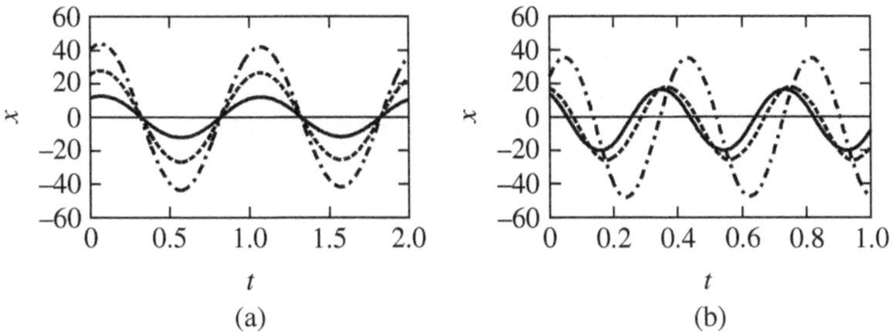

Figure 16.6 Complex modes in fluid conveying pipe: solid, broken, and dashed-dotted lines denote the displacements at points P_1, P_3, and P_5, respectively. (a) is a free oscillation without flow. (b) is a self-excited oscillation with flow.

and in the case when the fluid is conveyed with a high speed, flutter-type self-excited oscillations are produced. Figure 16.6a,b are the time histories at the three points indicated in Figure 16.5; Panels (a) and (b) are the cases without and with relatively high speed flow velocity, respectively. Videoclip 16.5(1) shows the case without flow velocity and Videoclip 16.5(2) is its slow motion. From these videos, the motions at each point on the pipe is synchronized. Videoclip 16.5(3) shows the case with flow velocity and Videoclip 16.5(4) is its slow motion. From these videos, the motion at each point on the pipe is nonsynchronized. The nonsynchronous motion is based on the circulatory force caused by the flow, that is a follower force.

16.6 Realization of Self-Excited Oscillation in a Practical Cantilever (Section 10.1.3)

Let us observe the self-excited oscillation in a cantilever beam by linear and nonlinear feedback control as discussed in Section 10.1.3. Videoclip 16.6(1) shows the free oscillation of a cantilever without feedback control, where the left and right ends of the beam are free and fixed, respectively. The time history at the point near the free end of the cantilever, which is measured by the laser sensor monitoring the displacement, is Figure 16.7. The

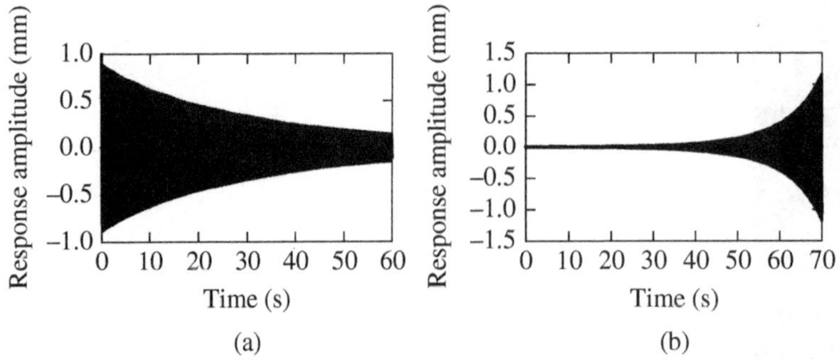

Figure 16.7 Free oscillation and self-excited oscillation with linear feedback control: Panel (a) is then time history of the free oscillation shown in Videoclip 16.6(1); Panel (b) is the time history of the self-excited oscillation shown in Videoclip 16.6(2).

displacement is periodically decayed with the frequency of about 3 Hz that is the first natural frequency of the cantilever. The envelope of the time history corresponds to the time variation of the amplitude theoretically expressed by Eq. (10.22) under the conditions, $\gamma > 0$ and $\Gamma = 0$. Videoclip 16.6(2) shows the self-excited oscillation of the cantilever under the linear velocity feedback. By differentiating the output signal of the sensor through an analog differentiator, we obtain the velocity at the point near the free end of the cantilever. The actuation for the control is applied to the fixed end by the piezo actuator according to Eq. (10.40); c_{lin} is set so that $c - c_{lin}$, i.e. γ in Eq. (10.43), becomes negative, and $c_{non} = 0$. The measured time history is Figure 16.7b. The envelope grows with time due to the negative damping until the limit of the power supply to the actuator.

Videoclips 16.6(3), 16.6(4), and 16.6(5) show behaviors of the cantilever under linear and nonlinear feedback, where the nonlinear feedback in Eq. (10.40) is $c_{non} < 0$ and then $\beta_3 > 0$ and $\Gamma > 0$. Then, as theoretical analysis, the supercritical Hopf bifurcation is produced. Videoclip 16.6(3) reports the behavior in the case when the linear feedback $c_{lin} > 0$ is applied, but it is not so large that $c - c_{lin}$, i.e. γ in Eq. (10.43), is still positive. As seen from the time history shown in Figure 16.8a, the deflection is oscillatory decayed as Figure 16.7a. Next, we observe the case when the linear feedback $c_{lin} > 0$ is applied, and it is so large that $c - c_{lin}$, i.e. γ in Eq. (10.43), is negative. Videoclips 16.6(4) and 16.6(5) are the behaviors in the cases with small and large initial disturbances, whose time histories are Figure 16.8b,c, respectively. In both cases, the amplitude converges to the same nontrivial steady-state amplitude after their transient states. This indicates that the behavior converges to the nontrivial stable limit cycle regardless of whether the initial condition is in or out of the stable limit cycle shown in Figure 7.6b.

Videoclips 16.6(6), 16.6(7), and 16.6(8) show behaviors of the cantilever under linear and nonlinear feedback, where the nonlinear feedback in Eq. (10.40) is $c_{non} > 0$ and then $\beta_3 < 0$ and $\Gamma < 0$. Then, as theoretical analysis, the subcritical Hopf bifurcation is produced. Videoclip 16.6(6) reports the behavior in the case when the linear feedback $c_{lin} > 0$ is applied and it is so large that $c - c_{lin}$, i.e. γ in Eq. (10.43), is negative. From the time history shown in Figure 16.9a, the behavior oscillatory grows with time as Figure 16.7b. Next, we observe the case when the linear feedback $c_{lin} > 0$ is applied, and it is not so large that $c - c_{lin}$, i.e. γ in Eq. (10.43), is positive. Videoclips 16.6(7) and 16.6(8) display the behaviors in the cases with

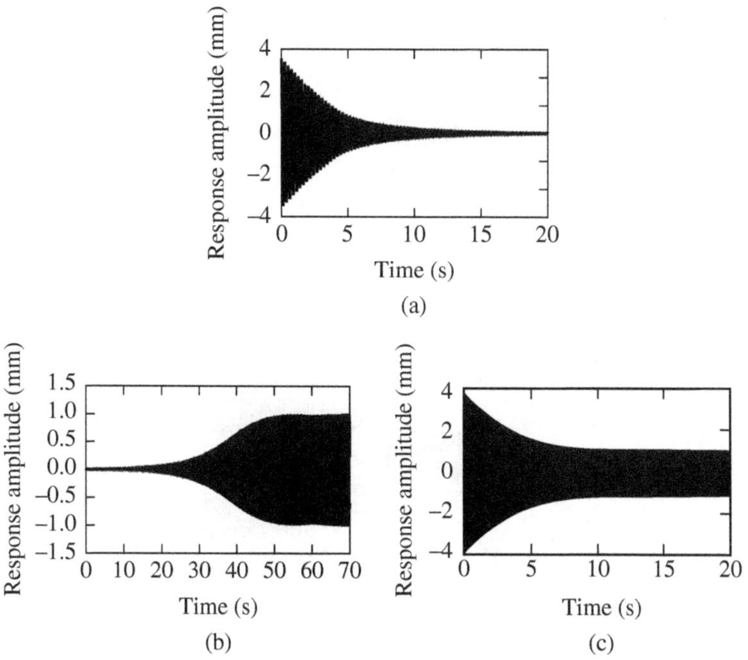

Figure 16.8 Self-excited oscillation through supercritical Hopf bifurcation ($\beta_3 > 0$). Panel (a) is the time history of Videoclip 16.6(3) in the case of $\gamma > 0$, i.e. below the supercritical Hopf bifurcation point. (b) and (c) are the time histories of Videoclips 16.6(3) and (4), respectively. They are in the case of $\gamma < 0$, i.e. above the supercritical Hopf bifurcation point. Initial disturbances of (a) and (b) are relatively small and large, respectively.

small and large initial disturbances, whose time histories are Figure 16.9b,c, respectively. In Figure 16.9b, the time history is periodically decayed as the case without nonlinear damping, i.e. $\Gamma = 0$. This phenomenon indicates that the initial disturbance is so small that the initial condition is located in the unstable limit cycle shown in Figure 7.7a. On the other hand, in the case of large initial disturbance, the time history periodically grows with time as shown in Figure 16.9c. This phenomenon indicates that the initial disturbance is so large that the initial condition is located out the unstable limit cycle shown in Figure 7.7a.

16.7 Parametric Resonance (Chapter 11)

We consider the cantilever in Videoclip 16.7. The clamped end is attached to the shaker and subject to vertically periodic excitation.

1. Videoclip 16.7(1) reports a free oscillation of the cantilever in case without excitation. From this result, we can find that the first natural frequency is 2.8 Hz
2. We apply the vertical excitation with frequency 2 Hz. Since the excitation frequency is far from twice the natural frequency, no resonance occurs as shown in Videoclip 16.7(2).
3. Videoclips 16.7(3) shows the motion of the cantilever when the excitation frequency is 5.6 Hz. Regardless of no initial disturbance, the parametric resonance occurs since the

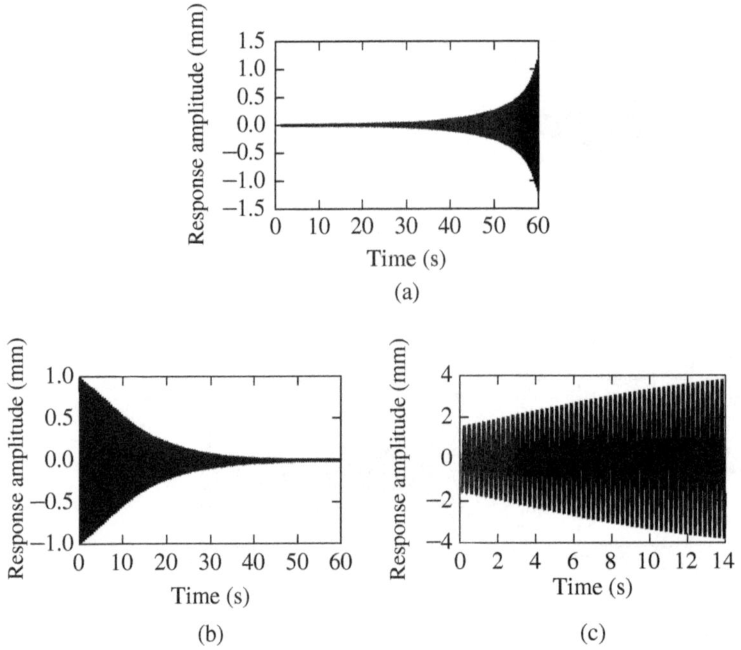

Figure 16.9 Self-excited oscillation through subcritical Hopf bifurcation ($\beta_3 < 0$). (a) is the time history of Videoclip 16.6(6) in the case of $\gamma < 0$, i.e. above the subcritical Hopf bifurcation point. (b) and (c) are the time histories of Videoclips 16.6(7) and (8), respectively. They are in the case of $\gamma > 0$, i.e. below the subcritical Hopf bifurcation point. Initial disturbances of (a) and (b) are relatively small and large, respectively.

combination of the excitation frequency and amplitude is in the unstable region shown in Figure 11.2; the detuning parameter σ of the excitation frequency is located between the points, Sup-PF and Sub-PF in Figure 11.3. The response amplitude reaches the nontrivial stable steady-state amplitudes after the transient state for any initial disturbance as shown in Figure 11.4b. The response frequency is 2.8 Hz that is half the excitation frequency.

4. We change the excitation frequency to 5.65 Hz. In the case without initial disturbance, the parametric resonance does not occur. As shown in Videoclip 16.7(4), since the combination of the excitation frequency and amplitude is out of the unstable region shown in Figure 11.2; the detuning parameter σ of the excitation frequency is located above Sub-PF in Figure 11.3.

   Also in the case with relatively small initial disturbance, corresponding to the case when the initial disturbance is located out of the hatched region in Figure 11.4c, the oscillation decays with time and no resonance occurs as shown in Videoclip 16.7(5).

   However, under the relatively large initial disturbance, corresponding to the case when the initial disturbance is located in the hatched region in Figure 11.4c, the parametric resonance can be produced as shown in Videoclip 16.7(6). In this excitation condition, since the detuning parameter σ of the excitation frequency is located above Sub-PF in Figure 11.3, there is the unstable nontrivial steady-state amplitude in addition to the stable nontrivial and stable trivial ones.

16.8 Stabilization of an Inverted Pendulum Under High-Frequency Vertical Excitation (Chapter 12)

Let us observe the stabilization of an inverted pendulum by high-frequency vertical excitation. The upright position of the pendulum is the unstable equilibrium state. It is well known that the state is stabilized by state feedback control (for example, Ogata (1997)). On the other hand, the high-frequency vertical excitation applied to the supporting point produces a subcritical pitchfork bifurcation and stabilizes the upright position without state feedback control as theoretically clarified in Chapter 12. The phenomenon is experimentally investigated in many studies (for example, Smith and Blackburn (1992)). Here, we see the stabilization phenomenon through Videoclip 16.8(1) corresponding to a study (Yabuno et al. (2004)). The natural frequency of the pendulum around the downward vertical stable steady state is 1.9 Hz. The excitation frequency is 31 Hz.

The high-frequency excitation improves the stability of elastic systems (Chelomei (1956); Jensen (2000); Jensen et al. (2000); Thomsen (2003)). Videoclip 16.8(2) shows the stabilization of a buckled beam. First, we applied the axial compressive force to a clamped–clamped beam by using two linear motors and observe the buckling. While keeping the compressive force, by using the shaker, we apply the axial excitation to the buckled beam with high-frequency. Then, the buckled beam returned to the original straight state; see (Yabuno and Tsumoto (2007)) for more detail.

Videoclip 16.8(3) shows the stabilization of a beam buckled under its own weight. Also in this case, the high-frequency excitation stabilizes the buckled beam to the original state. This stabilization is well known as Indian rope trick (Champneys and Fraser (2000); Fraser and Champneys (2002); Mullin et al. (2003)).

16.9 Self-Excited Coupled Cantilever Beams for Ultrasensitive Mass Sensing (Section 14.6)

Let us observe the motion of self-excited coupled cantilever beams in Videoclips 16.9(1) and (2). Videoclip 16.9(1) shows a response of weakly coupled macrocantilevers without measured mass. Mass of a cantilever is 1.9×10^{-2} kg. In case without a measured mass, the response amplitudes of the cantilevers are the same. On the other hand, when a measured mass 4.3×10^{-4} kg is attached to the upper beam, the amplitude ratio is 1 : 4.1 as shown in Videoclip 16.9(2).

Videoclip 16.9(3) shows the experimentally measured behavior of the coupled microcantilevers. The dimensions of the cantilever are 500 μm × 100 μm × 10 μm, the mass is 306 ng, and the first natural frequency is 55.3 kHz. The dimensionless coupling stiffness experimentally identified is $\kappa = 1.44 \times 10^{-2}$ (see Chapter 14 and (Endo et al. (2015))). In case without measured mass, the self-excited coupled cantilevers oscillates in phase motion with the same amplitude as the first half of the videoclip. On the other hand, when a measured mass (1.8 ng) is attached to the left cantilever, the amplitude ratio between the cantilevers are about 1 : 1.23 in the latter half of the videoclip.

16.10 Motion Control of an Underactuated Manipulator by Bifurcation Control (Chapter 15)

Let us observe the motion control of an underactuated manipulator in Videoclip 16.10. An actuator is equipped at the first joint, but no actuator at the second joint. Therefore, the second link is freely rotated. The number of degree of freedom is 2, but that of the actuator is 1. By using a high-frequency excitation and controlling the bifurcations produced in the second link, we swing up the second link to the upright position and stabilize the state. We set the angle of the first joint by only one actuator as

$$\theta_1 = \theta_{1off} + a_e \cos \nu t. \tag{16.4}$$

First, we set $\theta_{1off} = 0$ and the second link is swung up with a finite angle through a supercritical pitchfork bifurcation. Next, increasing θ_{1off} until $\pi/2$, the pitchfork bifurcation is perturbed and the magnitude of the perturbation increases. Along the stable equilibrium state changing with θ_{1off}, the free link is swung up to the upright position; see (Endo and Yabuno (2013); Yabuno et al. (2004, 2005)), for more detail.

References

Champneys, A. R. and W. B. Fraser (2000). The 'indian rope trick' for a parametrically excited flexible rod: linearized analysis. In *Proceedings of the Royal Society of London A: Mathematical, Physical and Engineering Sciences*, Volume 456, pp. 553–570. The Royal Society.

Chelomei, V. (1956). On the possibility of the increase of elastic system stability by means of vibrations. In *Dokl. AN SSSR*, Volume 110, pp. 345–347.

Endo, D., H. Yabuno, K. Higashino, Y. Yamamoto, and S. Matsumoto (2015). Self-excited coupled-microcantilevers for mass sensing. *Applied Physics Letters* 106(22), 223105.

Endo, K. and H. Yabuno (2013). Swing-up control of a three-link underactuated manipulator by high-frequency horizontal excitation. *Journal of Computational and Nonlinear Dynamics* 8(1), 011002.

Fraser, W. B. and A. R. Champneys (2002). The 'indian rope trick' for a parametrically excited flexible rod: nonlinear and subharmonic analysis. *Proceedings of the Royal Society of London. Series A: Mathematical, Physical and Engineering Sciences* 458(2022), 1353–1373.

Hetzler, H. (2014). Bifurcations in autonomous mechanical systems under the influence of joint damping. *Journal of Sound and Vibration* 333(23), 5953–5969.

Jensen, J. S. (2000). Buckling of an elastic beam with added high-frequency excitation. *International Journal of Non-Linear Mechanics* 35(2), 217–227.

Jensen, J. S., D. Tcherniak, and J. J. Thomsen (2000). Stiffening effects of high-frequency excitation: experiments for an axially loaded beam. *Journal of applied mechanics* 67(2), 397–402.

Mullin, T., A. Champneys, W. B. Fraser, J. Galan, and D. Acheson (2003). The 'indian wire trick' via parametric excitation: a comparison between theory and experiment. In *Proceedings of the Royal Society of London A: Mathematical, Physical and Engineering Sciences*, Volume 459, pp. 539–546. The Royal Society.

Ogata, K. (1997). *Modern Control Engineering*. Prentice Hall.

Sedighi, H. M. and K. H. Shirazi (2012). A survey of hopf bifurcation analysis in nonlinear railway wheelset dynamics. *Journal of Vibroengineering* 14(1).

Smith, H. and J. A. Blackburn (1992). Experimental study of an inverted pendulum. *American Journal of Physics* 60(10), 909–911.

Thomsen, J. J. (2003). Theories and experiments on the stiffening effect of high-frequency excitation for continuous elastic systems. *Journal of Sound and Vibration* 260(1), 117–139.

Wei, W. and H. Yabuno (2019). Subcritical hopf and saddle-node bifurcations in hunting motion caused by cubic and quintic nonlinearities: experimental identification of nonlinearities in a roller rig. *Nonlinear Dynamics* 98(1), 657–670.

Yabuno, H., K. Goto, and N. Aoshima (2004). Swing-up and stabilization of an underactuated manipulator without state feedback of free joint. *IEEE Transactions on Robotics and Automation* 20(2), 359–365.

Yabuno, H., T. Matsuda, and N. Aoshima (2005). Reachable and stabilizable area of an underactuated manipulator without state feedback control. *IEEE/ASME Transactions on Mechatronics* 10(4), 397–403.

Yabuno, H., M. Miura, and N. Aoshima (2004). Bifurcation in an inverted pendulum with tilted high-frequency excitation: analytical and experimental investigations on the symmetry-breaking of the bifurcation. *Journal of sound and vibration* 273(3), 493–513.

Yabuno, H., R. Oowada, and N. Aoshima (1999). Effect of coulomb damping on buckling of a simply supported beam. In *Proc. of ASME Design Engineering Technical Conferences*, Number VIB-8056, pp. 1–12.

Yabuno, H., H. Takano, and H. Okamoto (2008). Stabilization control of hunting motion of railway vehicle wheelset using gyroscopic damper. *Journal of Vibration and Control* 14(1-2), 209–230.

Yabuno, H. and K. Tsumoto (2007). Experimental investigation of a buckled beam under high-frequency excitation. *Archive of Applied Mechanics* 77(5), 339–351.

Yoshino, H., T. Hosoya, H. Yabuno, S. Lin, and Y. Suda (2015). Theoretical and experimental analyses on stabilization of hunting motion by utilizing the traction motor as a passive gyroscopic damper. *Proceedings of the Institution of Mechanical Engineers, Part F: Journal of Rail and Rapid Transit* 229(4), 395–401.

A

Cubic Nonlinear Characteristics

We consider a smooth function $F^{\#}$ of ξ with C^{∞}. $F^{\#}$ is expanded at $\xi = \xi_0$ by Taylor series as

$$F^{\#}(\xi) = F(\xi)^{\#}\big|_{\xi=\xi_0} + \left.\frac{dF^{\#}}{d\xi}\right|_{\xi=\xi_0}(\xi-\xi_0) + \frac{1}{2!}\left.\frac{d^2F^{\#}}{d\xi^2}\right|_{\xi=\xi_0}(\xi-\xi_0)^2 + \frac{1}{3!}\left.\frac{d^3F^{\#}}{d\xi^3}\right|_{\xi=\xi_0}(\xi-\xi_0)^3 + \cdots + \frac{1}{n!}\left.\frac{d^nF^{\#}}{d\xi^n}\right|_{\xi=\xi_0}(\xi-\xi_0)^n + \cdots, \tag{A.1}$$

or equivalently

$$F^{\#}(\xi) = F(\xi)^{\#}\big|_{\xi=\xi_0} + \left.\frac{dF^{\#}}{d\xi}\right|_{\xi=\xi_0}\Delta x + \frac{1}{2!}\left.\frac{d^2F^{\#}}{d\xi^2}\right|_{\xi=\xi_0}\Delta x^2 + \frac{1}{3!}\left.\frac{d^3F^{\#}}{d\xi^3}\right|_{\xi=\xi_0}\Delta x^3 + \cdots + \frac{1}{n!}\left.\frac{d^nF^{\#}}{d\xi^n}\right|_{\xi=\xi_0}\Delta x^n + \cdots, \tag{A.2}$$

where $\Delta x \overset{\text{def}}{=} \xi - \xi_0$. As shown in Figure A.1, we regard ξ_0, Δx, and $F^{\#}(\xi)$ as the natural length, the elongation of a spring, and the applied force, respectively. Letting $F^{\#}(\xi)\big|_{\xi=\xi_0} = 0$, Eq. (A.2) expresses the general nonlinear spring force with elongation Δx, i.e.

$$F^{\#}(\Delta x) = k_1\Delta x + k_2\Delta x^2 + k_3\Delta x^3 + \cdots + k_n\Delta x^n + \cdots \overset{\text{def}}{=} F_s^{\#}(\Delta x), \tag{A.3}$$

where $k_1 \overset{\text{def}}{=} \left.\frac{dF^{\#}}{d\xi}\right|_{\xi=\xi_0}$, $k_2 \overset{\text{def}}{=} \frac{1}{2!}\left.\frac{d^2F^{\#}}{d\xi^2}\right|_{\xi=\xi_0}$, $k_3 \overset{\text{def}}{=} \frac{1}{3!}\left.\frac{d^3F^{\#}}{d\xi^3}\right|_{\xi=\xi_0}$, and $k_n \overset{\text{def}}{=} \frac{1}{n!}\left.\frac{d^nF^{\#}}{d\xi^n}\right|_{\xi=\xi_0}$, are linear, quadratic nonlinear, cubic nonlinear, and nth-order nonlinear stiffness, respectively, with n being an integer.

A.1 Symmetric and Nonsymmetric Nonlinearities

If there are only odd powers of Δx, i.e. $F_s^{\#}(\Delta x) = -F_s^{\#}(-\Delta x)$, the nonlinearity of $F_s^{\#}(\Delta x)$ is called *symmetric* and Eq. (A.3) is rewritten as

$$F_s^{\#}(\Delta x) = k_1\Delta x + k_3\Delta x^3 + \cdots + k_n\Delta x^n + \cdots, \tag{A.4}$$

where n is odd integer. On the other hand, if there are even powers of Δx, i.e. $F_s^{\#}(\Delta x) \neq -F_s^{\#}(-\Delta x)$, the nonlinearity of $F_s^{\#}(\Delta x)$ is called *nonsymmetric*. The characteristic of a nonlinear spring is described as the solid line in Figure A.2(a). The spring becomes stiffer as $|x|$

Linear and Nonlinear Instabilities in Mechanical Systems: Analysis, Control and Application, First Edition. Hiroshi Yabuno.

Companion website: www.wiley.com/go/yabuno/instabilitiesinmechanicalsystems

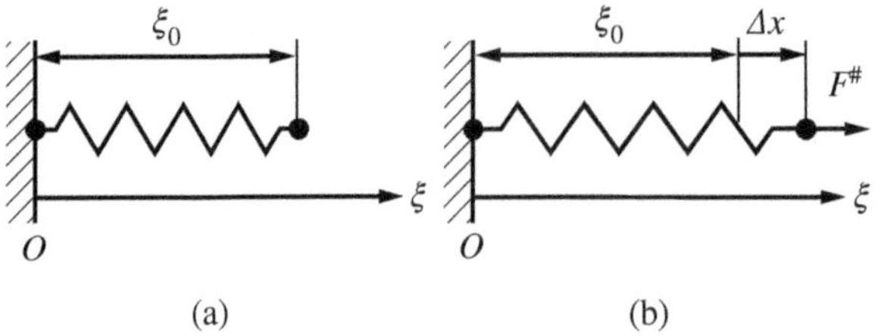

Figure A.1 Spring with natural length ξ_0. (a) The origin O of the coordinate ξ is located at the left end of the spring. (b) The elongation of the spring is Δx.

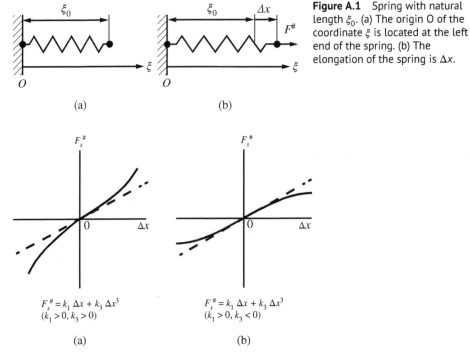

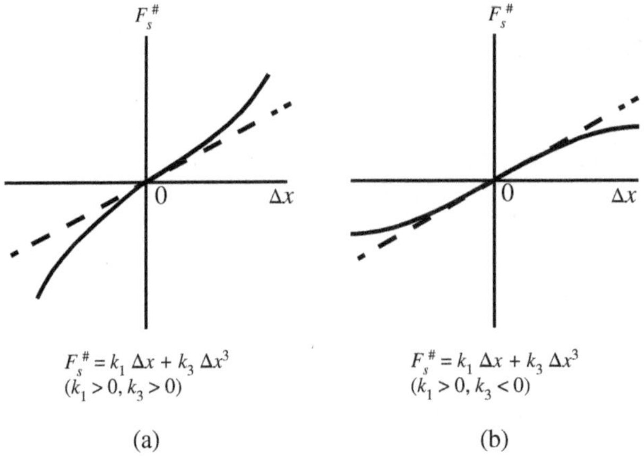

Figure A.2 Cubic nonlinear characteristics of a spring with positive linear stiffness: solid line denotes the nonlinear force. The dashed line tangent to the solid line at the origin is the linear component. (a) Hardening cubic nonlinearity: slope of solid line is larger than that of dashed line. (b) Softening cubic nonlinearity: slope of solid line is less than that of dashed line.

increases, while the stiffness of a linear spring that is described by the dashed line tangent to the solid line is constant independent of $|\Delta x|$. Such a nonlinear characteristic is called a *hardening* nonlinear stiffness. The characteristic of a nonlinear spring described by the solid line in Figure A.2(b) becomes less stiffer as $|\Delta x|$ increases. Such a nonlinear characteristic is called a softening nonlinear stiffness. Figure A.3 shows the nonlinear characteristics of a spring exhibiting a negative linear stiffness component so that the slope of the dashed line is negative at the origin. Nonlinear characteristics of Figure A.3(a,b) are hardening and softening, respectively.

A.2 Nonsymmetric Nonlinearity Due to the Shift of the Equilibrium State

In this section, we consider two spring-mass systems in which the mass is supported by a spring with a symmetric cubic nonlinear characteristic that is expressed as

$$F_s^{\#}(\Delta x) = k_1 \Delta x + k_3 \Delta x^3, \tag{A.5}$$

where higher order terms than quintic nonlinearity in Eq. (A.4) are truncated. One is the system as shown in Figure A.4(a). The other one is subject to gravity as shown in Figure 1.1,

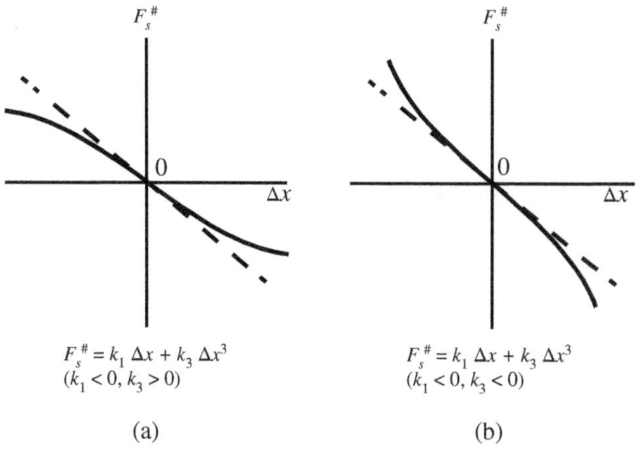

Figure A.3 Nonlinear characteristics of a spring with negative linear stiffness ($k_1 < 0$): (a) hardening ($k_3 > 0$); (b) softening ($k_3 < 0$).

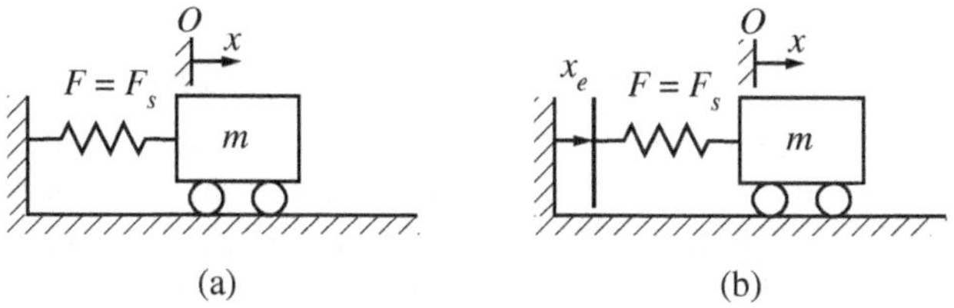

Figure A.4 Spring-mass system: (a) without excitation; (b) with excitation.

but the linear spring is replaced with a symmetric nonlinear spring whose characteristic is described as Figure A.2(a) or Figure A.2(b). The equation of motion of the first system is expressed as

$$m\frac{d^2x}{dt^2} = -F_s^{\#}(\Delta x) = -k_1\Delta x - k_3\Delta x^3 = -k_1x - k_3x^3, \tag{A.6}$$

where it is accounted that the displacement of the mass x is equal to the elongation of the spring Δx. In the neighborhood of the equilibrium state, i.e., $x = 0$, the spring force acting on the mass is symmetric. On the other hand, the equation of motion of the system under gravity in Figure 1.1 is expressed as

$$m\frac{d^2x'}{dt^2} = -F_s^{\#}(x') + mg = -k_1x' - k_3x'^3 + mg, \tag{A.7}$$

where x' is equal to the elongation of the spring Δx. The equilibrium state $x' = x_{st}$ satisfies

$$0 = -F_s^{\#}(x_{st}) + mg = -k_1x_{st} - k_3x_{st}^3 + mg. \tag{A.8}$$

By introducing $x(= x' - x_{st})$ and taking into account the equilibrium equation (A.8), Eq. (A.7) leads to

$$m\frac{d^2x}{dt^2} = -(k_1 + 3k_3x_{st}^2)x - 3k_3x_{st}x^2 - k_3x^3. \tag{A.9}$$

It is noticed that the quadratic nonlinear term with respect to x appears. Therefore, the nonsymmetric nonlinear spring force in the equilibrium state can act on the mass regardless if the spring is symmetric.

A.3 Effect of Harmonic External Excitation

We consider the effect of the displacement excitation x_e as shown in Figure A.4(b). By replacing Δx with $x - x_e$, Eq. (A.6) is changed to

$$m\frac{d^2x}{dt^2} = -F_s^{\#}(\Delta x) = -k_1(x - x_e) - k_3(x - x_e)^3. \tag{A.10}$$

In particular, considering the harmonic excitation as $x_e = a_e \cos \nu t$, we have the equation of motion as

$$\begin{aligned} m\frac{d^2x}{dt^2} + \left(k_1 + \frac{3}{2}a_e^2k_3 + \frac{3}{2}a_e^2k_3 \cos 2\nu t\right)x - 3a_ek_3 \cos \nu t x^2 + k_3x^3 \\ = \left(a_ek_1 + \frac{3}{4}a_e^3k_3\right)\cos \nu t + \frac{a_e^3}{4}k_3 \cos 3\nu t. \end{aligned} \tag{A.11}$$

The terms including k_3 are due to the cubic nonlinearity of the spring. The linear natural frequency is shifted from $\omega = \sqrt{\frac{k_1}{m}}$ to $\omega' = \sqrt{\frac{k_1 + 3a_e^2k_3/2}{m}}$. When ν is in the neighborhood of the shifted linear natural frequency ω', in addition to the primary resonance due to the first term on the right-hand side, the parametric resonance can be produced due to the term of $\frac{3}{2}a_e^2k_3 \cos 2\nu t \cdot x$. When the excitation frequency ν is in the neighborhood of triple the shifted natural frequency ω', the $\frac{1}{3}$-order subharmonic resonance can occur. The occurrence of such resonances will be dealt with in Appendix C.

B

Nondimensionalization and Scaling Nonlinearity

In this appendix, using the equation of motion of a spring–mass–damper system as that in Figure B.1(a), we consider a method of nondimensionalization and its necessity. Relating the method to choose the representative quantity in the nondimensionalization, the concept of *scaling* is introduced using a nonlinear spring as an example. Furthermore, the nondimensional equation of motion for a nonlinear spring mass damper system under a harmonic excitation is shown.

B.1 Nondimensionalization of Equations of Motion

Referring to Chapter 1, we can express the dimensional equation of motion of spring-mass-damper system shown in Figure B.1(a) as

$$m\frac{d^2x}{dt^2} + c\frac{dx}{dt} + kx = 0, \tag{B.1}$$

where all terms being summed on the left-hand side must have the same dimension of force by the *principle of dimensional homogeneity* (Witelski and Bowen (2015)). Letting the dimension of force be $[F]$, we can express the dimension of the first term on the left-hand side, i.e. of the inertia term, as

$$[F] = \left[\frac{ML}{T^2}\right], \tag{B.2}$$

where $[M]$, $[L]$, and $[T]$ express the dimensions of mass, length, and time, respectively. From this fact, we can determine the dimensions of the damping coefficient and the spring constant, c and k, which are

$$\left[\frac{F}{L/T}\right] = \left[\frac{ML/T^2}{L/T}\right] = \left[\frac{M}{T}\right] \tag{B.3}$$

and

$$\left[\frac{F}{L}\right] = \left[\frac{M}{T^2}\right], \tag{B.4}$$

respectively. Using the representative length X_r with dimension $[L]$ and the representative time T_r with dimension $[T]$, which are concretely selected depending on the problem, we

Linear and Nonlinear Instabilities in Mechanical Systems: Analysis, Control and Application, First Edition. Hiroshi Yabuno.

Companion website: www.wiley.com/go/yabuno/instabilitiesinmechanicalsystems

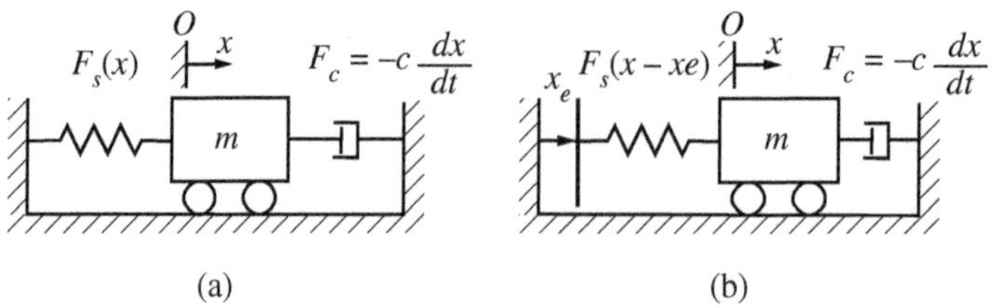

Figure B.1 Spring-mass-damper system: (a) without external excitation; (b) with external excitation.

set x and t as

$$x = X_r x^*, \quad t = T_r t^*. \tag{B.5}$$

Then, x^* and t^* are the nondimensional displacement and time, respectively. Substituting Eq. (B.5) into Eq. (B.1) yields

$$\frac{d^2x^*}{dt^{*2}} + T_r \frac{c}{m}\frac{dx^*}{dt^*} + T_r^2 \frac{k}{m} x^* = 0. \tag{B.6}$$

Choosing the inverse of the natural frequency $\sqrt{\frac{k}{m}}$ with dimension $[T]$ as the representative time T_r, i.e.

$$T_r = 1 \Big/ \sqrt{\frac{k}{m}} \tag{B.7}$$

leads to

$$\frac{d^2x^*}{dt^{*2}} + \frac{c}{\sqrt{mk}}\frac{dx^*}{dt^*} + x^* = 0. \tag{B.8}$$

Because each term is nondimensional, the coefficient of the second term $\frac{c}{\sqrt{mk}} \overset{\text{def}}{=} 2\gamma$ must be nondimensional; the nondimensional quantity γ is called as the *damping ratio*. Based on the *principle of dimensional homogeneity*, we know directly from Eq. (B.8) that γ is nondimensional.

Problem B.1 Show that γ is nondimensional.

Ans:

Using Eqs. (B.3) and (B.4),

$$\left[\frac{c}{\sqrt{mk}}\right] = \frac{[M]/[T]}{\sqrt{[M] \cdot [M]/[T]^2}} = [1]. \tag{B.9}$$

As a result, the original equation of motion with three parameters, m, c, and k leads to the nondimensionalized equation with only one parameter γ as

$$\frac{d^2x^*}{dt^{*2}} + 2\gamma\frac{dx^*}{dt^*} + x^* = 0. \tag{B.10}$$

This remaining nondimensional parameter γ is essential for the dynamics of the system and by the parameter study only on γ, the dynamics of Eq. (B.1) can be revealed. This is an important reason to nondimensionalization. Therefore, before employing a mathematical method for analyzing the governing equation, the nondimensionalization is generally performed (Nayfeh 1981).

Let us consider the analytical result for the case without damping to investigate more in detail the merit of the nondimensionalization. When the initial displacement and velocity are x_0 and v_0, respectively, let us examine the dynamics of the spring-mass system in Figure A.4(a) with $F_s = -kx$. The equation of motion in the dimensional form is

$$m\frac{d^2x}{dt^2} + kx = 0. \tag{B.11}$$

Choosing the representative time as $T_r = \frac{1}{\omega}$, where $\omega = \sqrt{\frac{k}{m}}$, we obtain the nondimensional equation of motion as

$$\frac{d^2x^*}{dt^{*2}} + x^* = 0. \tag{B.12}$$

Under the nondimensional initial conditions, $x_0^* = \frac{x_0}{X_r}$ and $v_0^* = \frac{v_0}{X_r\omega}$, the solution is

$$x^* = x_0^* \cos t^* + v_0^* \sin t^*, \tag{B.13}$$

equivalently,

$$x^* = \sqrt{x_0^{*2} + v_0^{*2}} \cos(t^* + \phi), \tag{B.14}$$

where ϕ satisfies

$$\begin{cases} \cos\phi = \dfrac{x_0^*}{\sqrt{x_0^{*2} + v_0^{*2}}}, & \text{(B.15)} \\[2ex] \sin\phi = -\dfrac{v_0^*}{\sqrt{x_0^{*2} + v_0^{*2}}}. & \text{(B.16)} \end{cases}$$

By choosing the representative length as $X_r = \sqrt{x_0^2 + \left(\frac{v_0}{\omega}\right)^2}$, Eq. (B.14) is simplified as

$$x^* = \cos(t^* + \phi). \tag{B.17}$$

Letting $t^* + \phi$ be $t^{*'}$, we have the time history as the graph of Figure B.2 with the ordinate of x^* and the abscissa $t^{*'}$. The time history of Eq. (B.14) is the same graph with the ordinate of x^* and the abscissa $t^*(= t^{*'} - \phi)$. Furthermore, the time history of the solution of Eq. (B.1) is the same graph with the ordinate of $x\left(= X_r x^* = \sqrt{x_0^2 + \left(\frac{v_0}{\omega}\right)^2} x^*\right)$ and the abscissa $t\left(= T_r t^* = \frac{1}{\omega} t^*\right)$.

If we want to find the dimensional time history, we just shift the abscissa and change the scales of both axes depending on the representative quantities while keeping the graph of Eq. (B.17). Hence, the analysis for the nondimensional equation of motion captures an essential characteristic of the dynamics.

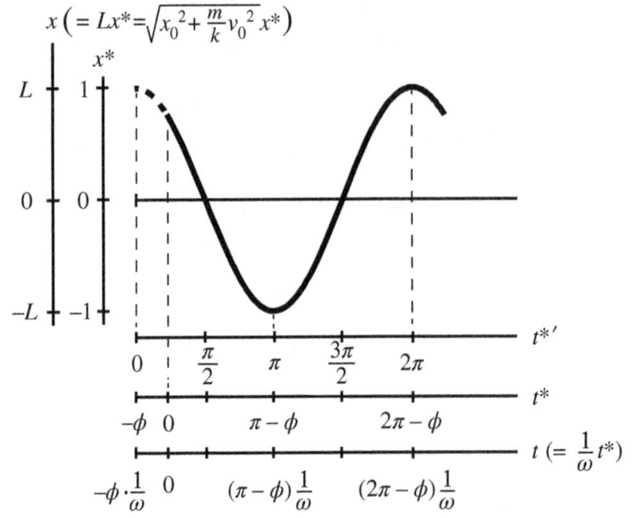

Figure B.2 Nondimensional and dimensional time histories.

B.2 Scaling of Nonlinearity

We consider an analytic function $F(x')$. In the neighborhood of an interested point, x_0, we can have the expression of Taylor series as

$$F(x') = k_0 + k_1(x' - x_0) + k_2(x' - x_0)^2 + \cdots + k_n(x' - x_0)^n + \cdots , \tag{B.18}$$

or using $x' - x_0 \overset{\text{def}}{=} x$,

$$F(x) = k_0 + k_1 x + k_2 x^2 + \cdots + k_n x^n + \cdots , \tag{B.19}$$

where $k_n = \frac{1}{n!} \frac{d^n F}{dx'^n}\Big|_{x'=x_{st}} = \frac{1}{n!} \frac{d^n F}{dx^n}\Big|_{x=0}$, with n being an integer. The radius of convergence R is calculated from

$$R^{-1} = \limsup_{n\to\infty} |k_n|^{1/n}. \tag{B.20}$$

In the case of $R \neq 0, f(x)$ converges uniformly and absolutely for $|x| < R$. Then, we can find a certain finite number $r(0 < r \leq R)$ such that ((see for example Yoshida (2010))

$$\sup |k_n r^n| < 1. \tag{B.21}$$

Rescaling or nondimensionalizing x with r as

$$x^* = \frac{x}{r} \tag{B.22}$$

yields

$$F(x^*) = \alpha_0 + \alpha_1 x^* + \alpha_2 x^{*2} + \cdots + \alpha_n x^{*n} + \cdots , \tag{B.23}$$

where $|\alpha_n| = |k_n r^n| < 1$ because of Eq. (B.21). Within the range of $|x^*| < 1$, Eq. (B.23) uniformly converges. Therefore, the number r $(0 < r \leq R)$ is a reference for x and Eq. (B.23) is a representative form of the analytic function $F(x')$ or $F(x)$.

B.3 Nondimensionalization of the Governing Equation of a Nonlinear Oscillator

Let us consider the spring-mass-damper system as shown in Figure B.1(a). It is assumed that the spring force $F_s^{\#}(x)$ is smooth with respect to the elongation and nonsymmetric, i.e. $F_s^{\#}(x) \neq -F_s^{\#}(-x)$. Then, $F_s^{\#}(x)$ is expressed by Taylor expansion as

$$F_s^{\#}(x) = k_1 x + k_2 x^2 + k_3 x^3 + \cdots + k_n x^n + \cdots , \tag{B.24}$$

which is rewritten as

$$F_s^{\#}(x^*) = k_1 X_r x^* + k_2 X_r^2 x^{*2} + k_3 X_r^3 x^{*3} + \cdots + k_n X_r^n x^{*n} + \cdots , \tag{B.25}$$

where $x^* = \frac{x}{X_r}$ is the nondimensional elongation. By the *principle of dimensional homogeneity*, the dimensions of the coefficients k_n in Eq. (B.24) are different, but the dimensions of the coefficients $k_n X_r^n$ in Eq. (B.25) are the same. By setting the representative length X_r to be less than or equal to the radius of convergence of $F_s^{\#}(x)$, $k_n X_r^n$ becomes less than 1. Equation (B.25) can be truncated at any order depending on a desired accuracy of analysis.

Taking into account the spring and damping forces acting on the mass, $F_s(= -F_s^{\#})$ and $F_d(= -c\frac{dx}{dt})$, respectively, we obtain the equation of motion in the dimensional form as

$$m\frac{d^2x}{dt^2} = F_s + F_d = -c\frac{dx}{dt} - (k_1 x + k_2 x^2 + k_3 x^3 + \cdots + k_n x^n + \cdots). \tag{B.26}$$

Using the nondimensional displacement x^* and the nondimensional time $t^* = \sqrt{\frac{k_1}{m}}t$, we obtain the following nondimensionalized equation:

$$\frac{d^2x^*}{dt^{*2}} = -2\gamma\frac{dx^*}{dt^*} - (x^* + \alpha_2 x^{*2} + \cdots + \alpha_n x^{*n} + \cdots), \tag{B.27}$$

where $2\gamma = \frac{c}{\sqrt{mk_1}}$ and α_n is a nth-order nonlinear stiffness

$$\alpha_n = \frac{k_n X_r^{n-1}}{k_1}. \tag{B.28}$$

B.4 Effect of Harmonic External Excitation

We consider the effect of the displacement excitation x_e as Figure B.1(b). By replacing x including the spring force with $x - x_e$, i.e. replacing x^* by $x^* - x_e^*$, where $x_e^* = x_e/X_r$, Eq. (B.27) is changed to

$$\frac{d^2x^*}{dt^{*2}} = -2\gamma\frac{dx^*}{dt^*} - (x^* - x_e^*) - \alpha_2(x^* - x_e^*)^2 - \alpha_3(x^* - x_e^*)^3 - \cdots . \tag{B.29}$$

In particular, we consider a harmonic excitation $x_e = a_e \cos \nu t$. Let the nondimensional excitation displacement x_e^* be

$$x_e^* = a_e^* \cos \nu^* t^*, \tag{B.30}$$

where $a_e^* \stackrel{\text{def}}{=} \frac{a_e}{X_r}$ is the nondimensional excitation amplitude and $\nu^* \stackrel{\text{def}}{=} \nu T_r$ is the nondimensional excitation frequency. Then, Eq. (B.29) leads to

$$\begin{aligned}
&\frac{d^2x^*}{dt^{*2}} + 2\gamma\frac{dx^*}{dt^*} + \left(1 + \frac{3a_e^2}{2}\alpha_3 - 2a_e^*\alpha_2 \cos\nu^*t^* + \frac{3a_e^2}{2}\alpha_3 \cos 2\nu^*t^*\right)x^* \\
&+ (\alpha_2 - 3a_e^*\alpha_3 \cos\nu^*t^*)x^{*2} + \alpha_3x^{*3} + \cdots \\
&= -\frac{a_e^{*2}}{2}\alpha_2 + \left(a_e^* + \frac{3a_e^{*3}}{4}\alpha_3\right)\cos\nu^*t^* - \frac{a_e^{*2}}{2}\alpha_2\cos 2\nu^*t^* + \frac{a_e^{*3}}{4}\alpha_3 \cos 3\nu^*t^* + \cdots .
\end{aligned} \tag{B.31}$$

Similar to the system of Eq. (A.11), the natural frequency is shifted to $\omega' = 1 + \frac{3a_e^2}{2}\alpha_3$ due to the cubic nonlinear effect. However, because of taking into account also the nonsymmetric nonlinearity, this equation has more nonlinear terms than those in Eq. (A.11); the terms including α_2 are additional. First of all, the DC component (direct current component) that is the component independent of time is produced in the response by means of the quadratic nonlinearity through the term α_2x^{*2}. When the excitation frequency ν^* is in the neighborhood of twice the shifted natural frequency ω', the *parametric* resonance, and *1/2-order subharmonic* resonance can be simultaneously produced due to the term $-2a_e^*\alpha_2\cos\nu^*t^* \cdot x^*$, and both terms α_2x^{*2} and $\left(a_e^* + \frac{3a_e^{*2}}{4}\alpha_3\right)\cos\nu^*t^*$, respectively. When the excitation frequency ν^* is in the neighborhood of triple the shifted natural frequency ω', the *1/3-order subharmonic* resonances can be produced due to both terms α_3x^{*3} and $\left(a_e^* + \frac{3a_e^{*2}}{4}\alpha_3\right)\cos\nu^*t^*$. Furthermore, when the excitation frequency ν^* is in the neighborhood of a half the shifted natural frequency ω', the *2nd-order superharmonic* resonance can be produced due to both terms α_2x^{*2} and $\left(a_e^* + \frac{3a_e^{*2}}{4}\alpha_3\right)\cos\nu^*t^*$. When the excitation frequency ν^* is in the neighborhood of one-third the shifted natural frequency ω', the *3rd-order superharmonic* resonance can be produced due to both terms α_3x^{*3} and $\left(a_e^* + \frac{3a_e^{*2}}{4}\alpha_3\right)\cos\nu^*t^*$. The occurrence predictions about these resonances will be dealt with in Appendix C.

References

Nayfeh, A. H. (1981). *Introduction to Perturbation Technique.* Wiley.

Witelski, T. and M. Bowen (2015). *Methods of mathematical modelling: Continuous systems and differential equations.* Springer.

Yoshida, Z. (2010). *Nonlinear science: the challenge of complex systems.* Springer Science & Business Media.

C

Occurrence Prediction for Some Types of Resonances

First, we summarize the dynamics of a linear oscillator subject to harmonic excitation and qualitatively examine the mechanism of resonance from physical and mathematical points of view. Using this resonance mechanism, we predict the occurrence of various types of resonances in a nonlinear oscillator such as subharmonic, superharmonic, and parametric resonances. To clarify the resonance mechanism of oscillators with the occurrence prediction is very important to theoretically analyze the nonlinear resonances by the perturbation method because the order of the leading term in the analytical solution can be determined based on the resonance mechanism.

C.1 Dynamics of a Linear Spring-Mass-Damper System Subject to Harmonic External Excitation

C.1.1 Case with Viscous Damping

For the linear oscillator subject to a harmonic external excitation as shown in Figure B.1(b), the nondimensional equation of motion is expressed from Eq. (B.31) with $\alpha_2 = \alpha_3 = 0$ as

$$\frac{d^2x}{dt^2} + 2\gamma\frac{dx}{dt} + x = a_e \cos \nu t, \tag{C.1}$$

where the first, second, and third terms are the inertia, damping, and restoring forces, respectively. Equation (C.1) and hereafter, the symbol "*" denoting the nondimensional variables is omitted for simplicity. The mass, stiffness, and natural frequency are normalized to 1, and the damping ratio γ is assumed to be positive. A complete solution consists of the homogeneous solution $\tilde{x}$ and a particular solution x_p as

$$x = \tilde{x} + x_p, \tag{C.2}$$

where $\tilde{x}$ is a homogeneous solution of Eq. (C.1) whose right-hand side is replaced with 0, i.e. the solution of

$$\frac{d^2\tilde{x}}{dt^2} + 2\gamma\frac{d\tilde{x}}{dt} + \tilde{x} = 0. \tag{C.3}$$

Linear and Nonlinear Instabilities in Mechanical Systems: Analysis, Control and Application, First Edition. Hiroshi Yabuno.

Companion website: www.wiley.com/go/yabuno/instabilitiesinmechanicalsystems

In the case when there is any positive damping, because of $\gamma > 0$, $\tilde{x}$ decays with time from the result of Chapter 2. A particular solution is

$$x_p = a_{pc} \cos \nu t + a_{ps} \sin \nu t, \tag{C.4}$$

where a_{pc} and a_{ps} are

$$a_{pc} = \frac{1-\nu^2}{(1-\nu^2)^2 + 4\gamma^2\nu^2} a_e, \tag{C.5}$$

$$a_{ps} = \frac{2\gamma\nu}{(1-\nu^2)^2 + 4\gamma^2\nu^2} a_e. \tag{C.6}$$

The particular solution Eq. (C.4) is also expressed as

$$x_p = a_p \cos(\nu t + \phi), \tag{C.7}$$

where

$$a_p = \frac{a_e}{\sqrt{(1-\nu^2)^2 + 4\gamma^2\nu^2}} \tag{C.8}$$

and the phase difference ϕ satisfies

$$\cos\phi = \frac{a_{pc}}{a_p}, \quad \sin\phi = -\frac{a_{ps}}{a_p} \tag{C.9}$$

or

$$\phi = -\arctan\frac{a_{ps}}{a_{pc}} \quad (-\pi < \phi \le 0). \tag{C.10}$$

Because $a_{ps} > 0$ is independent of ν, the phase difference ϕ is always negative and the response delays compared with the external excitation $a_e \cos \nu t$. The amplitude and phase difference depending on the excitation frequency ν are described as Figure C.1(a,b), respectively (for example, see (Balachandran and Magrab, 2008)).

C.1.2 Case Under No Viscous Damping

The dimensionless equation of motion is

$$\frac{d^2x}{dt^2} + x = a_e \cos \nu t. \tag{C.11}$$

When the excitation frequency is not equal to the natural frequency, i.e. $\nu \neq 1$, a particular solution is

$$x_p = a_p \cos(\nu t + \phi), \tag{C.12}$$

where

$$a_p = \left|\frac{a_e}{1-\nu^2}\right| \tag{C.13}$$

and

$$\phi = \begin{cases} 0 & (\nu < 1), \\ -\pi & (\nu > 1), \end{cases} \tag{C.14}$$

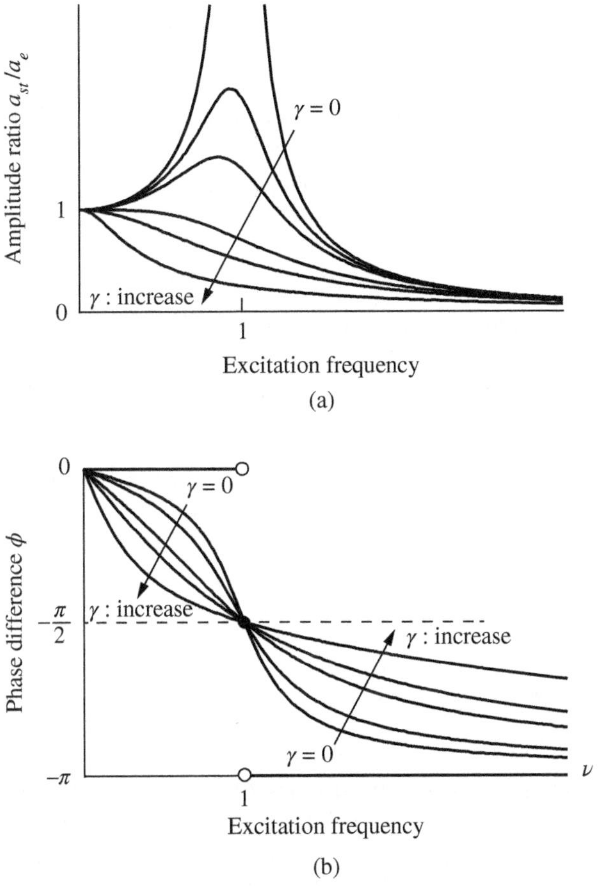

Figure C.1 Frequency response curve: (a) variation of response amplitude ratio a_{st}/a_e depending on the excitation frequency ν; (b) variation of phase difference depending on the excitation frequency ν. Source: Modified from Yabuno (2019).

respectively. The amplitude and phase difference depending on the excitation frequency ν are described as shown in Figure C.1. Then, the homogeneous solution is

$$\tilde{x} = a_c \cos t + a_s \sin t. \tag{C.15}$$

From Eq. (C.2), the complete solution of Eq. (C.11) is expressed for $\nu \neq 1$ as

$$x = a_c \cos t + a_s \sin t + \frac{a_e}{1 - \nu^2} \cos \nu t. \tag{C.16}$$

Under the initial conditions $x|_{t=0} \stackrel{\text{def}}{=} x_0$, and $\left.\frac{dx}{dt}\right|_{t=0} \stackrel{\text{def}}{=} v_0$, a_c and a_s are determined, and the complete solution of Eq. (C.11) is expressed as

$$x = \left(x_0 - \frac{a_e}{1 - \nu^2}\right) \cos t + v_0 \sin t + \frac{a_e}{1 - \nu^2} \cos \nu t, \tag{C.17}$$

or

$$x = Ae^{it} + A_p e^{i\nu t} + \mathrm{CC}, \tag{C.18}$$

where $A = \frac{1}{2}\left(x_0 - \frac{a_e}{1-\nu^2}\right) + \frac{v_0}{2i}$, $A_p = \frac{a_p}{2}e^{i\phi}$, and CC denotes the complex conjugates of the preceding terms.

When the excitation frequency is equal to the natural frequency, i.e. under the primary resonance condition $\nu = 1$, a particular solution is

$$x_p = \frac{a_e}{2} t \sin t, \tag{C.19}$$

which can be confirmed by the substitution of Eq. (C.19) into Eq. (C.11). Thus, the term proportional to t is called a *secular term*. The growth with time is a characteristic of resonance phenomena. Then, by the initial conditions, x_0 and v_0, the complete solution of Eq. (C.11) is expressed as

$$x = x_0 \cos t + v_0 \sin t + \frac{a_e}{2} t \sin t. \tag{C.20}$$

By the way, Eq. (C.11) is rewritten in the state equation form as

$$\frac{d}{dt}\begin{bmatrix} y_1 \\ y_2 \end{bmatrix} = B \begin{bmatrix} y_1 \\ y_2 \end{bmatrix} + \begin{bmatrix} 0 \\ a_e \end{bmatrix} \cos \nu t, \tag{C.21}$$

where $y_1 = x$, $y_2 = \frac{dx}{dt}$, and

$$B = \begin{bmatrix} 0 & 1 \\ -1 & 0 \end{bmatrix}. \tag{C.22}$$

The left-hand side and the first term on the right-hand side describe the dynamics of the spring–mass system and the second term on the right-hand side is the harmonic excitation.

The above mentioned resonance condition ($\nu = 1$) is also interpreted as follows. Equation (C.22) corresponds to the matrix B of Eq. (2.80) with the eigenvalues $\pm i$. The external harmonic excitation is expressed in the complex expression as

$$\begin{bmatrix} 0 \\ a_e \end{bmatrix} \cos \nu t = \begin{bmatrix} 0 \\ \frac{a_e}{2} \end{bmatrix} (e^{i\nu t} + e^{-i\nu t}). \tag{C.23}$$

When the coefficient of t in the exponent of the exponential function in an external excitation, i.e. $\pm i\nu$ in Eq. (C.23), are equal to the eigenvalues of matrix B, i.e. $\pm i$ in Eq. (C.22), the system of Eq. (C.21) satisfies the primary resonance condition and is in the primary resonant state.

C.2 Occurrence Prediction of Some Types of Resonances in a Nonlinear Spring-Mass-Damper System

Toward the occurrence prediction of resonances, the following two characteristics in the responses of the linear system under the external harmonic excitation should be noted from the above analytical result:

1. The response frequency is always equal to the excitation frequency ν.

2. The resonance is produced when the excitation frequency is equal to the natural frequency, which corresponds to the eigenvalues of matrix B.

We shall predict resonances produced in the nonlinear oscillator governed by Eq. (B 31) as an illustrative example. First, we rewrite this equation as

$$\frac{d^2x}{dt^2} + x = a_e \cos \nu t + 2\alpha_2 a_e \cos \nu t \cdot x - \alpha_2 x^2 - \alpha_3 x^3$$
$$-\frac{a_e^2}{2}\alpha_2 - \frac{a_e^2}{2}\alpha_2 \cos 2\nu t - 2\gamma \frac{dx}{dt} + \cdots, \tag{C.24}$$

where the terms representing the inertia and linear spring forces are left on the left-hand side and the other terms move on the right-hand side. The left-hand side of Eq. (C.24) can be regarded as the equation expressing the dynamics of the linear spring–mass system with the unit natural frequency, i.e. with the eigenvalues $\pm i$ of the matrix B in Section 2.4.4. The first term on the right-hand side is the actual external excitation and the other terms on the right-hand side of Eq. (C.24) can be regarded as "equivalent" external excitations which are applied to the linear spring–mass system on the left-hand side of Eq. (C.24). We shall predict the occurrence of the resonances due to these equivalent external excitations.

Before that, we examine more in details the solution of Eq. (C.11), which corresponds to Eq. (C.24) in the case without the nonlinear or damping effects including γ, α_2, α_3, and so on. Rewriting the solution Eq. (C.17) for $\nu \neq 1$ as

$$x = x_0 \cos t + v_0 \sin t + \frac{a_e(\cos \nu t - \cos t)}{1 - \nu^2} \tag{C.25}$$

and taking the limit $\nu \to 1$ by the aid of L'Hospital's Rule, we notice that Eq. (C.17) results in Eq. (C.20) for $\nu = 1$. Therefore, we can regard Eq. (C.17) or (C.18) as the solution for any ν.

Let us predict the occurrence of resonances due to each equivalent external excitation by substituting (C.18) into each term on the right-hand side of Eq. (C.24) individually.

C.2.1 Effect of the Linear Term with a Periodic Time-Varying Coefficient $2\alpha_2 a_e \cos \nu t \cdot x$

This term is rewritten as

$$2\alpha_2 a_e \cos \nu t x = \alpha_2 a_e (e^{i(\nu-1)t}\overline{A} + e^{i(\nu+1)t}A + a_p\, e^{2i\nu t} + a_p + \mathrm{CC}). \tag{C.26}$$

When the excitation frequency is equal to twice the natural frequency of the spring mass system, i.e. $\nu = 2$, the coefficient of t appearing in the first exponential term of Eq. (C.26) becomes equal to i that is an eigenvalue of the B matrix related to the left-hand side of Eq. (C.24). Hence, the system described by Eq. (C.24) satisfies the resonance condition mentioned in Section C.1 and produces a resonance with a resonance mechanism different from the primary resonance in the linear external excitation. This kind of resonance is induced from the term with the periodic time varying coefficient, which is called *parametric resonance* (see Chapter 11).

C.2.2 Effect of the Quadratic Nonlinear Term $-\alpha_2 x^2$

This term is rewritten as

$$-\alpha_2 x^2 = -\alpha_2 \left(2a_p e^{i(\nu-1)t}\overline{A} + a_p^2 e^{2i\nu t} + 2a_p e^{i(\nu+1)t} A + e^{2it} A^2 \right.$$
$$\left. + |A|^2 + a_p^2 + \mathrm{CC} \right) . \tag{C.27}$$

Other two types of resonances depending on the excitation frequency ν can occur.

The first one is the case when the excitation frequency is equal to twice the natural frequency of the spring mass system, i.e. $\nu = 2$. Since the coefficient of t in the first exponential function shown in Eq. (C.27) becomes equal to i that is the eigenvalue of the B matrix related to the left-hand side, a resonance can occur, which is called a *a half-order subharmonic resonance*.

The second one is the case when the excitation frequency is equal to a half the natural frequency of the spring mass system, i.e. $\nu = \frac{1}{2}$. Since the coefficient of t appearing in the second exponential function of Eq. (C.27) becomes equal to i that is the eigenvalue of the B matrix related to the left-hand side, a resonance can occur, which is called a *2nd-order superharmonic resonance.*

C.2.3 Effect of the Cubic Nonlinear Term $-\alpha_3 x^3$

This term is rewritten as

$$-\alpha_3 x^3 = -\alpha_3 \left(3a_p e^{i(\nu-2)t}\overline{A}^2 + a_p^3 e^{3i\nu t} + e^{3it} + 3a_p e^{i(2+\nu)t} A^2 + 3|A|^2 A e^{it} \right.$$
$$\left. + 5a_p e^{i\nu t}|A|^2 + 3a_p^2 e^{i(2\nu+1)t} A + 5a_p^2 e^{it} A + 3a_p^2 e^{i(2\nu-1)t} + 3a_p^3 e^{i\nu t} + \mathrm{CC} \right) . \tag{C.28}$$

Also in this case, it is possible to produce two types of resonances depending on the excitation frequency ν. The first one is the case when the excitation frequency is equal to triple the natural frequency of the spring mass system, i.e. $\nu = 3$. Since the coefficient of t in the first exponential function of Eq. (C.28) becomes i, that is the eigenvalue of the matrix B related to the left-hand side, a resonance can occur, which is called a *one-third-order subharmonic resonance.*

The second one is the case when the excitation frequency is equal to one-third the natural frequency of the spring mass system, i.e. $\nu = \frac{1}{3}$. Because the coefficient of t appearing in the second exponential function of Eq. (C.28) becomes i, that is the eigenvalue of the matrix B related to the left-hand side, a resonance can occur. This type of resonance is called a *third-order superharmonic resonance.*

In the above predictions, it is the resonance condition that the excitation frequency ν strictly equals to some values. It should be noted that the response frequency is equal to the natural frequency 1 in the above resonances. In general, if such conditions for the excitation frequency ν are approximately satisfied, these resonances can occur. The quantitative regime of ν for the resonances is clarified by the analytical approaches such as perturbation methods for the governing equation. Then, the response frequency is deviated from the natural frequency 1 depending on the *detuning* of excitation frequency ν from the resonance conditions.

References

Balachandran, B. and E. B. Magrab (2008). *Vibrations*. Cengage Learning.

Yabuno, H. (2019). Review of applications of self-excited oscillations to highly sensitive vibrational sensors. *ZAMM-Journal of Applied Mathematics and Mechanics/Zeitschrift für Angewandte Mathematik und Mechanik*, e201900009.

D

Order Estimation of Responses

In Appendix C, we have predicted the occurrence of some types of nonlinear resonances and the parametric resonance due to nonlinear effects. In the next step, we are interested in whether the response in the resonances is finite or not and if it is finite, how large the magnitude of the response is. In this appendix, we estimate the magnitude of the response by introducing the concept of balance among each term in governing equations. The estimation is utilized to determine the leading term in the *asymptotic expansion* form for the solution.

D.1 Order Symbol

We begin with the introduction of the order symbol "O," which is very useful to describe the solution of differential equations in the asymptotic expansion form and its error. The definition is as follows:

Definition D.1 When $f(\epsilon)$ and $g(\epsilon)$ satisfy

$$\lim_{\epsilon\to 0}\frac{f(\epsilon)}{g(\epsilon)} = A, \tag{D.1}$$

where A is finite, the relationship is expressed as "$f = O(g)$ as $\epsilon \to 0$," where "$\epsilon \to 0$" is often omitted.

Another symbol "o" is defined as follows:.

Definition D.2 When $f(\epsilon)$ and $g(\epsilon)$ satisfy

$$\lim_{\epsilon\to 0}\frac{f(\epsilon)}{g(\epsilon)} = 0, \tag{D.2}$$

the relationship is expressed as "$f = o(g)$ as $\epsilon \to 0$," where "$\epsilon \to 0$" is often omitted.

There is a related symbol "$\sim$" defined as follows:

Linear and Nonlinear Instabilities in Mechanical Systems: Analysis, Control and Application,
First Edition. Hiroshi Yabuno.

Companion website: www.wiley.com/go/yabuno/instabilitiesinmechanicalsystems

Definition D.3 When $f(\epsilon)$ and $g(\epsilon)$ satisfy

$$\lim_{\epsilon \to 0} \frac{f(\epsilon)}{g(\epsilon)} = 1, \tag{D.3}$$

the relationship is expressed as "$f \sim g$ as $\epsilon \to 0$," where "$\epsilon \to 0$" is often omitted.

Example D.1

- $\sin\theta = O(\theta)$ as $\theta \to 0$ and $\sin\theta \sim \theta$ as $\theta \to 0$, since

$$\lim_{\theta \to 0} \frac{\sin\theta}{\theta} = 1. \tag{D.4}$$

- $1 - \cos\theta = O(\theta^2)$ as $\theta \to 0$, since

$$\lim_{\theta \to 0} \frac{1 - \cos\theta}{\theta^2} = \frac{1}{2}. \tag{D.5}$$

D.2 Asymptotic Expression of Solution

In the approximate analytical approach as perturbation methods including the method of multiple scales, the solution is assumed using the following *asymptotic expansion*:

$$x(t, \epsilon) = \delta_0(\epsilon)x_0(t) + \delta_1(\epsilon)x_1(t) + \delta_2(\epsilon)x_2(t) + \cdots, \tag{D.6}$$

where $x_n = O(1)$ and $\delta_n(\epsilon)$ called a *gauge function* satisfies

$$\delta_{n+1}(\epsilon) = o(\delta_n(\epsilon)). \tag{D.7}$$

The first term on the right-hand side of Eq. (D.6) is called the *leading term*. It is very important to estimate the order of the leading term, i.e. $\delta_0(\epsilon)$ as first step in the application of the perturbation methods.

D.3 Linear Oscillator Under Harmonic External Excitation

We consider the nondimensional linear oscillator systems Eq. (C.1) with small damping and harmonic excitation, $\gamma = \epsilon\hat{\gamma} > 0$ ($\hat{\gamma} = O(1)$) and $a_e = \epsilon\hat{a}_e > 0$ ($\hat{a}_e = O(1)$), respectively. The equation is rewritten as

$$\underbrace{\frac{d^2x}{dt^2} + x}_{(A)} + \underbrace{2\epsilon\hat{\gamma}\frac{dx}{dt}}_{(B)} = \underbrace{\epsilon\hat{a}_e \cos\nu t}_{(C)}, \tag{D.8}$$

where (A), (B), and (C) are the sum of the inertia and spring forces, the viscous damping force, and the external force, respectively. As mentioned in Section C.2, the response frequency in the linear oscillator systems consists of the excitation frequency and the natural frequency. Here, focusing on the response with the excitation frequency, we can express the response as

$$x = \delta_0(\epsilon)\hat{a}\cos(\nu t - \phi), \tag{D.9}$$

where $\delta_0(\epsilon) = \epsilon^i$ $(0 \le i \in \mathbb{Z})$ is a gauge function. Substituting this expression into (A) and (B) of Eq. (D.8), we obtain

$$(\text{A}):\ \frac{d^2x}{dt^2} + x = \delta_0(\epsilon)(-\nu^2 + 1)\hat{a}\cos(\nu t - \phi), \tag{D.10}$$

$$(\text{B}):\ 2\epsilon\hat{\gamma}\frac{dx}{dt} = -2\epsilon\hat{\gamma}\delta_0(\epsilon)\nu\hat{a}\sin(\nu t - \phi) = O(\epsilon\delta_0(\epsilon)), \tag{D.11}$$

$$(\text{C}):\ \epsilon\hat{a}_e\cos\nu t = O(\epsilon). \tag{D.12}$$

We find the terms that balance the external excitation (C). For the balance among the terms, it is necessary that their orders, their phases, in particular frequencies, are the same.

D.3.1 Nonresonant Case

In the nonresonant case, when the excitation frequency is far from the natural frequency, i.e. $-\nu^2 + 1 = O(1)$, because of $\hat{a}\cos(\nu t - \phi) = O(1)$, the order of Eq. (D.10) is expressed by using $\delta_0(\epsilon)$ as follows:

$$\delta_0(\epsilon)(-\nu^2 + 1)\hat{a}\cos(\nu t - \phi) = O(\delta_0(\epsilon)). \tag{D.13}$$

Then, since the order of Eq. (D.10) is greater than that of Eq. (D.11), i.e. $O(\delta_0(\epsilon)) \gg O(\epsilon\delta_0(\epsilon))$, the external excitation (C) necessarily balances the sum of the inertia and spring forces (A). Therefore, since the order of Eq. (D.10) is equal to that of the external excitation $O(\epsilon)$, i.e. $\delta_0(\epsilon) = \epsilon$, we can determine the gauge function of the leading term $\delta_0(\epsilon)$, as $\delta_0(\epsilon) = \epsilon$. In other words, since the order of the damping effect (B) is $O(\epsilon\delta_0(\epsilon)) = O(\epsilon^2)$, which is less than the sum of the inertia and spring forces (A), the response is governed mainly by the inertia and spring forces, but not by the damping effect.

D.3.2 Resonant Case

We consider the resonant case, that is, when the excitation frequency is equal to the natural frequency, i.e., $\nu = 1$. Since the inertia and spring forces are cancelled out each other, i.e. Eq. (D.10) is zero, the external excitation (C) necessarily balances the remaining damping effect (B) in the system. From Eqs. (D.11) and (D.12), we have the following order estimation:

$$O(\epsilon\delta_0(\epsilon)) = O(\epsilon) \Rightarrow \delta_0(\epsilon) = 1. \tag{D.14}$$

Therefore, we can determine the order of the leading term in the asymptotic solution, i.e. the order of the response, as $\delta_0(\epsilon) = 1$. Then, the order at which the external excitation balances the damping effect $O(\epsilon)$ is different from that of the response, that is, $O(1)$; The former and latter, $O(1)$ and $O(\epsilon)$, are later called the *dominant level* and *subdominant one*, respectively. This shows a characteristic of resonance and leads to the concept of *hierarchical levels* which will be mentioned in Section 9.2.

D.3.3 Near-Resonant Case

We consider the case when the excitation frequency is in the neighborhood of the natural frequency, i.e. $-\nu^2 + 1 = O(\epsilon)$. Since the order of Eqs. (D.10) and (D.11) is $O(\epsilon\delta_0(\epsilon))$, the external excitation (C) balances the sum of inertia and spring forces (A) in addition to the damping effect (B). Then, we can determine the order of the response as $\delta_0(\epsilon) = 1$; this order estimation is the same as that in the case of the resonant condition. Since the order of the response, $O(1)$, is greater than that of the excitation, $O(\epsilon)$, the system under the condition $-\nu^2 + 1 = O(\epsilon)$ can be regarded as in a near-resonant state. The value of $-\nu^2 + 1$ expresses the deviation of the excitation frequency from the natural frequency, which is called the *detuning* of the excitation frequency with respect to the natural frequency. In the method of multiple scales, the analysis of dynamics in the near-resonant state is performed by suitably setting the order of the detuning.

D.4 Cubic Nonlinear Oscillator Under External Harmonic Excitation

Next, we consider the following system that includes a cubic nonlinear stiffness:

$$\underbrace{\frac{d^2x}{dt^2} + x}_{\text{(A)}} + \underbrace{2\gamma\frac{dx}{dt}}_{\text{(B)}} + \underbrace{\alpha x^3}_{\text{(C)}} = \underbrace{\epsilon\hat{a}_e \cos \nu t}_{\text{(D)}}, \tag{D.15}$$

where (A), (B), (C), and (D) are the sum of the inertia and spring forces, the viscous damping force, nonlinear force and the external harmonic excitation with $O(\epsilon)$, respectively. We consider the dynamics in the resonant case when the inertia and spring forces are cancelled out each other. We examine the dependency of the magnitude of the damping force on the order of the response according to the order estimation for the linear system in Section D.3. Then, we clarify the effect of the nonlinear restoring force on the resonance.

We assume the response as

$$x = \delta_0(\epsilon)\hat{a}\cos(\nu t + \phi), \tag{D.16}$$

where $\delta_0(\epsilon)$ is a gauge function. After examining the gauge function depending on the magnitude of the damping, we clarify the effect of the nonlinear restoring force on the resonance.

D.4.1 Large Damping Case ($\gamma = O(1)$)

First, we consider the large damping case ($\gamma = O(1)$). The orders of the damping force and external excitation, which are balanced out in the resonance of the linear oscillator, are

$$\text{(B)}: \ 2\gamma\frac{dx}{dt} = -2\gamma\delta_0(\epsilon)\nu\hat{a}\sin(\nu t + \phi) = O(\delta_0(\epsilon)), \tag{D.17}$$

$$\text{(D)}: \ \epsilon\hat{a}_e \cos \nu t = O(\epsilon). \tag{D.18}$$

Since the damping force should be the same order as that of the external excitation $O(\epsilon)$, we can determine the magnitude of the response, i.e., the order of the leading term in the solution, as $\delta_0(\epsilon) = \epsilon$. Then, the order of the cubic nonlinear term is estimated as follows:

$$\text{(C)}: \; \alpha_3 x^3 = \delta_0(\epsilon)^3 \alpha_3 \hat{a}^3 \left\{ \frac{1}{4}\cos(3\nu t + 3\phi) + \frac{3}{4}\cos(\nu t + \phi) \right\} = O(\delta_0(\epsilon)^3) = O(\epsilon^3). \tag{D.19}$$

Since this order is less than that of the external excitation, $O(\epsilon)$, and the external excitation does not balance the cubic nonlinear restoring force, the cubic nonlinear restoring force does not contribute to the leading term of the asymptotic solution. In the large damping case, there is not any hierarchical level, which characterizes the resonance mechanism. The dynamics is qualitatively the same as that in the nonresonant state even at the resonant condition $\nu = 1$.

D.4.2 Relatively Small Damping Case ($\gamma = O(\epsilon^{2/3})$)

In the case when the damping is relatively small, i.e., $\gamma = O(\epsilon^{2/3})$ and $\gamma = \epsilon^{2/3}\hat{\gamma}$ ($\hat{\gamma} = O(1)$), the order of damping term is

$$\text{(B)}: \; 2\gamma \frac{dx}{dt} = -2\epsilon^{2/3}\delta_0(\epsilon)\hat{\gamma}\nu\hat{a}\sin(\nu t + \phi) = O(\epsilon^{2/3}\delta_0(\epsilon)). \tag{D.20}$$

Since this damping term is balanced by the external excitation at $O(\epsilon)$, we can determine $\delta_0(\epsilon) = \epsilon^{1/3}$. Then, the order of the cubic nonlinear term is estimated as

$$\text{(C)}: \; \alpha_3 x^3 = O(\delta_0(\epsilon)^3) = O(\epsilon). \tag{D.21}$$

As a result, the magnitude of the response is $O(\epsilon^{1/3})$, which is realized by balancing at $O(\epsilon)$ the damping force and the nonlinear restoring force with the external excitation. Since the order in the balancing, ϵ, is less than that of the leading term in the asymptotic solution, $\delta_0(\epsilon) = \epsilon^{1/3}$, the nonlinear restoring force does contribute to the leading term of the asymptotic solution indirectly. In such a situation, the system of Eq. (D.15) is regarded as a weakly nonlinear system.

The order of the response $O(\epsilon^{1/3})$, which is the dominant level, and the order at the balance condition $O(\epsilon)$, which is the subdominant level, construct the *hierarchical levels*, which play an important role in the application process of the perturbation methods. The existence of such hierarchical levels requires the application of singular perturbation methods as the method of multiple scales to obtain the asymptotic solution.

D.4.3 Small Damping Case ($\gamma = O(\epsilon)$)

We consider that the damping is small, i.e. ($\gamma = O(\epsilon)$) and $\gamma = \epsilon\hat{\gamma}$, ($\hat{\gamma} = O(1)$), which is the same order as the external excitation. Then, the order of damping term is

$$\text{(B)}: \; 2\gamma \frac{dx}{dt} = -2\epsilon\delta_0(\epsilon)\hat{\gamma}\nu\hat{a}\sin(\nu t + \phi) = O(\epsilon\delta_0(\epsilon)). \tag{D.22}$$

Since this term balances the external excitation at $O(\epsilon)$, we can determine $\delta_0(\epsilon) = 1$. Then, the order of the cubic nonlinear term is estimated as

$$\alpha_3 x^3 = O(\delta_0(\epsilon)^3) = O(1). \tag{D.23}$$

Since the order of the cubic nonlinear term is the same as that of the leading term in the asymptotic solution, it contributes to the leading term of the asymptotic solution directly. Then, the system of Eq. (D.15) is regarded as a strongly nonlinear system. In general, strongly nonlinear problems cannot be solved by the perturbation methods except for some special case as mentioned in Chapters 12 and 15.

D.5 Linear Oscillator with Negative Damping

We consider the following linear oscillator with damping:

$$\frac{d^2x}{dt^2} + 2\gamma\frac{dx}{dt} + x = 0. \tag{D.24}$$

We focus on the neighborhood of the critical point for the dynamic instability obtained by the linear stability theory, $\gamma = 0$ ($k' = 1$ and $c' = 0$ in Figure 2.7). When $\gamma < 0$, the self-excited oscillation occurs (see Section 2.6). By setting $\gamma = \epsilon\hat{\gamma}$ ($\hat{\gamma} = O(1)$), Eq. (D.24) is rewritten as

$$\underbrace{\frac{d^2x}{dt^2} + x}_{\text{(A)}} + \underbrace{2\epsilon\hat{\gamma}\frac{dx}{dt}}_{\text{(B)}} = 0, \tag{D.25}$$

where (A) and (B) express the sum of the inertia and spring forces and the linear damping force, respectively. Let us determine the order of the leading term $\delta_0(\epsilon)$ of the asymptotic solution,

$$x(t,\epsilon) = \delta_0(\epsilon)x_0(t) + \delta_1(\epsilon)x_1(t) + \delta_2(\epsilon)x_2(t) + \cdots . \tag{D.26}$$

Substituting $x = \delta_0(\epsilon)x_0$ into Eq. (D.25) yields

$$\underbrace{\delta_0(\epsilon)\left(\frac{d^2x_0}{dt^2} + x_0\right)}_{\text{(A)}} + \underbrace{2\epsilon\delta_0(\epsilon)\hat{\gamma}\frac{dx_0}{dt}}_{\text{(B)}} = 0. \tag{D.27}$$

The order of the sum of the inertia and spring forces, $O(\delta_0(\epsilon))$, is greater than that of the damping force, $O(\epsilon\delta_0(\epsilon))$. Picking up the terms of order $O(\delta_0(\epsilon))$ yields

$$\underbrace{\delta_0(\epsilon)\left(\frac{d^2x_0}{dt^2} + x_0\right)}_{\text{(A)}} = 0 \tag{D.28}$$

or

$$\frac{d^2x_0}{dt^2} + x_0 = 0, \tag{D.29}$$

which is the equation expressing the dominant level. On the other hand, at the subdominant level whose order is $O(\epsilon\delta_0(\epsilon))$, there is only the damping term (B). Therefore, since there is no terms to balance the negative damping term (B), the response grows infinitely with time. In general, the linear system is often an approximation of the original nonlinear system. If the response grows with time in the theoretical solution for the linear system, we have to reanalyze the equation that takes into account the nonlinear terms neglected in the linearization, i.e. the mathematical model that is more accurately derived from the physical problem. In Section D.6, we perform the order estimation of the response for the system in which the nonlinear damping is taken into account.

D.6 Van der Pol Oscillator

We consider the van der Pol oscillator given in Eq. (10.5), which takes into account the nonlinear damping force in addition to the linear damping one. We again focus on the dynamics in the neighborhood of the critical point obtained by the linear stability theory, $\gamma = 0$, and set $\gamma = \epsilon\hat{\gamma}$ ($\hat{\gamma} = O(1)$). Then, Eq. (10.5) is rewritten as

$$\underbrace{\frac{d^2x}{dt^2} + x}_{\text{(A)}} + \underbrace{2\epsilon\hat{\gamma}\frac{dx}{dt}}_{\text{(B)}} + \underbrace{\beta_3 x^2\frac{dx}{dt}}_{\text{(C)}} = 0, \tag{D.30}$$

where (A), (B), and (C) express the sum of the inertia and spring forces, the linear damping force, and the nonlinear damping force (see Section 10.1.1), respectively. Let us determine the order of the leading term $\delta_0(\epsilon)$ of the asymptotic solution,

$$x(t,\epsilon) = \delta_0(\epsilon)x_0(t) + \delta_1(\epsilon)x_1(t) + \delta_2(\epsilon)x_2(t) + \cdots . \tag{D.31}$$

Substituting $x = \delta_0(\epsilon)x_0$ into Eq. (D.30) yields

$$\underbrace{\delta_0(\epsilon)\left(\frac{d^2x_0}{dt^2} + x_0\right)}_{\text{(A)}} + \underbrace{2\epsilon\delta_0(\epsilon)\hat{\gamma}\frac{dx_0}{dt}}_{\text{(B)}} + \underbrace{\delta_0(\epsilon)^3\beta_3 x_0^2\frac{dx_0}{dt}}_{\text{(C)}} = 0. \tag{D.32}$$

D.6.1 Large Response Case ($\delta_0(\epsilon) = 1$)

First, we consider the case when the order of the response is $O(1)$, i.e. $\delta_0(\epsilon) = 1$. Since the order of the cubic nonlinear force (C) is the same as that of the leading term in the asymptotic solution $O(1)$, it contributes to the leading term of the asymptotic solution directly. Then, the system of Eq. (D.30) is regarded as a strongly nonlinear system.

D.6.2 Small But Finite Response Case ($\delta_0(\epsilon) = o(1)$)

Secondly, we consider the case when the order of the response is less than $O(1)$, i.e., $\delta_0(\epsilon) = o(1)$. Then, since $\delta_0(\epsilon)^3 = o(\delta_0(\epsilon))$, the order of the cubic nonlinear force is less than that of the sum of the inertia and spring forces and the cubic nonlinear force is not included in the dominant level. Picking up the leading terms with $O(\delta_0(\epsilon))$ in Eq. (D.32) yields

$$\underbrace{\delta_0(\epsilon)\left(\frac{d^2x_0}{dt^2} + x_0\right)}_{\text{(A)}} = 0 \tag{D.33}$$

or

$$\frac{d^2x_0}{dt^2} + x_0 = 0. \tag{D.34}$$

In the dominant part consisting of the leading terms, the inertia and spring forces are balanced out. The solution is

$$x_0 = Ae^{it} + \bar{A}e^{-it}, \tag{D.35}$$

where A is a complex amplitude. The order of $\delta_0(\epsilon)$, i.e. the magnitude of the response amplitude, cannot be determined at this stage yet. Namely, the order of the leading term of the asymptotic solution Eq. (D.31), $\delta_0(\epsilon)$, is not determined at the dominant level, but at the following subdominant level with lower order.

Let us determine $\delta_0(\epsilon)$. Substituting Eq. (D.35) into Eq. (D.32) yields

$$\underbrace{2i\epsilon\hat{\gamma}\delta_0(\epsilon)Ae^{it}}_{\text{(B)}} + \underbrace{i\beta_3\delta_0(\epsilon)^3(A^3e^{3it} + |A|^2Ae^{it})}_{\text{(C)}} + \mathrm{CC} = 0. \tag{D.36}$$

Focusing on the frequency component 1 of the solution Eq. (D.35) in the dominant level expressed by Eq. (D.34), we consider the balance of the terms proportional to e^{it} in the subdominant level expressed by Eq. (D.36). Then, from the following relationship:

$$O(\epsilon\delta_0(\epsilon)) = O(\delta_0(\epsilon)^3) \Rightarrow \epsilon\delta_0(\epsilon) = \delta_0(\epsilon)^3 \Rightarrow \delta_0(\epsilon) = \epsilon^{1/2}, \tag{D.37}$$

we can determine the leading order of the asymptotic solution as $\delta_0(\epsilon) = \epsilon^{1/2}$.

D.7 Parametrically Excited Oscillator

Next, we consider the following system with periodically varying stiffness and positive damping ($\gamma > 0$):

$$\frac{d^2x}{dt^2} + 2\gamma\frac{dx}{dt} + (1 + \epsilon\hat{a}_e \cos \nu t)x = 0 \tag{D.38}$$

or

$$\underbrace{\frac{d^2x}{dt^2} + x}_{\text{(A)}} + \underbrace{\epsilon\hat{a}_e \cos \nu t \cdot x}_{\text{(B)}} + \underbrace{2\gamma\frac{dx}{dt}}_{\text{(C)}} = 0. \tag{D.39}$$

Let us determine the order of the leading term $\delta_0(\epsilon)$ of the asymptotic solution,

$$x(t,\epsilon) = \delta_0(\epsilon)x_0(t) + \delta_1(\epsilon)x_1(t) + \delta_2(\epsilon)x_2(t) + \cdots . \tag{D.40}$$

Substituting the leading term of the approximate solution $x = \delta_0(\epsilon)x_0$ into Eq. (D.39) yields

$$\underbrace{\delta_0(\epsilon)\left(\frac{d^2x_0}{dt^2} + x_0\right)}_{\text{(A)}} + \underbrace{\epsilon\delta_0(\epsilon)x_0\hat{a}_e \cos \nu t}_{\text{(B)}} + \underbrace{2\delta_0(\epsilon)\gamma\frac{dx_0}{dt}}_{\text{(C)}} = 0. \tag{D.41}$$

D.7.1 Large Damping Case ($\gamma = O(1)$)

First, we consider that the order of the damping is $O(1)$, i.e. $\gamma = O(1)$. Then, neglecting the term (B) that is smaller than the other terms, we rewrite Eq. (D.41) as

$$\underbrace{\delta_0(\epsilon)\left(\frac{d^2x_0}{dt^2}+x_0\right)}_{(A)} + \underbrace{2\delta_0(\epsilon)\gamma\frac{dx_0}{dt}}_{(C)} = 0 \tag{D.42}$$

or

$$\frac{d^2x_0}{dt^2}+2\gamma\frac{dx_0}{dt}+x_0=0. \tag{D.43}$$

Because of $k' = 1$ and $c' = 2\gamma > 0$ in Figure 2.7, x_0 is decayed with time. Therefore, regardless of the periodically varying stiffness, the parametric resonance is not produced due to the large damping.

D.7.2 Small Damping Case ($\gamma = O(\epsilon)$)

Next, we consider that the order of the damping force is $O(\epsilon)$, i.e. $\gamma = \epsilon\hat{\gamma}$. Then, Eq. (D.41) is expressed as

$$\underbrace{\delta_0(\epsilon)\left(\frac{d^2x_0}{dt^2}+x_0\right)}_{(A)} + \underbrace{\epsilon\delta_0(\epsilon)x_0\hat{a}_e\cos\nu t}_{(B)} + \underbrace{2\epsilon\delta_0(\epsilon)\hat{\gamma}\frac{dx_0}{dt}}_{(C)} = 0. \tag{D.44}$$

Picking up the leading terms with $O(\delta_0(\epsilon))$ in Eq. (D.44) yields the equation expressing the dominant part:

$$\underbrace{\delta_0(\epsilon)\left(\frac{d^2x_0}{dt^2}+x_0\right)}_{(A)} = 0 \tag{D.45}$$

or

$$\frac{d^2x_0}{dt^2}+x_0=0. \tag{D.46}$$

In the dominant part consisting of the leading terms, the inertia and spring forces are balanced out. The solution is

$$x_0 = Ae^{it}+\bar{A}e^{-it}, \tag{D.47}$$

where A is a complex amplitude.

The order of $\delta_0(\epsilon)$, i.e. the magnitude of the response amplitude, cannot be determined at this stage yet. Namely, the order of the leading term of the asymptotic solution Eq. (D.40), $\delta_0(\epsilon)$, is not determined at the dominant level, but at the following subdominant level with lower order.

Substituting Eq. (D.47) into Eq. (D.44) yields

$$\underbrace{\frac{1}{2}\epsilon\delta_0(\epsilon)\hat{a}_e\{Ae^{i(\nu+1)t}+\bar{A}e^{i(\nu-1)t}\}}_{(B)} + \underbrace{2i\epsilon\delta_0(\epsilon)\hat{\gamma}Ae^{it}}_{(C)} + CC = 0. \tag{D.48}$$

According to the prediction in Section C.2, in the case when the excitation frequency is twice the natural frequency, i.e., $\nu = 2$, the parametric resonance can be produced. Then, the second term of (B), which is related to the periodically changing stiffness, has the frequency 1. Focusing on the frequency component 1 of the solution Eq. (D.47) in the dominant level expressed by Eq. (D.46), we consider the balance of the second term of (B) and the term of (C) which are proportional to e^{it} in the subdominant level expressed by Eq. (D.48). Then, we have the estimation

$$O(\epsilon\delta_0(\epsilon)) = O(\epsilon\delta_0(\epsilon)) \Rightarrow \delta_0(\epsilon) = \delta_0(\epsilon). \tag{D.49}$$

Therefore, different from the resonant case in the linear oscillator under harmonic external excitation in Section D.3, the order of the response, $\delta_0(\epsilon)$, is undetermined even in the subdominant level and from a physical point of view, the viscous damping effect does not govern the magnitude of the response in the parametric resonance, i.e. the response amplitude. In other words, in the state where the parametric resonance occurs, the response amplitude is governed by another effect which is not taken into account in the system Eq. (D.38). Thus, we take into account the nonlinear components in the system, which are usually included in the system, but neglected due to the smallness compared with the linear component. As an example, we take into account the cubic nonlinear component of the restoring force in Section D.7.3 (Nayfeh and Mook, 2008).

D.7.3 Case with Cubic Nonlinear Component of Restoring Force

We consider the following system with the cubic nonlinear component of the restoring force as follows:

$$\frac{d^2x}{dt^2} + 2\gamma\frac{dx}{dt} + (1 + \epsilon\hat{a}_e \cos \nu t)x + \alpha x^3 = 0 \tag{D.50}$$

or

$$\underbrace{\frac{d^2x}{dt^2} + x}_{(A)} + \underbrace{\epsilon\hat{a}_e \cos \nu t \cdot x}_{(B)} + \underbrace{2\gamma\frac{dx}{dt}}_{(C)} + \underbrace{\alpha x^3}_{(D)} = 0. \tag{D.51}$$

Then, Eq. (D.44) is changed as

$$\underbrace{\delta_0(\epsilon)\left(\frac{d^2x_0}{dt^2} + x_0\right)}_{(A)} + \underbrace{\epsilon\delta_0(\epsilon)x_0\hat{a}_e \cos \nu t}_{(B)} + \underbrace{2\epsilon\delta_0(\epsilon)\hat{\gamma}\frac{dx_0}{dt}}_{(C)} + \underbrace{\delta_0(\epsilon)^3\alpha x_0^3}_{(D)} = 0. \tag{D.52}$$

Because the equation governing the dominant level is the same as Eq. (D.46), substituting Eq. (D.47) into Eq. (D.52) yields the following equation adding the cubic terms to Eq. (D.48):

$$\underbrace{\frac{1}{2}\epsilon\delta_0(\epsilon)\{Ae^{i(\nu+1)t} + \bar{A}e^{i(\nu-1)t}\}}_{(B)} + \underbrace{2i\epsilon\delta_0(\epsilon)\hat{\gamma}Ae^{it}}_{(C)}$$
$$\underbrace{+i\delta_0(\epsilon)^3\alpha(A^3e^{3it} + 3|A|^2Ae^{it})}_{(D)} + CC = 0, \tag{D.53}$$

and expresses the subdominant part. Focusing on the frequency component 1 as the Section D.7.2, we consider the balance of the term proportional to e^{it} in (B) related to the parametric excitation and that in (D) related to the cubic nonlinear component of the restoring force. Then, from the following relationship:

$$O(\epsilon\delta_0(\epsilon)) = O(\delta_0(\epsilon)^3) \Rightarrow \epsilon\delta_0(\epsilon) = \delta_0(\epsilon)^3 \Rightarrow \delta_0(\epsilon) = \epsilon^{1/2}, \tag{D.54}$$

we can determine the order of the response $\delta_0(\epsilon)$ as $\epsilon^{1/2}$ thus, from a physical point of view, the cubic nonlinear component of the restoring force governs the magnitude of the response, i.e. the response amplitude. In other words, the balancing of the cubic nonlinear component of the restoring force to the parametric excitation can realize a finite response amplitude in the parametric resonant state.

D.7.4 Near-Resonant Case

We consider the case when the excitation frequency is not exactly equal to twice the natural frequency. We set the excitation frequency as

$$\nu = 2 + \sigma, \tag{D.55}$$

where σ is the detuning parameter ($|\sigma| \ll 1$) and is expressed as

$$\sigma = \delta_\sigma(\epsilon)\hat{\sigma} = o(1), \quad \hat{\sigma} = O(1), \tag{D.56}$$

where $\delta_\sigma(\epsilon)$ shows the nearness of the excitation frequency to twice the natural frequency and will be suitably set for the application of the method of multiple scales later. Substituting Eq. (D.55) into Eq. (D.53) yields

$$\underbrace{\frac{1}{2}\epsilon\delta_0(\epsilon)\{Ae^{i(3+\sigma)t} + \bar{A}e^{i(1+\sigma)t}\}}_{\text{(B)}} + \underbrace{2i\epsilon\delta_0(\epsilon)\hat{\gamma}Ae^{it}}_{\text{(C)}}$$
$$\underbrace{+i\delta_0(\epsilon)^3\alpha(A^3e^{3it} + 3|A|^2Ae^{it})}_{\text{(D)}} + \text{CC} = 0. \tag{D.57}$$

Therefore, different from the case of $\nu = 2$, the cubic nonlinear component of the restoring force, i.e. the second term of (D), cannot balance the effect of the parametric excitation, i.e. the second term of (B), since these terms do not have the same frequency. Here, instead of Eq. (D.47), we employ the following expression:

$$x_0 = Ae^{i\frac{\nu}{2}t} + \bar{A}e^{-i\frac{\nu}{2}t}, \tag{D.58}$$

where $\frac{\nu}{2} = 1 + \frac{\delta_\sigma(\epsilon)\hat{\sigma}}{2} = 1 + o(1)$. Substituting Eq. (D.58) into Eq. (D.52) yields

$$\underbrace{\delta_0(\epsilon)\left(1 - \frac{\nu^2}{4}\right)Ae^{i\frac{\nu}{2}t}}_{\text{(A)}} + \underbrace{\frac{1}{2}\epsilon\delta_0(\epsilon)\hat{a}_e\left\{Ae^{i\frac{3\nu}{2}t} + \bar{A}e^{i\frac{\nu}{2}t}\right\}}_{\text{(B)}} + \underbrace{i\epsilon\delta_0(\epsilon)\nu\hat{\gamma}Ae^{i\frac{\nu}{2}t}}_{\text{(C)}}$$
$$\underbrace{+\delta_0(\epsilon)^3\alpha\left(A^3e^{i\frac{3\nu}{2}t} + 3|A|^2Ae^{i\frac{\nu}{2}t}\right)}_{\text{(D)}} + \text{CC} = 0. \tag{D.59}$$

In this case, as the case of $\nu = 2$, the cubic nonlinear component of the restoring force, i.e. the second term of (D) can balance the effect of the parametric excitation, i.e. the second term of (B), since these terms have the same frequency $\frac{\nu}{2}$. Then, we can determine the leading order of the asymptotic solution as $\delta_0(\epsilon) = \epsilon^{1/2}$. Equation (D.59) expresses the subdominant level with the order of $O(\epsilon^{3/2})$, at which the terms of (B), (C), and (D) are balanced. If we set the terms of (A) as this order of $O(\epsilon^{3/2})$, since the sum of the inertia and linear restoring forces can be included in the subdominant level, we can analyze the contribution of the inertia and spring forces on the parametric resonance; such a state corresponds to the near-resonant state. Therefore, in the dynamics in near-resonant state, the order of (A), is estimated as

$$O\left(\delta_0(\epsilon)\left(1-\frac{\nu^2}{4}\right)Ae^{i\frac{\nu}{2}t}\right) = O(\epsilon^{1/2}\delta_\sigma(\epsilon)) = O(\epsilon^{3/2}) \tag{D.60}$$

and to analyze the parametric resonance in the near-resonant state, the order of the detuning is set as $\delta_\sigma(\epsilon) = \epsilon$.

References

Nayfeh, A. H. and D. T. Mook (2008). *Nonlinear oscillations*. John Wiley & Sons.

E

Free Oscillation of Spring-Mass System Under Coulomb Friction and Its Dead Zone

The purpose of this appendix is to explore the effect of the Coulomb friction on the behavior of a spring-mass system and, more in detail, on pitchfork bifurcation in Section 7.6. We will here consider the equilibrium states in the case when the stiffness of the spring is small, whereas the dynamics in the case when the stiffness is large is widely discussed in several textbooks (for example, Meirovitch (1975)). However, at first we will briefly review the transient state and the final rest position after the transient state on a spring-mass system with large stiffness. Next, in the case when the stiffness is very small, we will infer that the Coulomb friction, even if it is slight, has a significant influence on the final rest position that is equilibrium state.

E.1 Characteristics of Friction

There is a large variety of models to characterize friction. Here, we briefly introduce some which are used in the analysis of dynamical systems (for example, see Leine and Van de Wouw (2007); Leine and Nijmeijer (2013) for modeling Coulomb friction in detail). We consider the friction between Bodies I and II in Figure E.1. The friction is characterized by the relationship between the relative velocity dx/dt of Body I with respect to Body II and the external force $F_c^{\#}$ in Figure E.1. Figure E.2a represents the most fundamental model of friction, where the thick vertical solid line denotes the range of the static friction. When the absolute value of $F_c^{\#}$ is less than the maximum static friction force $F_{cmax}(> 0)$, i.e. $|F_c^{\#}| < F_{cmax}$, the relative velocity is kept zero and this state is called *stick*. When the absolute value of the external force reaches F_{cmax}, the relative velocity can be produced. Then, in the case when the external force is set as $F_c^{\#} = F_{cmax}$ and $F_c^{\#} = -F_{cmax}$, Body I can have any positive and negative relative velocities, respectively, and this state is called *slip*; F_{cmax} and $-F_{cmax}$ are both denoted as dynamic friction.

Figure E.2b is the model which takes into account that the magnitude of the dynamic friction is less than that of the maximum static friction. The characteristic shows hysteresis. For example, we consider the state without relative velocity, $\frac{dx}{dt} = 0$; increasing the external force $F_c^{\#}$, the external force reaches the maximum static friction F_{cmax} and then the relative velocity is produced. Hence, when the relative velocity is changed from zero to nonzero, i.e. from stick to slip state, the external force must exceed the maximum static friction force.

Linear and Nonlinear Instabilities in Mechanical Systems: Analysis, Control and Application, First Edition. Hiroshi Yabuno.

Companion website: www.wiley.com/go/yabuno/instabilitiesinmechanicalsystems

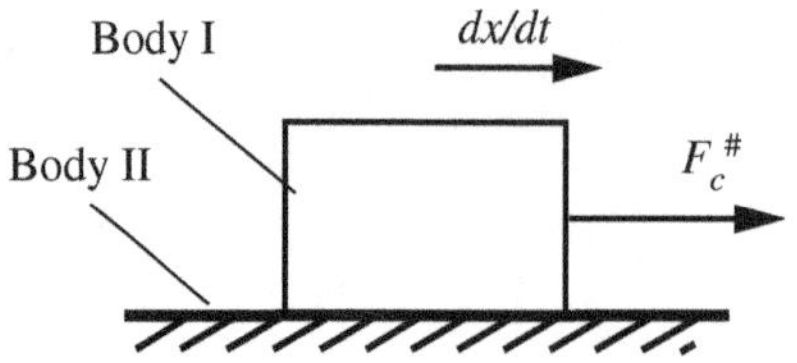

Figure E.1 Relative velocity dx/dt between Bodies I and II due to the external force $F_c^{\#}$.

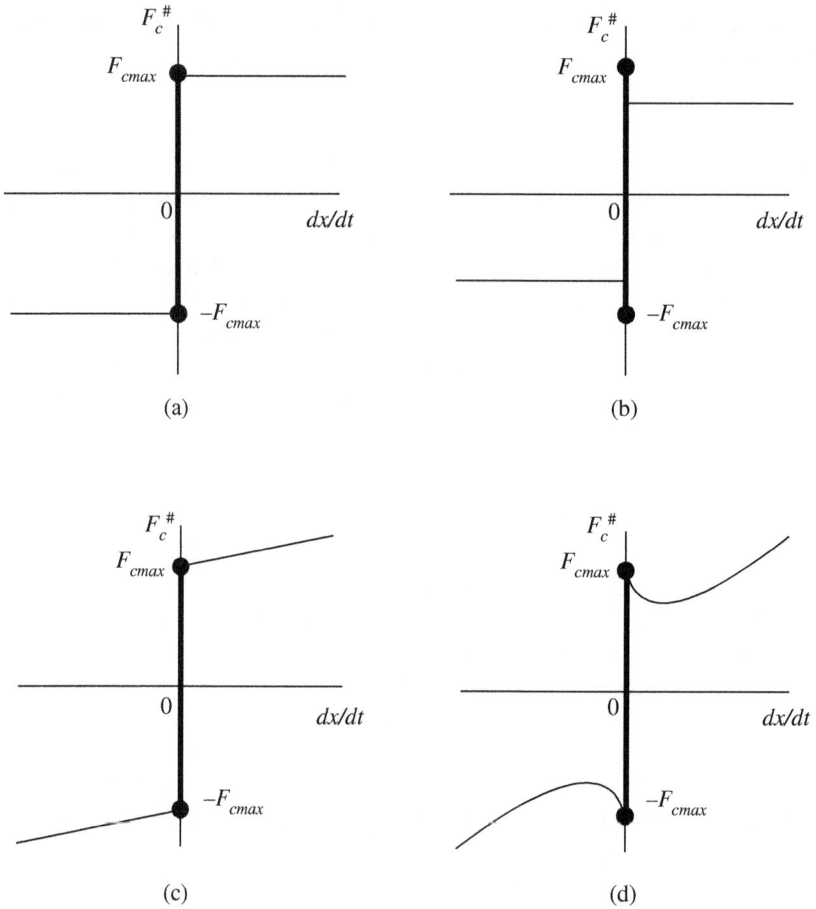

Figure E.2 Models to characterize friction.

This situation is schematically described as Figure E.3a; to pass (1) is allowable while to pass (2) is not allowable in the switch from the stick to the slip state. On the other hand, when the relative velocity is changed from nonzero to zero, the state is turned from slip into stick. At the instant when the relative velocity becomes zero, the static friction is not necessarily the maximum static friction, but at the instant when the state becomes stick, the static friction can be any value in $-F_{cmax} < F_c^{\#} < F_{cmax}$. This situation is schematically described by the arrows in Figure E.3b.

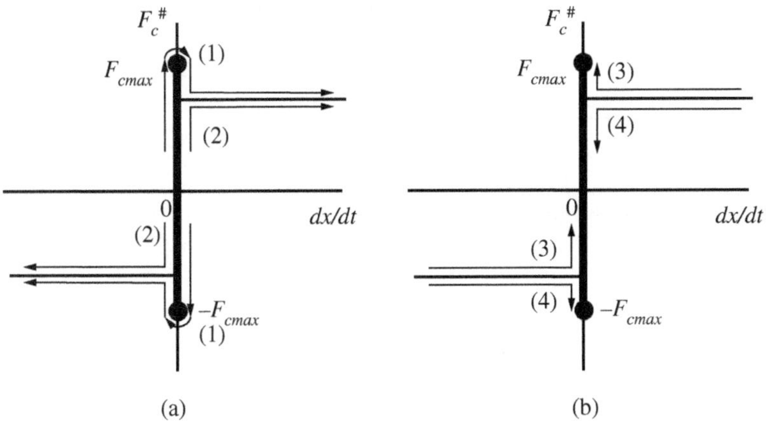

Figure E.3 Hysteresis of the friction characteristic referred to Figure E.2b.

Figure E.2c represents the model taking into account the viscous effect on the dynamic friction which is proportional to the relative velocity. Figure E.2d is the characteristic of the so-called Stribeck effect (for example, see Leine and Van de Wouw (2007)). When the magnitude of the relative velocity is relatively small and large, the friction exhibits negative slope (negative damping) and positive slope (positive damping), respectively.

E.2 Free Oscillation Under Coulomb Friction

Let us consider a spring-mass system with Coulomb friction as shown in Figure E.4. The governing equation of motion is expressed as follows:

$$m\frac{d^2x}{dt^2} + kx = F_c. \tag{E.1}$$

The m and k are the mass and the stiffness of the spring, respectively. The x is the displacement of the mass and the origin, O, is the position of the mass where the spring is unstretched; hereafter this position is called the original position. The F_c expresses the Coulomb friction acting on the mass. The relationship between the friction force and the velocity of the mass relative to the base can be expressed as Figure E.5a except for the case when the relative velocity is zero, where $F_c^\# = -F_c$. At the rest position, the friction force takes values between $-F_{cmax}$ and F_{cmax} and is equal to the restoring force for equilibrium states as Figure E.5b. In the transient state until the final rest position, F_c is assumed as $-F_{cmax}\text{sign}(dx/dt)$, where F_{cmax} is a positive constant value. In the final rest position after the transient state, F_c satisfies $-F_{cmax} < F_c < F_{cmax}$.

Let us summarize the transient state under the initial conditions $x|_{t=0} = x(0) > 0$ and $\left.\frac{dx}{dt}\right|_{t=0} = 0$ (Liang and Feeny, 1998). If we introduce the quantity, $\omega^2 = k/m$, Eq. (E.1) becomes

$$\frac{d^2x}{dt^2} + \omega^2 x = \frac{F_c}{m}. \tag{E.2}$$

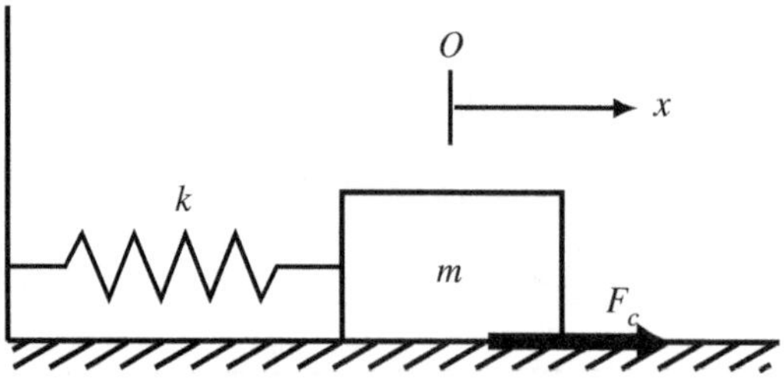

Figure E.4 Spring-mass system with Coulomb friction.

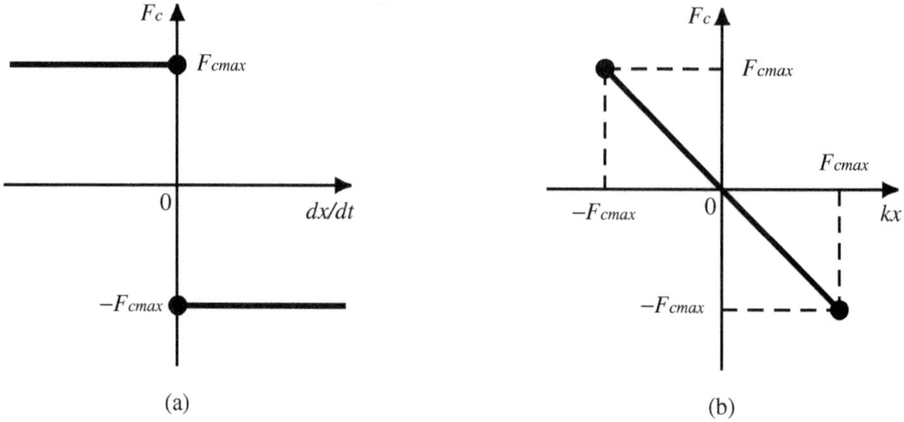

Figure E.5 Coulomb friction in the cases of (a) $dx/dt \neq 0$ and (b) $dx/dt = 0$.

In the interval $0 \leq t \leq \pi/\omega \overset{\text{def}}{=} t_1$, the solution is expressed as

$$x = \left(x(0) - \frac{F_{cmax}}{k}\right)\cos\omega t + \frac{F_{cmax}}{k}, \tag{E.3}$$

$$x|_{t=t_1} = \frac{2F_{cmax}}{k} - x(0) \overset{\text{def}}{=} x(t_1), \tag{E.4}$$

because the motion starts with $\left.\frac{d^2x}{dt^2}\right|_{t=0+0} < 0$ and $\left.\frac{dx}{dt}\right|_{t=0+0} < 0$.

In the next half cycle, $t_1 \leq t \leq t_1 + \pi/\omega \overset{\text{def}}{=} t_2$, the initial condition is regarded as $x|_{t=t_1} = x(t_1)$ and $\left.\frac{dx}{dt}\right|_{t=t_1} = 0$. The response in this half cycle is expressed as follows:

$$x = \left(x(t_1) + \frac{F_{cmax}}{k}\right)\cos\omega(t - t_1) - \frac{F_{cmax}}{k}, \tag{E.5}$$

$$x(t_2) = -\frac{2F_{cmax}}{k} - x(t_1) = -\frac{4F_{cmax}}{k} + x(0), \tag{E.6}$$

because the motion starts with $\left.\frac{d^2x}{dt^2}\right|_{t=t_1+0} > 0$ and $\left.\frac{dx}{dt}\right|_{t=t_1+0} > 0$.

The free oscillation decays through the repeat of the above procedure as shown in Figure E.6a, where $m = 1$, $k = 0.8$, and $F_{cmax} = 0.05$. The phase plane is also described in Figure E.6b. When the two conditions, $\frac{dx}{dt} = 0$ and $|\omega^2 x| < \frac{F_{cmax}}{m}$, i.e. $|x| < \frac{F_{cmax}}{k}$, are satisfied

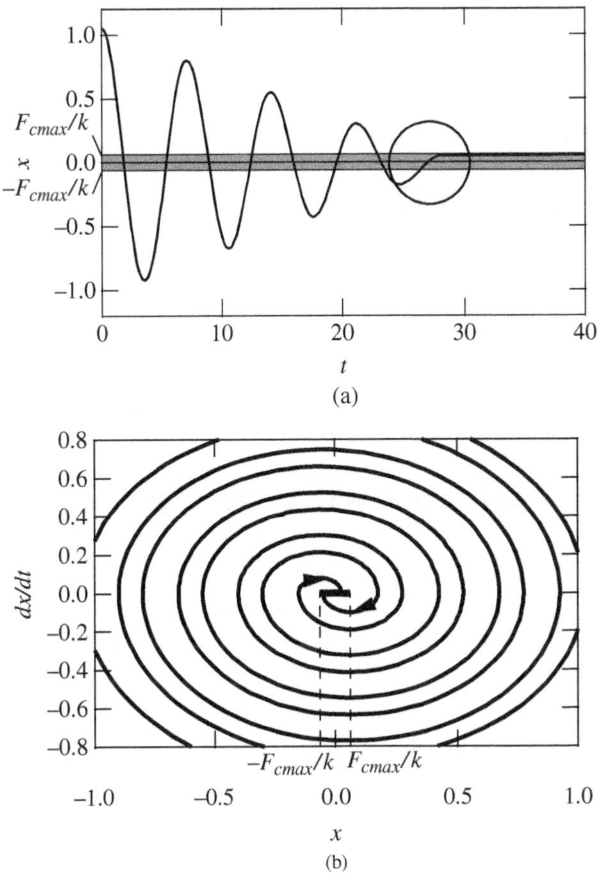

Figure E.6 Free oscillation of the system with large stiffness ($m = 1$, $k = 0.8$, and $F_{cmax} = 0.05$): (a) Time history; (b) Phase portrait. The hatched region in (a), $|x| < F_{cmax}/k$, is the dead zone, which is denoted the thick solid line in (b).

at the first time, the transient state ends and turns out into the final rest state. At the final rest position, as shown in Figure E.5b, the Coulomb friction balances with the restoring force of the spring. The latter condition is shown as the hatched region in Figure E.6a; the expansion in the part of the circle is shown in Figure E.7. This region is called *Dead Zone* (Beards, 1981) and also expressed as a thick solid line in the phase plane Figure E.6b. The final rest position is on this thick solid line. Since the dead zone is very small when the stiffness is not so small (i.e. k is not very small), the final rest position is in the neighborhood of the original position, $x = 0$, regardless of the initial position. In addition, when the initial position is in the narrow hatched region in Figure E.6a whose expansion is Figure E.7, i.e. on the thick line in Figure E.6b, and the initial velocity is zero, the mass remains at rest at the initial position.

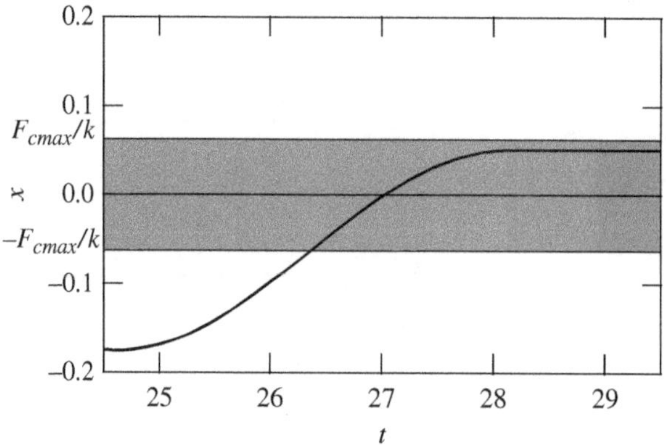

Figure E.7 Expansion of the time history near the final rest position.

E.3 Variation of the Final Rest Position with Decrease in the Stiffness

We consider the initial condition as in Section E.2, $x|_{t=0} = x(0) > 0$ and $\left.\frac{dx}{dt}\right|_{t=0} = 0$. The position with zero velocity in the transient state is deduced from Eqs. (E.4) and (E.6) as follows:

$$x(t_i) = (-1)^{i+1}\frac{2iF_{cmax}}{k} + (-1)^i x(0). \tag{E.7}$$

Letting n be the number of the half-cycle just prior to the cessation of the transient motion, n is the smallest integer, i, satisfying the inequality

$$|x(t_i)| = \left|(-1)^{i+1}\frac{2iF_{cmax}}{k} + (-1)^i x(0)\right| < \frac{F_{cmax}}{k}. \tag{E.8}$$

From the above result, we illustrate the relationship between the stiffness k, and the final rest position x_{st}. The curves of Eq. (E.7) are described for some values of i in $k - x$ plane as Figure E.8. The points on the curve with smallest i in the region, $|x| < \frac{F_{cmax}}{k}$, provide the final rest positions for each k; the set of the positions are described by the thick solid curve. It is

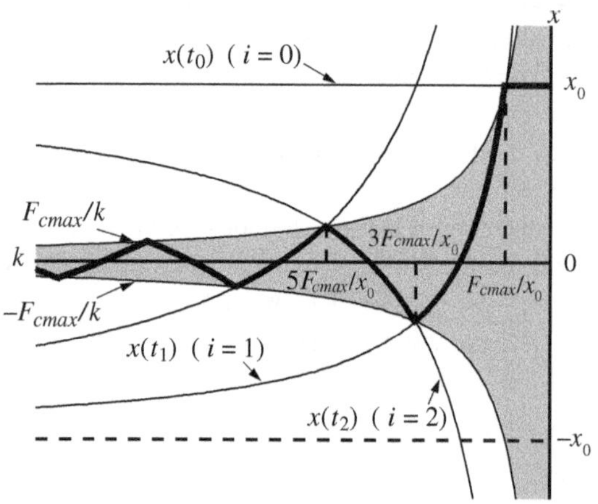

Figure E.8 Variation of the final rest position with stiffness.

Figure E.9 Dead zone depending on the stiffness k. The dead zone is the hatched region and becomes wider as decreasing the stiffness.

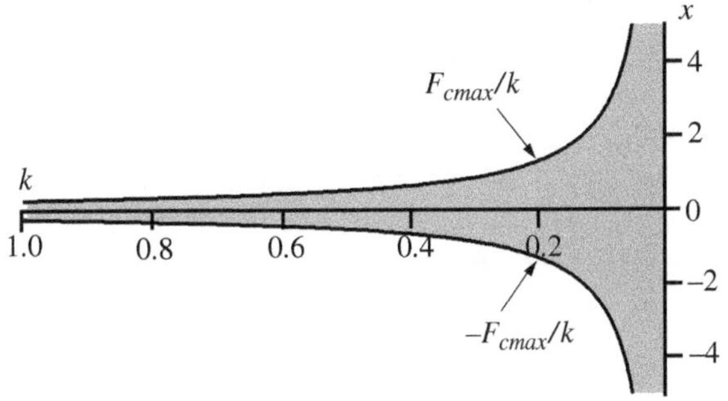

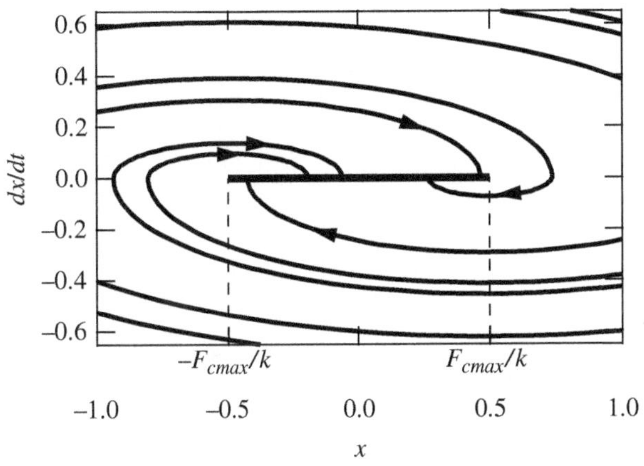

Figure E.10 Phase plane of the system with small positive stiffness ($m = 1$, $k = 0.1$, and $F_{cmax} = 0.05$); the thick solid line is the dead zone, i.e. the equilibrium region. If the initial condition is on this line, the system remains at rest.

noted that, when the stiffness is very small, the final rest position departs widely from the original position $x = 0$, since the dead zone, $|x| < \frac{F_{cmax}}{k}$, becomes wide. In the case when the initial position $x = x(0)$ is in the dead zone, the mass remains at rest at the initial position; this state corresponds to the straight part of the thick solid line.

By the way, the state under the small stiffness corresponds to the neighborhood of the buckling point, i.e. the pitchfork bifurcation point, at $k = 0$. Then, as discussed above, the slight Coulomb friction has a significant influence on the behavior of the system, because the dead zone is very wide as shown in Figure E.9. We can regard this figure as a bifurcation diagram with a control parameter, k, in which all points in the dead zone are equilibrium states. Also, in the phase plane, the thick line expressing the dead zone is stretched as Figure E.10 comparing with Figure E.6b.

References

Beards, C. F. (1981). Vibration analysis and control system dynamics. Chichester, Sussex, England, Ellis Horwood, Ltd.; New York, Halsted Press, 1981. 169 p.

Leine, R. I. and H. Nijmeijer (2013). *Dynamics and bifurcations of non-smooth mechanical systems*, Volume 18. Springer Science & Business Media.

Leine, R. I. and N. Van de Wouw (2007). *Stability and convergence of mechanical systems with unilateral constraints*, Volume 36. Springer Science & Business Media.

Liang, J. and B. Feeny (1998). Identifying coulomb and viscous friction from free-vibration decrements. *Nonlinear Dynamics* 16(4), 337–347.

Meirovitch, L. (1975). *Elements of vibration analysis*. McGraw-Hill.

F

Projection by Adjoint Vector

In the case when the basis $\boldsymbol{\Phi}_i$ $(i = 1, 2)$, where $|\boldsymbol{\Phi}_i| = 1$, is not orthogonal, we consider the projection of a vector $\boldsymbol{x}$ along the basis $\boldsymbol{\Phi}_i$. It is discussed using the two dimensional system in Section 4.4.1 as an example. There exist two unit basis, $\boldsymbol{\Phi}_1$ and $\boldsymbol{\Phi}_2$, which are not orthogonal each other as shown in Figure F.1. As defined in Section 4.4, the adjoint vectors $\boldsymbol{\Psi}_i$ $(i = 1, 2)$ for $\boldsymbol{\Phi}_i$ $(i = 1, 2)$ are expressed as Eq. (4.64) or

$$\boldsymbol{\Psi}_j \cdot \boldsymbol{\Phi}_i = \delta_{ji}. \tag{F.1}$$

Vector $\boldsymbol{x}$ spanned by $\boldsymbol{\Phi}_1$ and $\boldsymbol{\Phi}_2$ is expressed as

$$\boldsymbol{x} = \boldsymbol{\Phi}_1 \xi_1 + \boldsymbol{\Phi}_2 \xi_2. \tag{F.2}$$

We can geometrically find that ξ_1 is obtained from the projection of $\boldsymbol{x}$ along $\boldsymbol{\Psi}_1$, i.e. the adjoint vector of $\boldsymbol{\Phi}_1$. Recalling Eq. (F.1), we notice that $|\boldsymbol{\Psi}_1|$ is $\frac{1}{\cos\theta}$ from Figure F.1, where θ is the angle between $\boldsymbol{\Phi}_1$ and $\boldsymbol{\Psi}_1$. By considering that θ' is the angle between $\boldsymbol{x}$ and $\boldsymbol{\Psi}_1$ in this figure, we obtain the following equation:

$$\boldsymbol{x} \cdot \boldsymbol{\Psi}_1 = |\boldsymbol{x}| \cos\theta' |\boldsymbol{\Psi}_1| = \xi_1 \cos\theta \cdot \frac{1}{\cos\theta} = \xi_1. \tag{F.3}$$

Similarly, ξ_2 is obtained by the following projection of $\boldsymbol{x}$ along $\boldsymbol{\Psi}_2$:

$$\boldsymbol{x} \cdot \boldsymbol{\Psi}_2 = \xi_2. \tag{F.4}$$

The matrix form for the above result is Eq. (4.76).

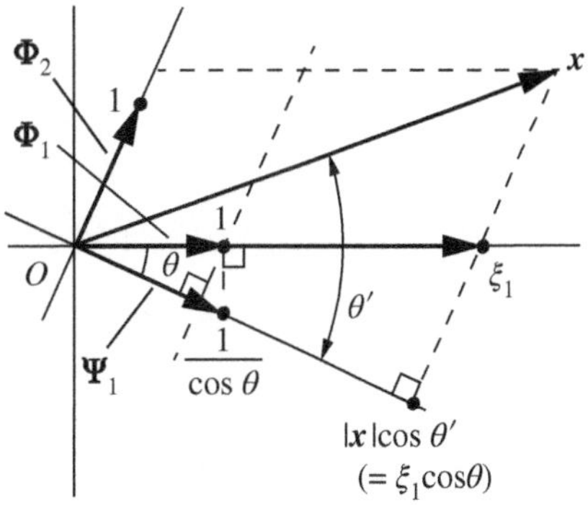

Figure F.1 Component ξ_1 of vector $\boldsymbol{x}$ defined in a nonorthogonal basis, $\boldsymbol{\Phi}_1$ and $\boldsymbol{\Phi}_2$ ($\boldsymbol{x} = \xi_1 \boldsymbol{\Phi}_1 + \xi_2 \boldsymbol{\Phi}_2$); derivation by projection of $\boldsymbol{x}$ along $\boldsymbol{\Psi}_1$.

Linear and Nonlinear Instabilities in Mechanical Systems: Analysis, Control and Application, First Edition. Hiroshi Yabuno.

Companion website: www.wiley.com/go/yabuno/instabilitiesinmechanicalsystems

G

Solvability Condition

G.1 Kernel and Image of Linear Transformation

We consider a linear transformation $L : V \to W$ between two vector spaces V and W defined by $\boldsymbol{x} \in \mathbb{R}^n \mapsto L\boldsymbol{x} \in \mathbb{R}^n$, where the linear operator L is a $n \times n$ matrix. The set of all elements $\boldsymbol{x}$ of V for which $L\boldsymbol{x} = \mathbf{0}$ is called the *kernel* of the linear transformation L, which is expressed as

$$\mathrm{Ker}(L) = \{\boldsymbol{x}_0 \in \mathbb{R}^n | L\boldsymbol{x}_0 = \mathbf{0}\}. \tag{G.1}$$

The set of $L\boldsymbol{x}$ for any elements $\boldsymbol{x}$ of V is called the *image* of L, which is expressed as

$$\mathrm{Image}(L) = \{L\boldsymbol{x} | \boldsymbol{x} \in \mathbb{R}^n\}. \tag{G.2}$$

The conjugate transpose of matrix L is called an adjoint matrix, which is expressed as

$$L^* = \overline{L}^T. \tag{G.3}$$

If $L^* = L$, the operator L is said to be selfadjoint. On the other hand, if $L^* \neq L$, the operator L is said to be non-selfadjoint. The kernel of L^* is expressed as

$$\mathrm{Ker}(L^*) = \{\tilde{\boldsymbol{x}}_0 \in \mathbb{R}^n | L^*\tilde{\boldsymbol{x}}_0 = \mathbf{0}\}, \tag{G.4}$$

where $\tilde{\boldsymbol{x}}_0$ is called the adjoint vector of $\boldsymbol{x}_0$.

Regardless of selfadjointness, the inner product of the adjoint vector of $\boldsymbol{x}_0$ and the image of L, i.e. $\tilde{\boldsymbol{x}}_0$ and $\boldsymbol{y} = L\boldsymbol{x}$, is zero since

$$\boldsymbol{y} \cdot \tilde{\boldsymbol{x}}_0 = \overline{\tilde{\boldsymbol{x}}}_0^T L\boldsymbol{x} = (\overline{\tilde{\boldsymbol{x}}}_0^T L\boldsymbol{x})^T = \boldsymbol{x}^T L^T \overline{\tilde{\boldsymbol{x}}}_0 = \boldsymbol{x}^T \overline{(L^* \tilde{\boldsymbol{x}}_0)} = 0. \tag{G.5}$$

To gain insight on the geometrically adjoint vector, let us discuss a simple example of linear transformation defined by 2×2 matrix

$$L = \begin{bmatrix} 0 & 1 \\ 0 & 1 \end{bmatrix}. \tag{G.6}$$

The kernel of L is

$$\boldsymbol{x}_0 = \mathrm{Ker}(L) = c_0 \begin{bmatrix} 1 \\ 0 \end{bmatrix}, \quad c_0 \in \mathbb{R}. \tag{G.7}$$

Linear and Nonlinear Instabilities in Mechanical Systems: Analysis, Control and Application,
First Edition. Hiroshi Yabuno.

Companion website: www.wiley.com/go/yabuno/instabilitiesinmechanicalsystems

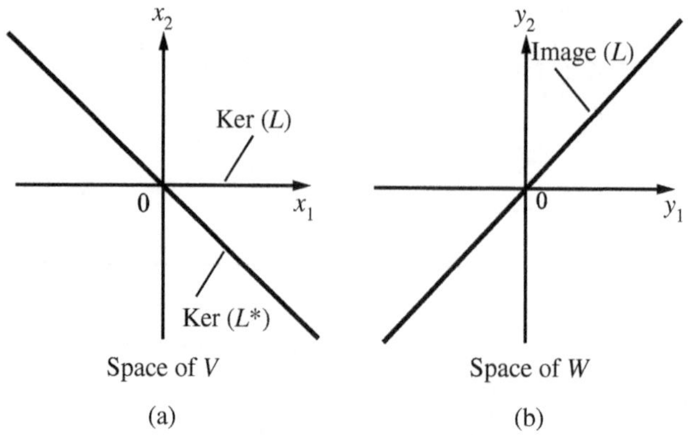

Figure G.1 Geometrical relationship between Ker (L), Ker (L^*) and Image (L). Ker (L^*) and Image (L) are orthogonal.

The image of L is

$$\boldsymbol{y} = \text{Image}(L) = c\begin{bmatrix}1\\1\end{bmatrix}, \quad c \in \mathbb{R}. \tag{G.8}$$

On the other hand, the adjoint matrix of L is

$$L^* = \begin{bmatrix}0 & 0\\1 & 1\end{bmatrix}. \tag{G.9}$$

The kernel of L^* is

$$\tilde{\boldsymbol{x}}_0 = \text{Ker}(L^*) = \tilde{c}\begin{bmatrix}1\\-1\end{bmatrix}, \quad \tilde{c} \in \mathbb{R}. \tag{G.10}$$

As seen from Figure G.1, the inner product of $\tilde{\boldsymbol{x}}_0$ and $\boldsymbol{y}$ is 0 as

$$\tilde{\boldsymbol{x}}_0 \cdot \boldsymbol{y} = c\tilde{c}\begin{bmatrix}1\\1\end{bmatrix}^T \begin{bmatrix}1\\-1\end{bmatrix} = 0. \tag{G.11}$$

G.2 Solvability Condition

We consider the following coupled equations as

$$\begin{cases} L\boldsymbol{x}_0 = 0, & \text{(G.12)}\\ L\boldsymbol{x}_1 = \boldsymbol{f} & \text{(G.13)}\end{cases}$$

$\boldsymbol{x}_0$ is the kernel of L. Equation (G.13) does not have a solution for every $\boldsymbol{f}$, but has a solution for specified $\boldsymbol{f}$. Let us consider the adjoint vector of $\boldsymbol{x}_0$ that satisfies

$$L^*\tilde{\boldsymbol{x}}_0 = 0. \tag{G.14}$$

We take the inner product of both sides of Eq. (G.13) with theadjoint vector $\tilde{\boldsymbol{x}}_0$, i.e.

$$\tilde{\boldsymbol{x}}_0 \cdot L\boldsymbol{x}_1 = \tilde{\boldsymbol{x}}_0 \cdot \boldsymbol{f}. \tag{G.15}$$

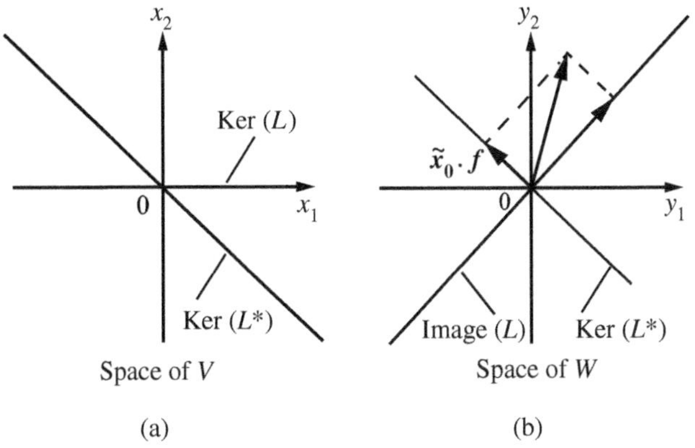

Figure G.2 Solvability condition: If there exists a solution of $L\boldsymbol{x}_1 = \boldsymbol{f}$, $\boldsymbol{f}$ must be in Image (L) and the component of $\boldsymbol{f}$ parallel to Ker (L^*), i.e. $\tilde{\boldsymbol{x}}_0 \cdot \boldsymbol{f}$, must not exist, where $|\boldsymbol{x}_0| = 1$.

Because of Eq. (G.5), the left-hand side is zero. Therefore, the condition that $\boldsymbol{x}_1$ exists, i.e. Eq. (G.13) is solvable, is $\tilde{\boldsymbol{x}}_0 \cdot \boldsymbol{f} = 0$, which is called *solvability condition* for $\boldsymbol{x}_1$. Figure G.2 shows the condition from a geometrical point of view. If Eq. (G.13) is solvable, $\boldsymbol{f}$ has to be included in Image (L). The condition to vanish the component of $\boldsymbol{f}$ orthogonal to Image (L), i.e. the component parallel to Ker (L^*), is the solvability condition.

Coupling Eqs. (G.12) and (G.13), we may consider the following equation:

$$L\boldsymbol{x} = \epsilon \boldsymbol{f}(\boldsymbol{x}), \tag{G.16}$$

where $0 < \epsilon \ll 1$. Assuming the solution to be

$$\boldsymbol{x} = \boldsymbol{x}_0 + \epsilon \boldsymbol{x}_1 + \cdots, \tag{G.17}$$

we have

$$\begin{cases} O(\epsilon^0) : L\boldsymbol{x}_0 = 0, & \text{(G.18)} \\ O(\epsilon) : L\boldsymbol{x}_1 = \boldsymbol{f}(\boldsymbol{x}_0). & \text{(G.19)} \end{cases}$$

Then, the solvability condition for $\boldsymbol{x}_1$ is $\tilde{\boldsymbol{x}}_0 \cdot \boldsymbol{f}(\boldsymbol{x}_0) = 0$.

Next, using the above method based on solvability condition, let us obtain an approximate solution of the following Duffing equation:

$$\frac{d^2x}{dt^2} + x + \alpha_3 x^3 = 0. \tag{G.20}$$

We consider the case when α_3 is much less than 1, $\alpha_3 = \epsilon\hat{\alpha}_3$ $(0 < \epsilon \ll 1)$, and assume the solution as

$$x = x_0 + \epsilon x_1 + \cdots. \tag{G.21}$$

Introducing the multiple time scales, $t_0 = t$ and $t_1 = \epsilon t$, we have

$$\frac{d^2}{dt^2} = D_0^2 + 2\epsilon D_0 D_1 + \cdots, \tag{G.22}$$

where $D_0 = \frac{\partial}{\partial t_0}$ and $D_1 = \frac{\partial}{\partial t_1}$. Substituting Eqs. (G.21) and (G.22) into Eq. (G.20) and equating the coefficients of like power of ϵ yields

$$\begin{cases} O(\epsilon^0) : D_0^2 x_0 + x_0 = 0, & \text{(G.23)} \\ O(\epsilon) : D_0^2 x_1 + x_1 = -2D_0 D_1 x_0 - \hat{\alpha}_3 x_0^3. & \text{(G.24)} \end{cases}$$

Substituting

$$x_0 = A e^{\lambda t_0}, \tag{G.25}$$

into Eq. (G.23) yields

$$L(\lambda)A = 0, \tag{G.26}$$

where

$$L(\lambda) = \lambda^2 + 1. \tag{G.27}$$

When $\lambda = i$ and $\lambda = -i$, Eq. (G.23) has a nontrivial solution as

$$x_0 = A_{01}(t_1)e^{\lambda_1 t_0} + \bar{A}_{01}(t_1)e^{\bar{\lambda}_1 t_0}, \tag{G.28}$$

where $\lambda_1 = i$ and

$$L(\lambda_1) = \lambda_1^2 + 1 = 0. \tag{G.29}$$

$A_{01}(t_1)$ satisfying

$$L(\lambda_1)A_{01} = 0, \tag{G.30}$$

is a function of t_1 but not explicitly determined at this level.

Substituting Eq. (G.28) into the right-hand side on Eq. (G.24) yields

$$D_0^2 x_1 + x_1 = -(2\lambda_1 D_1 A_{01} + 3\hat{\alpha}_3 |A_{01}|^2 A_{01})e^{\lambda_1 t_0} - \hat{\alpha}_3 A_{01}^3 e^{\lambda_3 t_0} + CC, \tag{G.31}$$

where $\lambda_3 = 3\lambda_1$. Inspecting the right-hand side on Eq. (G.31), we assume the solution to be

$$x_1 = A_{11}e^{\lambda_1 t_0} + A_{13}e^{\lambda_3 t_0} + CC. \tag{G.32}$$

Substituting Eq. (G.32) into Eq. (G.31) and equating the coefficients of $e^{\lambda_1 t_0}$ and $e^{\lambda_3 t_0}$ on both sides yields

$$\begin{cases} e^{\lambda_1 t_0} : L(\lambda_1)A_{11} = -(2\lambda_1 D_1 A_{01} + 3\hat{\alpha}_3 |A_{01}|^2 A_{01}), & \text{(G.33)} \\ e^{\lambda_3 t_0} : L(\lambda_3)A_{13} = -\hat{\alpha}_3 A_{01}^3. & \text{(G.34)} \end{cases}$$

The combination of Eqs. (G.30) and (G.33) corresponds to that of Eqs. (G.18) and (G.19). Since $L(\lambda_1)$ in Eqs. (G.30) and (G.33) is a scalar, the solvability condition of A_{11} in Eq. (G.33) is

$$2\lambda_1 D_1 A_{01} + 3\hat{\alpha}_3 |A_{01}|^2 A_{01} = 0. \tag{G.35}$$

This is the same as Eq. (10.18) with $\gamma = \Gamma = 0$. Letting

$$A_{01} = \frac{1}{2}a(t_1)e^{i\phi(t_1)}, \tag{G.36}$$

we obtain the equations corresponding to Eqs. (10.22) and (10.23) as

$$\begin{cases} D_1 a = 0, & \text{(G.37)} \\ aD_1\phi = \frac{3}{8}\hat{\alpha}_3 a^3, & \text{(G.38)} \end{cases}$$

or

$$\begin{cases} \dfrac{da}{dt} = 0, & \text{(G.39)} \\ a\dfrac{d\phi}{dt} = \dfrac{3}{8}\alpha_3 a^3. & \text{(G.40)} \end{cases}$$

From Eq. (G.39), a is independent of t and assumed to be equal to a_{st}. Then, from Eq. (G.40), we obtain ϕ as

$$\phi = \frac{3}{8}\alpha_3 a_{st}^2 t + \phi_0, \tag{G.41}$$

where ϕ_0 is a constant determined by the initial condition. Therefore, A_{01} is expressed as

$$A_{01} = \frac{1}{2}a_{st}e^{i(\frac{3}{8}\alpha_3 a_{st}^2 t + \phi_0)}. \tag{G.42}$$

Next, we investigate the solution of Eq. (G.34). Because of $L(\lambda_3) \neq 0$, the equation is solvable without imposing any special conditions and A_{13} is expressed as

$$A_{13} = \frac{1}{8}\hat{\alpha}_3 A_{01}^3 = \frac{1}{64}\hat{\alpha}_3 a_{st}^3 e^{3i(\frac{3}{8}\alpha_3 a_{st}^2 t + \phi_0)}. \tag{G.43}$$

The term A_{11} is disregarded since Eqs. (G.30) and (G.33) become equivalent under the solvability condition Eq. (G.35). Then, we obtain an approximate solution of Eq. (G.20) as

$$\begin{aligned} x &= x_0 + \epsilon x_1 + O(\epsilon^2) \\ &= a_{st}\cos\left\{\left(1 + \frac{3}{8}\alpha_3 a_{st}^2\right)t + \phi_0\right\} \\ &\quad + \frac{1}{32}\alpha_3 a_{st}^3 \cos\left\{3\left(1 + \frac{3}{8}\alpha_3 a_{st}^2\right)t + 3\phi_0\right\} + O(\epsilon^3). \end{aligned} \tag{G.44}$$

The frequency ω_n in the first term

$$\omega_n = 1 + \frac{3}{8}\alpha_3 a_{st}^2, \tag{G.45}$$

is related to the amplitude a_{st} through the cubic nonlinearity and results in the linear natural frequency 1 when a_{st} tends to 0. The relationship between ω_n and a_{st} is called a backbone curve and described as Figure G.3a,b, where (a) and (b) are in the cases of $\alpha_3 > 0$, i.e. hardening cubic nonlinearity, and $\alpha_3 < 0$, i.e. softening cubic nonlinearity, respectively (see Section A.1).

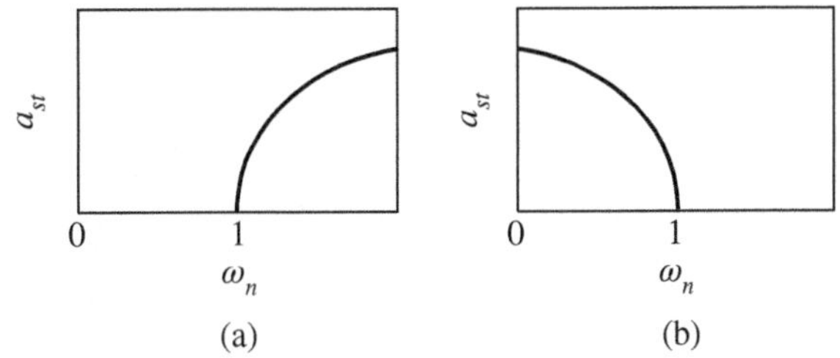

Figure G.3 Backbone curve of Duffing oscillator. (a) $\alpha_3 > 0$ (hardening cubic nonlinearity), the natural frequency ω_n increases as the response amplitude becomes larger; (b) $\alpha_3 < 0$ (softening cubic nonlinearity), the natural frequency ω_n increases as the response amplitude becomes smaller.

H

Effect of Contact Force on the Dynamics of Railway Vehicle Wheelset

Regarding the equations governing the lateral and yaw motions in a railway vehicle wheelset, Eq. (6.39), the matrix related to the positional force, i.e. K' in Eq. (3.1), is nonsymmetric. As clarified in Section 6.3, this characteristic produces the so-called hunting motion, which is the self-excited oscillation caused by the same mechanism as those in other systems mentioned in Sections 6.2, 6.4.4, and 6.5. In this appendix, we reveal the source of the nonsymmetric characteristic through the derivation of the equations governing the lateral and yaw motions in the railway vehicle wheelset, Eq. (6.39).

First, dealing with simple problem, we express the absolute speed at the contact point related to the friction force acting on the disk. By extending the way, we describe the absolute speed at the contact point of a wheel using the velocity of the center of the wheel and the angular velocity about the axle. It is deduced that the contact force depending on these velocities produces the nonsymmetric characteristic of the matrix related to the positional force in Eq. (6.39), i.e. is circulatory.

H.1 A Slip at the Contact Point of Rolling Disk on a Plane

First, we consider a simple model of a rolling disk with radius r on $x - y$ plane as shown in Figure H.1a. In addition to the unit vectors, $\boldsymbol{e}_x$, $\boldsymbol{e}_y$, and $\boldsymbol{e}_z$, along x-, y-, and z-axes, respectively, introducing the polar coordinate base vectors, $\boldsymbol{e}_r$ and $\boldsymbol{e}_\theta$, yields an expression of the position vector $\boldsymbol{q}$ of a point Q on the circumference as

$$\boldsymbol{q} = x\boldsymbol{e}_x + r\boldsymbol{e}_r, \tag{H.1}$$

where x is the displacement of the center of the disk, and the velocity vector is

$$\frac{d\boldsymbol{q}}{dt} = v\boldsymbol{e}_x + r\omega\boldsymbol{e}_\theta, \tag{H.2}$$

where $v = \frac{dx}{dt}$ and $\omega = \frac{d\theta}{dt}$ are the velocity of the center of the disk and the angular velocity, respectively. Because of $\boldsymbol{e}_\theta|_{\theta=\pi} = [-\sin\theta\boldsymbol{e}_z + \cos\theta\boldsymbol{e}_x]_{\theta=\pi} = -\boldsymbol{e}_x$, the velocity at contact point Q_c is

$$\frac{d\boldsymbol{q}_c}{dt} = (v - r\omega)\boldsymbol{e}_x, \tag{H.3}$$

Linear and Nonlinear Instabilities in Mechanical Systems: Analysis, Control and Application, First Edition. Hiroshi Yabuno.

Companion website: www.wiley.com/go/yabuno/instabilitiesinmechanicalsystems

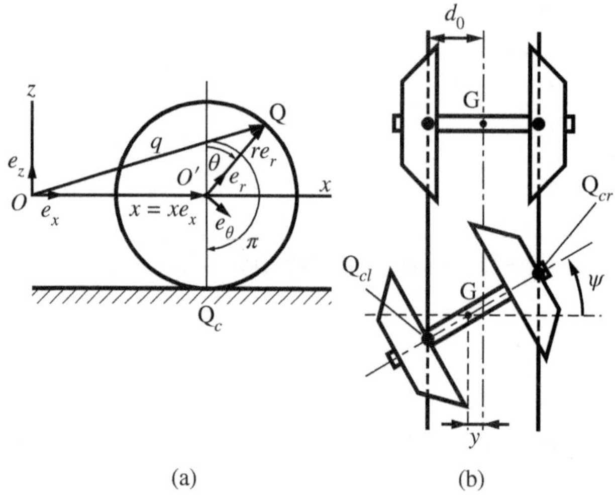

Figure H.1 Motion of rolling disk and configuration of wheels. (a) Coordinate system to obtain the speed of contact point Q_c. (b) Configuration of wheelset under lateral and yaw motions, y and ψ.

which is the difference between the actual forward speed and the peripheral one. Let us consider the situation that the wheelset moves in y-direction and rotates with yaw angle ψ as in Figure H.1b. By taking into account Figure H.2a, the actual speeds of the right and left wheels in x-direction, v_{rx} and v_{lx}, are

$$v_{rx} \approx V + d_0 \frac{d\psi}{dt}, \tag{H.4}$$

$$v_{lx} \approx V - d_0 \frac{d\psi}{dt}, \tag{H.5}$$

where V is the running speed of the vehicle. On the other hand, the actual speeds of the right and left wheels in y-direction, v_{ry} and v_{ly}, are

$$v_{ry} \approx \frac{dy}{dt}, \tag{H.6}$$

$$v_{ly} \approx \frac{dy}{dt}. \tag{H.7}$$

By the way, the contact point between a rail and a wheel describes a circle by the rotation of the wheel. The radii about the right and left wheels are expressed as

$$r_r = r_0 - \gamma_e y, \tag{H.8}$$

$$r_l = r_0 + \gamma_e y. \tag{H.9}$$

Figure H.2b shows the peripheral speeds of the right and left wheels. The peripheral speeds of the right and left wheels in x-direction are

$$u_{rx} \approx r_r \omega = (r_0 - \gamma_e y)\omega = V - \gamma_e \omega y, \tag{H.10}$$

$$u_{lx} \approx r_l \omega = (r_0 + \gamma_e y)\omega = V + \gamma_e \omega y, \tag{H.11}$$

and those in y-direction are

$$u_{ry} = r_r \omega \psi = (r_0 - \gamma_e y)\omega \psi \approx r_0 \omega \psi = V\psi, \tag{H.12}$$

$$u_{ly} = r_l\omega\psi = (r_0 + \gamma_e y)\omega\psi \approx r_0\omega\psi = V\psi. \tag{H.13}$$

The *creep ratio* related to the relative motion of the contact point is defined as

$$\nu = \frac{\upsilon - u}{V}, \tag{H.14}$$

where υ, u, and V are the actual forward speed, the peripheral speed of the wheel at the contact point, and the running speed, respectively. Then, the creep ratios of the right and left wheels in x-direction, ν_{rx} and ν_{lx}, are

$$\nu_{rx} = \frac{\upsilon_{rx} - u_{rx}}{V} = \frac{d_0\frac{d\psi}{dt} + \gamma_e\omega y}{V}, \tag{H.15}$$

$$\nu_{lx} = \frac{\upsilon_{lx} - u_{lx}}{V} = -\frac{d_0\frac{d\psi}{dt} + \gamma_e\omega y}{V}. \tag{H.16}$$

In addition, the creep ratios of the right and left wheels in y-direction, ν_{ry} and ν_{ly}, are

$$\nu_{ry} = \frac{\upsilon_{ry} - u_{ry}}{V} = \frac{\frac{dy}{dt} - V\psi}{V}, \tag{H.17}$$

$$\nu_{ly} = \frac{\upsilon_{ly} - u_{ly}}{V} = \frac{\frac{dy}{dt} - V\psi}{V}. \tag{H.18}$$

Finally, by taking into account $V = r_0\omega$ and Eqs. (H.8) and (H.9), the contact forces acting on the right and left wheels are expressed in x-direction as

$$F_{rx} = -\kappa_{xx}\nu_{rx} = -\frac{\kappa_{xx}d_0}{V}\frac{d\psi}{dt} - \frac{\kappa_{xx}\gamma_e}{r_0}y, \tag{H.19}$$

$$F_{lx} = -\kappa_{xx}\nu_{lx} = \frac{\kappa_{xx}d_0}{V}\frac{d\psi}{dt} + \frac{\kappa_{xx}\gamma_e}{r_0}y, \tag{H.20}$$

and in y-direction as

$$F_{ry} = -\kappa_{yy}\nu_{ry} = -\frac{\kappa_{yy}}{V}\frac{dy}{dt} + \kappa_{yy}\psi, \tag{H.21}$$

$$F_{ly} = -\kappa_{yy}\nu_{ly} = -\frac{\kappa_{yy}}{V}\frac{dy}{dt} + \kappa_{yy}\psi, \tag{H.22}$$

where κ_{xx} and κ_{yy} are called *creep coefficients* defined as constants depending on geometry and normal load. Then, the moment around z-axis due to F_{rx} and F_{lx} is

$$N = (d_0 + y)F_{rx} - (d_0 - y)F_{lx} \approx -\frac{2\kappa_{xx}d_0^2}{V}\frac{d\psi}{dt} - \frac{2\kappa_{xx}\gamma_e d_0}{r_0}y. \tag{H.23}$$

As a result, we obtain the equations governing the lateral and yaw motions:

$$m\frac{d^2y}{dt^2} + \frac{2\kappa_{yy}}{V}\frac{dy}{dt} + k_y y - 2\kappa_{yy}\psi = 0, \tag{H.24}$$

$$I\frac{d^2\psi}{dt^2} + \frac{2\kappa_{xx}d_0^2}{V}\frac{d\psi}{dt} + \frac{2\kappa_{xx}\gamma_e d_0}{r_0}y + k_x d_1^2\psi = 0, \tag{H.25}$$

where the third term in Eq. (H.24) is the lateral restoring force by the springs attached in y-direction and the fourth term in Eq. (H.25) is the restoring moment by the springs in x-direction. These equations are the same as Eq. (6.39), in which the matrix with respect to

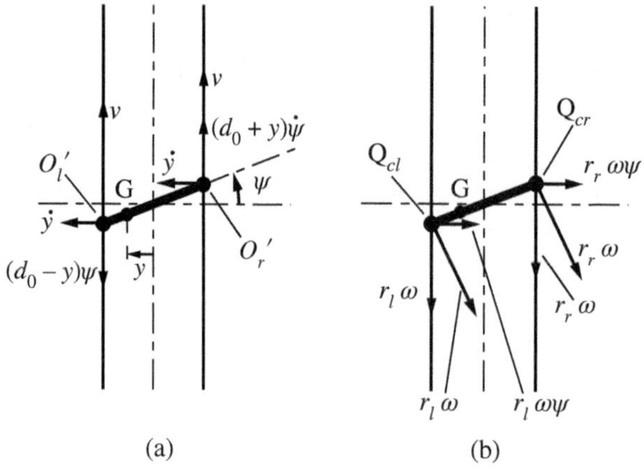

Figure H.2 Real speed and the peripheral speed at the contact points. (a) x and y components of the real speeds at the center of right and left wheels, O and O. (b) x and y components of the peripheral speed at the contact points Q_{cr} and Q_{cl} in Figure H.1b.

the positional force, i.e. K' in Eq. (3.1), is unsymmetric. As seen from the above derivation, the unsymmetric characteristic is caused from the contact force, which can produce the so-called hunting motion which is a flutter-type self-excited oscillation.

Index

Linear and Nonlinear Instabilities in Mechanical Systems: Analysis, Control and Application, First Edition. Hiroshi Yabuno.

Companion website: www.wiley.com/go/yabuno/instabilitiesinmechanicalsystems